SPEAKING MINDS

Speaking Minds

Interviews with Twenty Eminent Cognitive Scientists

EDITED BY PETER BAUMGARTNER
AND SABINE PAYR

Princeton University Press   Princeton, New Jersey

Library of Congress Cataloging-in-Publication Data

Speaking minds : interviews with twenty eminent cognitive scientists /
edited by Peter Baumgartner and Sabine Payr.
p.   cm.
Includes bibliographical references and index.
ISBN 0-691-03678-0 (cloth)
ISBN 0-691-02901-6 (paperback)
1. Cognitive science.   2. Scientists—Interviews.
I. Baumgartner, Peter, 1953– . II. Payr, Sabine, 1956– .
BF311.S657   1995
153—dc20   94-24797

This book has been composed in Adobe Sabon and Gill Sans

Princeton University Press books are printed on acid-free paper
and meet the guidelines for permanence and durability of the
Committee on Production Guidelines for Book Longevity of
the Council on Library Resources

Second printing, and first paperback printing, 1996

Printed in the United States of America
by Princeton Academic Press

10 9 8 7 6 5 4 3

Text designed by Laury A. Egan

CONTENTS

SPEAKING MINDS

Introduction

Why a Book of Interviews?

The idea for a collection of interviews with notable cognitive scientists originated as early as 1989, while we were in residence at the University of California at Berkeley, working as research scholars at the Institute of Cognitive Studies. Peter Baumgartner was preparing a book on educational philosophy dealing specifically with the background of knowledge (Baumgartner 1993), and Sabine Payr was working on her doctoral dissertation on knowledge-based machine translation (Payr 1992). So we both had already been dealing with the problems of Artificial Intelligence and cognitive science, and we shared a focus on the philosophical and social questions raised by these fields.

During our stay in Berkeley we had the invaluable opportunity to meet many scientists of different disciplines, more or less involved in cognitive science but none of them indifferent to the emergence, promises, and pitfalls of this young and lively research program on the Berkeley campus. We were deeply impressed by the commitment, intensity, and quality of the interdisciplinary discourse.

This dynamic personal discussion seemed to us—at least in some respects—much more concrete and powerful than the more general and often more moderate and balanced considerations in the written publications of the same scientists. We felt that in approaching a field of research by reading articles and textbooks alone, one leaves out an essential aspect: the personal commitment of the scientist, his or her connections and relationships with the scientific community, the importance of the claims for other thinkers—be they favorable or not.

So we began to think about ways in which the atmosphere of this interdisciplinary argument could be captured and conveyed to others. Different possibilities came to mind. In our opinion, the best way would have been a workshop, where all the notable people could meet in person and confront one another about the "hot" questions of the field. We saw two kinds of difficulties with such an endeavor:

First, organizational and technical difficulties: how to bring together, at the same time and in the same place, a respectable number of busy and active scientists, some of whom may be among the most sought-after personalities of American academic life? A videoconference might have been an alternative. But then we would have needed rich experience with this medium in order to make them forget the technological nature of their gathering and to impart to the conference a constructive character in which they would be ready to let down their defenses and welcome attacks and criticism.

The second reason seems quite trivial and pragmatic in comparison: as temporary guests from Europe, we were not in a position to organize (much less fund) such a major conference.

The next best thing, we thought, would be a personal talk with each of these scientists. If they were acquainted with the other people interviewed and their opinions, there would still be a chance to instigate some open and spontaneous scientific discourse. The interviewer could serve as a common point of reference whereby the scientists could—albeit very indirectly—refer to and react to one another and defend their position against attacks from the other members of this (virtual) panel. With this goal in mind, we set out to organize the interviews for this book.

It was clear that such an interview was not the setting either to develop and transmit entirely new, revolutionary ideas or to discuss single questions very thoroughly. Furthermore, a detailed discussion of the work of each single scientist would inevitably have led into his or her special discipline and away from the interdisciplinary discourse that had fascinated us in the first place and that we wanted to capture for the readers. On the other hand, an open interview always threatens to degenerate into small talk. So we tried to focus each conversation on a general outline of the views of the person interviewed and to stress the differences of that person's claims. In order to steer this middle course that we had chosen, we established and followed roughly the same procedure for all interviews. Peter, who was responsible for conducting the interviews, first wrote a letter stating his request for an interview, in which he had enclosed a list of questions. Of course, he also mentioned that everybody was free to elaborate or add their own questions. These are the original questions:

How did you get involved in cognitive science research?

How would you outline the interdisciplinary connection of sciences in the field of cognitive science? Which sciences are supposed to contribute to which question or point of view?

Which sciences play (or should play) an important role in cognitive science and Artificial Intelligence?

How would you characterize the importance of Artificial Intelligence in cognitive science?

What is, in your opinion, the very idea of Artificial Intelligence?

What are the most important periods in the history of Artificial Intelligence?

What were, in your opinion, the most important recent successes and failures of Artificial Intelligence?

Which role do theory and experiment play in cognitive science research?

What about the Turing Test: do you think it is a useful test (idea)?

At the moment, which is the biggest problem that has to be overcome for the future development of Artificial Intelligence?

Do you think that the question of commonsense is a special problem for Artificial Intelligence? And if so, how could it be solved?

In your opinion, will there arise any social problems in the further development or application of Artificial Intelligence research? If so, what are they, and what might their social impact be?

What are, in your opinion, the future prospects of Artificial Intelligence?

The choice of personalities was influenced by several different motives: without doubt, some were obviously the people that anyone who wanted to cover the field of cognitive science should talk to. Unfortunately (but also quite predictably), not all the "key" people were available for an interview within the limited timeframe and budget of our "interview tour." Such was the case with Noam Chomsky, who had to cancel because of his health, or with Roger C. Schank, who had then just moved to Chicago, which was, alas, far outside the planned route. Other people were recommended to us by members of the cognitive science group in Berkeley with whom we discussed the project and who mentioned one person or another who would, in their view, have some interesting aspect to add. Some of the scientists to whom Peter wrote in the first round also recommended further contacts. We were not able to pursue all these helpful hints, of course—again for lack of time, but also because of the limitations of book form and the demands on a reader. So the virtual meeting of notable researchers in cognitive science that the reader will find in this book is by no means complete.

About two weeks after his first letter Peter tried to establish an initial meeting to explain our plans in more detail. Surprisingly, the vast majority of the people to whom he had written were willing to grant an interview. In only one case was he unable to elicit any response. Except in this one instance, the responsibility for having left important names off the list is therefore entirely ours.

One of the hardest parts of the project was to coordinate the places and dates of the actual interviews. In order to stay within the (very tight) limits of our budget, Peter had to try to talk to everyone on the West Coast during one trip and everyone on the East Coast during another.

Each interview took approximately two hours and was tape-recorded. In one case (George Lakoff) the interview had to be repeated because the social scientist's constant nightmare came true and the tape recorder failed without Peter noticing it. We want to thank George in particular for undergoing the tedious procedure twice! Two other interviews (with Allen Newell and Hilary Putnam) had to be cut somewhat before the end because they had other urgent obligations.

Our real work began back in Austria, transforming the raw material—the tapes—into the finished product, this book. The demands on our time and energy that the return to our jobs made forced us to interrupt work on this book several times. The reasons for the repeated interruptions were not only professional but also of a more personal and psychological order: we had frankly underestimated how difficult it would be, back in Europe and temporarily "cut off" from American academic life, to keep alive the spirit of the scientific debate that had gotten the project started in the first place. Our return to Europe added another very real problem to this psychological obstacle: all the questions, corrections, and follow-up contacts concerning the transcription, revision, and authorization of the interviews had to be done by mail—another fact that postponed considerably the completion of the book. Sabine Payr took over most of this work, from the transcription and initial "smoothing out" of the interview texts, to the correspondence with the interviewees, to the final correction, annotation, and authorization of the interviews.

From the first raw transcription a smoother second draft was made: whereas in some interview contexts it is important to convey hesitations, pauses, slips of the tongue, and so forth, this was not desirable for a book with easily readable texts. On the other hand, the interviews should still be recognizable as what they were: an open, spontaneous conversation with its characteristic style and language. Documentation of persons referred to and works quoted in the interviews was completed as far as possible. The wide range of fields and backgrounds represented forced us, however, to leave one reference or another incomplete.

This draft version of the interviews was sent back to the participants, usually accompanied by requests for clarification, comments, and so on. Because some of the participants were referred to and criticized by oth-

Because some of the participants were referred to and criticized by others, we sent these remarks along as well, in order to offer an opportunity to respond to them and thus to strengthen the relevance of the interviews to one another.

The participants then had ample opportunity to correct and change the interview in whatever way they wished. The majority, however, were content to make a few stylistic modifications and helpfully correct errors that we, as nonnative English speakers, had introduced or overlooked. In our opinion, this reveals not only that our interviewees are extraordinarily articulate, even in half-formal settings, but also that the interviews were conducted in an atmosphere where they could follow their train of thought without feeling pressured to say things they might not say, or might say differently, even a year later.

When the corrected interviews were returned to us, together with the participants' biographical notes and their lists of titles for our "Recommended Works" sections, the last stage of final correction began. By sending back the interview text, the scientists also gave—by prior agreement—their authorization for the interviews to be published.

Out of the twenty-five original interviews, twenty were sent back and therefore authorized for publication. Four interviews were withdrawn: these researchers felt that their interviews no longer accurately reflected their positions and would therefore need more modification that they were willing or able to invest. In addition, we lost contact with one of these researchers.

The publication of this book is overshadowed by a grievous loss to the entire cognitive science community: Allen Newell, one of the pioneers of Artificial Intelligence, died during the summer of 1993. He did not have the opportunity to correct his interview, but we felt that it should be our small contribution to his memory to publish it in this volume. We would especially like to thank Betsy Herk of the Soar Project for making the publication of his interview possible by giving us the authorization to print it here and by providing all the necessary materials.

Ways of Reading

The product that we had in mind during this laborious process was a book that is, primarily, a particular sort of introduction to the field of cognitive science: clearly, it is not a textbook whose aim is to lay out the history, the methodology, the key findings, and open problems of the field. The problem of textbooks, as has often been noted, is precisely this necessity to cover their subject as a whole. Often enough, the effect of doing so is a slight distortion: while the common ground of the field and its uncontested methods and results are stressed, the differences and open questions must remain somewhat underdeveloped. And while readers get a more or less thorough introduction to the "hard facts" of the field, they do not so easily get the "feel" of it as a heterogeneous, multifaceted, lively community of researchers, pursuing their work in a rich context of often heated and sometimes even passionate debate.

By these remarks, we do not want to reduce the need for and the usefulness of such textbooks. On the contrary, we feel that this book will be much more useful to newcomers and students in the disciplines associated with cognitive science when it accompanies a textbook. Equipped with basic knowledge of the field, the reader should be ready to enjoy the arguments unfolding across the different interviews presented here. This recommendation, however, should by no means exclude or discourage students who prefer a more "inductive" way to become acquainted with cognitive science: the interviews may also be read as a sort of directory of questions, inviting the reader to explore them in more depth.

In our view, there are, then, different ways to approach this book:

One can read every interview in its own right—that is, as a personal introduction to the work of the interviewees and to their main concerns and research strategies. This kind of focus could complement the more general introductions in textbooks as well as more specialized and difficult access via the works of the author. We have tried to facilitate such a "personal" approach by including a photo of each author, a short biographical note, and a list of his or her important works. In compiling these sections of recommended readings, we have paid special attention to the accessibility of the material: where an article has been reprinted in a reader or anthology, perhaps together with work of others interviewed in this book, we have preferred this publication to the original, but possibly less accessible, source. In addition, the glossary can be of help in this introductory reading by clarifying some of the technical terms that appear in the interviews.

Another point of interest could be offered by the central issues that turn up again and again in the interviews. Of course, these issues were, to a certain degree, chosen already by the proposed questionnaire and thus reflect our interests: the knowledge representation problem, particularly with regard to the so-called everyday or commonsense knowledge or background; the Turing Test; and the Chinese Room Argument. An issue that was not already present in the questionnaire but almost unavoidable, at least at the time when the interviews took place, is connectionism and the respective prospects of symbol manipulation and neural networks approaches. We were also interested in the development of cognitive science as an interdisciplinary field of research and the way in which the academic institutions reflect this project. We will discuss some of these key issues and debates in the third part of this introduction, so that readers interested in this type of "cross-reading" can familiarize themselves with these issues and their role in cognitive science.

Finally, we can imagine that some readers might be interested in the complex network of relationships among the interviewed researchers, their personal involvement in and attachment to the field, the roles of their mentors, their academic backgrounds, and their locations. We think history—in this case, the history of cognitive science—is not a straightforward process that can be explained by social factors on a macro level alone but is also influenced by circumstances (part of them quite accidental) on the micro or personal level. But, at the same time, history is not just the sum of individual biographies: a new scientific paradigm is more than the accumulated ideas and achievements of a single founding father or a handful of pioneers. Although a collection of interviews with individual researchers is not the best place to develop this aspect, the interviewees themselves make it clear in their interviews how much their development depends on and is a response to a multitude of factors.

With these different ways of reading in mind, we had to make some compromises in producing this book. On the one hand, by printing each interview in its entirety and presenting the views of each researcher in his or her own right, some parallelism and redundancy was unavoidable. On the other hand, only a common framework of questions and subjects could make comparisons and cross-reading possible. What resulted from these conflicts is a book that might as well—or better—have been conceived as a hypertext. It invites not so much a sequential reading from the first to the last page as a conscious choice of text according to persons, or themes, or disciplines. For this reason, we ultimately refrained from grouping the interviews into parts or chapters and let the readers choose their way of reading. Each interview was completed with the intended goal of a text that could stand alone, so we chose to repeat notes and references from interview to interview rather than force the reader to search back and forth for cross-references.

A Guide to Some Issues Raised in the Interviews

The various suggested ways of reading this book have served as a guideline for the choice of questions. Although, as is usual in open interviews, some questions are individual responses to the development of the conversation, such as requests for clarifications, arguments and counter-arguments from other researchers, and so on, there are certain subjects that we discussed with most of the participants. These more or less constant leitmotifs can be the point of departure for a thematic cross-reading and cross-referencing of the interviews. In the remaining part of this introduction, we will try to outline these themes and discuss the sometimes long-standing argument surrounding them. This should provide the reader some background for the way these subjects are approached in the interviews.

In what follows, we refrain deliberately from an exhaustive discussion of these topics and the development of our own arguments. There are two reasons for our doing so: first, it would not have been fair to the interviewed researchers. For whereas they had to give relatively spontaneous answers to our questions during the interview and then consented to having these conversations published, we would have had the advantage that a written article affords for elaborating the questions. Also, they did not see the other interviews and therefore could not respond directly to what others had said. Conversely, we had access to all the interviews and thus would have been in a better position to arbitrate than they. Another part of the second reason is that we are not members of the cognitive science community in the way that most of these researchers are. In our work since then, we have been concerned with some of the questions that are discussed in the interviews and have developed a position that we try to apply to our own research, mainly in the field of education (Baumgartner and Payr 1990; Baumgartner 1993; Baumgartner and Payr 1994). But the purpose of the interviews was *not* explicitly to discuss our views with American cognitive scientists. Our status was that of external observers of and newcomers to this field, and we think the atmosphere that is mediated by the interviews owes much to the fact that the researchers perceived us as such. Only in this way, we think, was it possible to present what often amounts to a private introductory lecture.

An Interdisciplinary Discipline?

While it is not possible here to give a comprehensive introduction to cognitive science, some words must be said about this term and its meaning. H. Gardner (1985) presents, in our view, an excellent history of the research programs and efforts that came to be known as "cognitive science." In historical perspective, cognitive science is often seen as a reaction against the dominant paradigm in psychology, behaviorism. Whereas behaviorism—so the story goes—refrained from dealing with "inner" or mental states, this new way of thinking proposed to explore them. And, because methodological behaviorism had given up mental states not by way of principle but only (and reasonably) in light of the impossibility of studying them directly, the "cognitive revolution," as this movement has often been called, had to be tied to a new methodology that could claim to make the study of the mind possible, albeit indirectly. This is where the close ties between cognitive science and the computer (better: Artificial Intelligence) originate. Historically, Artificial Intelligence as a branch of engineering concerned with the construction of "machine intelligence" probably came first. But almost from the beginning it was equally important, or more so, to some to model human intelligence.

Cognitive science as "the mind's new science" (Gardner 1985) is almost by definition interdisciplinary. Whenever authors try to define what cognitive science is, they point out that it is a joint effort of specific disciplines to answer long-standing questions about the working of the mind—particularly knowledge, its acquisition, storage, and use in intelligent activity. In most cases, the disciplines of psychology, philosophy, Artificial Intelligence (computer science), linguistics, and neuroscience are listed as the five key disciplines contributing to and involved in cognitive science. Gardner (1985, 7) adds anthropology in this list, but this inclusion remains rather isolated.

It is easier to claim interdisciplinarity than to realize it: what does it mean in everyday scientific life? Can cognitive science really be characterized as an interdisciplinary research program, in which people from different fields work together, or is it instead held together by certain common goals and problems on which the different disciplines continue working as they always did?

It was tempting to ask cognitive scientists themselves how they conceive of the field and what they think about interdisciplinary work in cognitive science—in theory and in practice. Do they actually work together with colleagues from other fields, or are they satisfied with understanding and using their findings? Did they have to go outside their own field to learn others, or is it enough for them to know about the other fields?

As was to be expected, the range of views is very broad on this point. There are indeed researchers who found it necessary actually to learn the methods and problems of another field, and there are others who choose only occasionally to use the results of other fields to build their own work on, while continuing in the sort of research work in which they were trained. The notion of "interdisciplinarity" itself is at once vague enough to describe this range of attitudes and too vague to differentiate among them. It seems legitimate to conclude that cognitive science is generally considered in theory to be an interdisciplinary field. In practice, whether the actual research done in cognitive science is also carried out cooperatively among disciplines depends on individual subfields and interests.

The institutional setting in which cognitive scientists work seems to play an important role in the development of practical interdisciplinarity. Some of the researchers interviewed in this book report their experiences with all sorts of formal and informal types of exchange among the fields that constitute cognitive science. It comes as no surprise that the possibility of having colleagues from different disciplines working next door and of meeting them informally is felt as a positive influence and a motivation to broaden one's perspective.

At this point it should be asked what role is played by institutionalization—that is, cognitive science departments, or cognitive science programs that have been established in some universities. Although this sort of development is seen mostly as the outcome of already-existing interests and research work, one would like to ask how much the very existence of institutions itself contributed to creating the field. Academic systems are generally considered to be slow to react to new demands and developments. But on the other hand, once such a reaction takes place, its weight is considerable and might itself influence and change the needs to which it answered in the first place.

Such a development could take place with the creation of cognitive science departments and studies. Whereas at the time of the interviews, a number of researchers were not sure whether there would be a need for cognitive scientists trained "from scratch" in the different disciplines (a degree in cognitive science, for instance, is not considered necessary or even useful by a number of them), the second generation of cognitive science researchers could well redefine the field in a way completely different from the way the first generation—trained as specialists in their respective fields and only later interested in interdisciplinary work— views cognitive science.

The exact opposite is also possible, however. Training in more than one discipline becomes ever more difficult in a time of ever-increasing specialization of scientists. The risk of interdisciplinary training is—as some correctly point out—that of educating "jacks of all trades and masters of none," because the demands on a good researcher in a single field are already difficult enough to satisfy by standard academic training. So it would also seem possible that cognitive science might disappear as a field, giving way to very specialized but possibly still interdisciplinary fields of research, perhaps centered around a certain specific aspect of today's cognitive science—such as vision, or language acquisition. According to this point of view, "cognitive science" as a whole would then emerge as an embryonic, as yet unspecified and undeveloped precursor to a whole range of sciences of cognition. Such a change would certainly also be reflected in the establishment of institutions and studies. Whereas today's cognitive scientists might, for the most part, have their students repeat the way in which they themselves approached cognitive science—from a special discipline to the interdisciplinary work—the next generation may see it the other way around: starting from a basis in general cognitive science that lays out the goals and problems of all the disciplines and leading to a strongly specialized field that, however, cuts across some of today's accepted disciplines.

Of course, these are only speculations. But in this context, we intend them as an invitation to the reader to search for the explicit and implicit statements about individual working styles and conceptions of cognitive science in the interviews.

The Computer Metaphor

A good bit of each interview is concerned with the role of Artificial Intelligence. This relative emphasis on Artificial Intelligence may seem odd, but we had reasons for giving it so much attention here—reasons that have, in our view, little to do with the importance that individual researchers do or do not attribute to Artificial Intelligence as a discipline of cognitive science. They go much deeper, to what has often been called the "computer metaphor."

The uses to which the digital computer has been put since World War II—namely, symbolic processing—have played a key role in bringing about a new way to think about the mind that is now a central premise of cognitive science: that it is possible to study the brain from the standpoint of information processing. The computer has become a powerful metaphor for the mind in all the sciences that cognitive science comprises.

The power of the computer metaphor does not depend on the degree to which a discipline actually uses the computer in modeling its theories. But the mere availability of a device that can carry out logical and arithmetic operations made it a candidate for a brain model in action. Sometimes, the attractiveness of the computer metaphor has been compared to the fascination with which each period in the history of science has adopted the then most advanced technology—clocks, steam engines, or switchboards—as an analogy for the functioning of man and mind. The difference might be seen in the fact that the computer is a working metaphor in a way that former machines were not. The distinction between computer hardware and software lent itself particularly well to analogies with the mind-brain.

The symbol processing approach to cognitive science and the new connectionism differ in numerous and important ways, but it appears that the computer metaphor is still functioning for both of them and unites them in spite of their differences and arguments. Not only are computer models at least as important for neural network researchers as they were for symbol processing approaches but part of the argument can also be seen as an argument about who has the "better," more correct interpretation of the computer metaphor—those who are concerned only with software or those who take into account the actual (or at least virtual) architecture of the hardware as well.

In this sense, both ways of thinking are undoubtedly part of cognitive science. The foundations and metaphors common to both approaches are really the only bases on which an argument such as the one carried out throughout this book is made possible. It is interesting, however, to note that the only relatively recent discussion of computer architecture, raised by the new connectionists, also leads to questions about the degree to which the computer metaphor has guided cognitive science research: were not even the problems that research in cognitive science and Artificial Intelligence set out to solve influenced by the working of computers? Was perhaps even the concept of "cognition" prompted by what computers could be programmed to do? To pursue this line of thought a bit further, the well-represented discussion about computer architectures in this book—von Neumann machines versus parallel distributed processing—might well lead to a more skeptical view about the computer as a (mental or experimental) model for studying the mind, and to more radical questions: are there questions about the mind that are completely ignored as long as the computer metaphor is dominant in the (branches of) disciplines that constitute cognitive science—both in symbol manipulation and in connectionist approaches?

The Turing Test and Intelligence

Alan Turing's work is generally considered to be a milestone in making the computer metaphor functional. The abstract machine he conceived—now called the Turing machine—is an idea that made it possible to think of the computer as a universal machine capable of simulating any other machine—including the brain.

To the question that was already raised in his time—the late forties and fifties—whether such a machine could be intelligent or should be called intelligent, Turing (1950) answered with his now famous imitation game, or "Turing Test." In short, he proposed the following scenario: an investigator is placed in one room, and a person and a computer each in two separate rooms, with which the investigator communicates via teletype. The investigator's task is to find out in which room is the human being. The computer's "task" is to pass as human, while the human being should be as helpful as possible. Turing suggested that a machine should be considered intelligent if it can pass as human in such a test. With this test, he wanted to end fruitless discussions about the "nature" and essence of human intelligence and to replace them with a judgment based on behavior.

From today's view, the concept of intelligence was thus preserved and could be taken over by the field of Artificial Intelligence, where the Turing Test became a criterion for success. But the vagueness of the notion was never clarified and returned to haunt the discipline, so that Artificial Intelligence researchers complained that whenever the computer achieved a certain capacity it could no longer be characterized as "intelligent." This was the case with, for example, optical character recognition, which has become a part of everyday office computing. It is no clearer today than it was forty years ago how intelligence should be defined.

It may be seen as a result of these long-standing problems with both artificial and human intelligence that today's cognitive scientists seem reluctant even to use the word *intelligence*. When they study vision, for example, it is no longer of interest whether this capacity involves intelligence or not. It can be seen as progress in knowledge in a field when general notions are replaced by more specific, fine-grained concepts. Cognition and cognitive skills in the view of today's cognitive science comprise much more than in the early days of Artificial Intelligence. No more are chess playing, problem solving, or theorem proving the prototypical achievements of the mind, but rather perception, language understanding, or motor control. It seems evident that Turing's simulation game can no longer be of much use, and cannot even serve as a theoretical goal, because many of these capacities do not depend on language and are not even restricted to human beings.

But this does not mean that there are now better tests for the success of a computer model than in the early days of cognitive science. It is still the behavior of, say, a neural network that can be studied and compared to that of living beings. But as neural network researchers are careful today not to equate simple network models with complex brains, a much more reserved attitude seems to reign over the explanatory power of computer models.

It may be this dilemma that prompts the wide variety of attitudes toward the Turing Test expressed by the people interviewed in this book.

While a few of them subscribe fully to its usefulness, others maintain that this question is far removed from their research work or even that it has always biased in an unwholesome way the method of doing cognitive science.

The Chinese Room and the First-Person Perspective

When John Searle published this parable in his article "Minds, Brains, and Programs" in 1980, it immediately raised a storm of criticism that has hardly diminished since then. The parable is well known, and we will present it here in a very short form: imagine yourself sitting in a room, provided with lists of Chinese symbols and rules to match them. Your only connection to the outside world is a slot. Through this slot, you receive input in the form of certain Chinese symbols. According to your lists, you replace the incoming symbols with others and pass them back through the slot as output. For the people outside the room, the input was a question in Chinese, and the output was a reasonable answer to it. They would be justified in supposing a Chinese speaker to be inside. But Searle's point is: would you yourself say that you understand Chinese?

The obvious answer is no. And because Searle proposes this way of processing information as an analogue for the functioning of a computer, we then could under no circumstances say that a computer "understands" anything. Of course, this conclusion provoked the criticism of members of the Artificial Intelligence community. It would lead too far even to try to resume the discussion that followed from it, but some of the criticisms are expressed in the interviews in this book.

The interesting point in this discussion is, from our view, that it must express some common ground even between adversaries: all parties must agree about how to conceive of the mind that makes this argument possible. To us, this common ground can be characterized by the orientation around the individual mind or system. The Chinese Room Argument revolves around the question of what it takes for this special system to have meaning (semantics). Whereas some of the criticisms and replies take into account the necessity of a causal relationship to the world (such as the robot reply) that can be translated into meaningful action, the model of the "lonely actor" confronted with a material outside world remains intact. Under the dominance of the cognitive science paradigm, this model is widely accepted. From this viewpoint, social competence and intersubjectivity are seen as the results of the individual's knowledge about a "world" that, in this perspective, makes no distinction between objects and other subjects. But mind need not be seen as a property of an individual at all (as cognitive science must in our opinion). One could also, like Mead (1934) or Bateson (1972), start out by viewing mind as a relation between subjects that is created by their interaction. The question of such possible alternative views, based on the embeddedness of cognition in situation, interaction, and society, could be among those that cognitive science, because of its basic assumptions, cannot even raise.

Commonsense and Background

It became clear relatively early on in the history of Artificial Intelligence that most behavior that could be called "intelligent"—even by the standards the field set for itself, such as the Turing Test—had to rely on the use of knowledge of the "world." The notion of a "world" was set in quotation marks by all careful researchers who were conscious that what they had to represent in a program was knowledge about a strictly limited and artificial model of the world ("microworld"). However, problems like natural language processing or the completing and summarizing of stories soon made clear that it is hard to limit the kinds and amounts of knowledge that human beings use in these tasks: they can rely on the vast resources of their everyday experience that they accumulate during their lives, and most of the time they seem to have instantly available precisely the skills and knowledge that are relevant in a particular situation.

It is the nature of these experiences and skills that raised the question of whether the problem of representing commonsense knowledge is simply one of scale and could therefore be solved by large storage capacities, a lot of hard work, and ingenious search algorithms or whether it is a problem that lies in the nature of human capacities—that is, that they cannot be "represented" symbolically in formal languages (list structures, frames, scripts, etc.). The doubts raised about commonsense have highlighted an assumption to which Artificial Intelligence subscribed and which cognitive science adopted: the assumption that, just as in von Neumann computer architecture, knowledge is stored and retrieved in memory very much like data in computer storage.

Some philosophers have criticized this assumption for a long time. The interviews collected here indicate that connectionism has taken this criticism seriously and is searching for other ways of dealing with knowledge. It remains to be seen whether connectionist models offer a better explanation for commonsense and whether it is a viable alternative to representationalism. An important part of the debate between the schools of symbol manipulation and connectionism that is illustrated in this book concerns the question of how much of cognition can be modeled with neural networks—that is, without symbol manipulation and symbolic knowledge representation. Another question that is also raised goes deeper and asks whether neural network models really constitute a departure from the Turing machine and all that is involved in this concept.

For reasons that we tried to outline at the beginning of this introduction, a considerable amount of time has elapsed between the conducting of the interviews and their publication in this book. Undoubtedly, this also gives the book a historical perspective that was not anticipated at the beginning. The original interviews were made in 1989, but some of them were revised by the interviewed researchers as late as 1993. But because most of the interviews remained almost unchanged from this book's genesis to its publication, it can be said to capture a certain moment in the development of cognitive science. This moment was not chosen deliberately, so that the picture that emerges from it should be completely arbitrary. However, the moment during which these interviews were made fell within a (longer) period that was characterized by a renewed excitement in the field, due to the lively debate between two very different approaches in cognitive science: symbol manipulation and connectionism. By capturing this debate, this book can be seen as an example for the way in which scientific discourse between the "old" and the "new" takes place. Different schools of thought and opinions are presented here, which might well be confusing for a newcomer to the field who searches for straightforward answers and uncontested facts. In this sense, this book also tells a story about the way science proceeds—that is, not along a straight road and rarely through big revelations but instead along a winding path where arguments, contradictions, and puzzles appear at every turn.

To conclude, we would like to thank all the people who helped make this book possible, foremost the researchers who granted us the interviews. It is clear that this sample of interviews could neither cover the complete interdisciplinary field nor be representative of its debates and issues. The choice of questions and the thematic guidelines of the interviews were influenced by a number of people at the University of California, Berkeley. In particular, our contacts with Hubert L. Dreyfus and John R. Searle at Berkeley "biased" the interviews with permanent references. In a sense, the positions with which we were confronted at Berkeley became the starting point for the whole enterprise. Consequently, they turn up again and again in the interviews. But we want to stress that what is presented here as these stances is, of course, the interviewer's—Peter's—understanding and interpretation of them at that time and that the interviewees are in no way responsible for any mistakes that we may have made in presenting their views.

We hardly consider this bias to be a disadvantage, however. We became interested in the discourse mainly through personal contacts, and this kind of involvement motivated the desire to establish contact with other famous people in the field. And we think this personal interest was also the main motive for their answering our sometimes naive questions. All interviewees contributed to the form this book has ultimately taken, and we would like to express our gratitude to them all for their response and cooperation.

Work on this book was supported by a Fulbright travel grant (Peter Baumgartner) and a research grant from the Austrian Federal Ministry of Science and Research (Sabine Payr), and by our institute, IFF, which allows and even encourages an impressive range of unusual research work to be carried out under its roof. Of course, it is impossible to express our gratitude to all the people who contributed in some way or other to the creation of this book. We would like to thank especially Markus Peschl, who helped us with some technical parts of the glossary; Matthias Goldmann, who supported us with his experiences in the publishing world; and Josef Mitterer, for his remarks and encouragement. We also thanks Arno Bammé, Laszlo Böszörmenyi, Walter De Mulder, Vinod Goel, Ernst Kotzmann, Ernie Lepore, Anthonie Meijers, Dieter Münch, Jan Nuyts, Franz Reichl, Georg Schwarz, and Angela Tuczek, whose contributions to this book were the fascinating discussions we had with them partly at Berkeley and partly later, on the occasion of a workshop in Austria. Our special thanks go to our editors at Princeton University Press, Malcolm DeBevoise and Lauren Oppenheim, who expertly saw the book through the final stages of production. Of course, any shortcomings or errors that remain are our responsibility alone.

Patricia Smith Churchland was born in 1943. She studied philosophy at the University of British Columbia (Canada), the University of Pittsburgh, and Oxford University. She worked at the University of Manitoba from 1969 to 1984, first as assistant professor and thereafter as associate and then full professor, until she became Professor of Philosophy at the University of California at San Diego in 1984. From early on she has specialized in the philosophy of mind, of science, and especially of neuroscience, in which latter field she has become most noted for her book, *Neurophilosophy*.

PATRICIA SMITH CHURCHLAND

Take It Apart and See How It Runs

When I was a graduate student at Pittsburgh in 1966, we worked through Quine's book *Word and Object*.[1] That book was one of the few things in philosophy that made any sense to me at that time, apart, of course, from Hume. Several other graduate students whom I came to know quite well were utterly contemptuous of the later Wittgenstein in particular and of so-called ordinary language philosophy in general. Their scoffing was typically directed toward specific claims, such as the private language argument, and they generally had very good arguments to support their criticism. And apart from seeming willfully obscure, Wittgenstein was at best insensitive to the possible role of scientific data in philosophical questions and, at worst, openly hostile to science. Thus the skepticism about conceptual analysis as a method for finding out how things are in the world, including that part of the world that is the mind, was germinating and taking root while I was at Pittsburgh. In Oxford from 1966 to 1969, by contrast, Wittgenstein was revered; and having taken a Quinean position on the a priori and on intentionality, I ended up making a nuisance of myself in Oxford by swimming in the other direction. Paul [Churchland] was most helpful around this time, because he was very straightforward. He thought we ought to approach mind questions in basically the same scientific spirit with which one approached other questions of fact. So instead of flailing around aimlessly analyzing concepts, I had some sense of direction. And so I tried to figure out what I could do in the philosophy of mind if it was not conceptual analysis. Because I am a materialist, it finally occurred to me that there might be a lot in neuroscience that might help answer the questions I was interested in—for example, questions about reasoning, decision making, and consciousness.

During the sixties and seventies most philosophers said, "No, no, no—studying the brain is a waste of time for philosophers. First, nothing much is known anyway; and second, even if a lot were known, it would be completely irrelevant." At the time I was deciding that philosophy seemed like a waste of time, Jerry Fodor's book *The Language of Thought* was published.[2] In this book, Fodor is very clear: neuroscience is simply irrelevant to understanding psychological functions. Neuro-anatomy and neurophysiology have nothing of significance to teach us about the nature of cognition.

I was convinced that Fodor was wrong, because, after all, the brain is the machine that does these things—the thinking and the feeling and the representing, and so on. It would be amazing if it turned out that the organization and mechanisms were not going to teach us anything about cognitive processes. My hunch in this direction also fit well with Paul's ideas about scientific realism and about the nonsentential nature of rep-resentations. So I decided to explore this hunch and find out whether Fodor was more likely to be wrong, or I was.

It was at this point that I realized I needed to study neuroscience sys-tematically. I knew that I needed to know anatomy, and it is next to impossible really to learn anatomy from a book. You must have the ac-tual 3-D object there, and you need to look at it from all angles. The faculty at the University of Manitoba Medical School were very welcom-ing. They thought it was enormously amusing to have a philosopher among the medical students, but they also took my general aims very seriously and wanted to talk about the philosophical issues. So I ended up doing experiments and dissections and observing human patients with brain damage in neurology rounds (from about 1975 to 1984). The first thing that really sunk in is how important it is to see something for yourself, not just read about what someone else says it looks like. Neu-rology rounds were fascinating because all the questions about con-sciousness and cognition were raised in this completely different con-text, and it made me think about everything in a different way.

So in a sense I was lucky. I was disgruntled intellectually so my mind was readied for a new slant, and I stumbled into neuroscience, which was exactly what I needed to give me that new slant.

Slowly I began to realize how much really *was* known about the brain. I was increasingly convinced that the way the brain is put together, the way individual cells operate and talk to one another, had to be impor-tant to macrofunctions like perception, imaging, thinking, and reason-ing. This, shall I say, "tilt" to the neurosciences by people interested in cognition was by no means unique to Paul and me. Later I came to know other people who were moving in the same direction—people like David Rumelhart, James McClelland, Terry Sejnowski, and John Anderson. It was kind of a heretical thing to be thinking at that time, because the standard view was that of Artificial Intelligence: thinking is symbol manipulation according to rules. Thinking is like running a program; and if you want to understand cognition, you want to understand the program. Of course, a program needs hardware to run on, but looking at the hardware will not tell you much about the software it is running.

I was also quite simply captured by neuroscience itself. I thought it was absolutely fascinating. Once I had done as much as the medical students do, I took graduate courses in neurophysiology, and ultimately I became attached to Larry Jordan's neurophysiology lab, which was focused primarily on the spinal cord mechanisms for locomotion. That was a tremendously rewarding experience for me for several reasons. First, it taught me some basic things about neuroscientific techniques, especially concerning what you can and cannot do, and how to interpret results, what to be skeptical about, and so forth. So that was invaluable. Second, it was not the lab I had expected to be in because, after all, it was aimed at studying locomotion, and I was not especially interested in locomotion. I thought the "juicy" topics were memory, perception, consciousness—why should I be interested in how animals move? The lab taught me why I should be. Gradually it dawned on me, mainly by talking to Jordan, that for evolutionary reasons, cognition is closely tied to motor control, since adaptive movement is critical in survival. You might learn a lot about cognitive functions, at least in their general character, by looking at cognition from the perspective of an organism needing to feed, fight, flee, and reproduce (Paul Maclean's "four Fs"). The lab thus shifted my thinking, so the conceptual profit from being in the lab was also valuable. Once I got rolling, I began to have the opportunity to talk to people like Rodolfo Llinas and Francis Crick,[3] who taught me simply an enormous amount—both about details and about how to think about the problems: which ones are interesting, which ones might be tackled successfully relative to experimental limitations.

What was the idea behind this attraction to neuroscience at a time when people were still stuck with the idea that the architecture did not matter?

I suppose I was bloody-minded, but it just seemed obvious to me that the brain and its mechanisms and structure were relevant. It seemed obvious to me that you can learn a lot about visual perception by understanding the eye, the retina, and so on. One of the books I read early on that I think had a major impact on me was Richard Gregory's book, *Eye and Brain.*[4] It was a very important book because on the one hand he discussed the psychophysics of visual perception—the phenomenology and behavioral parameters of perception of motion, color, depth, and so on—but he also tied it to the physiology, to the extent that it was known. In Gregory's book, you could begin to see—but only *begin*, because so much is still unknown—that the physiology can tell you a lot about how we see. Consequently, when I read the functionalists' claim that the brain is irrelevant, I thought it was a bit outrageous. It did not seem real to me, in a funny kind of a way. And of course I did not have any stake in the orthodoxy, because I was not part of the "inner circle" in Boston. So I could afford to say, "Ha! That's a bunch of baloney!" and strike out on my own. In Manitoba I had a truly marvelous freedom to move in whatever direction I wanted to.

From way, way back, my take on how the world works was rather mechanistic, in the sense that I generally assumed that if you wanted to understand how something worked, you should try to take it apart and see how it runs. Maybe it's because I grew up on a farm, and as a kid I had to solve, as a matter of daily life, a lot of practical problems. If an irrigation pump did not work or a cow was having trouble calving, we had to figure out how the thing worked and do what we could to fix it. It seemed reasonable to me that taking something apart and finding out the causal connection between the parts would reveal a lot about how the thing as a whole worked. This seemed true not only for artifacts such as engines but also for the hearts and gizzards that I took apart after cleaning the chickens. I also knew that simply taking a gizzard apart was not enough—that the inner structure and all the little stones could give you a clue, but you also got clues from seeing what it was connected to, the relation to the gut and to the mouth, what the chicken ate, and so forth. What Jerry Fodor's book did for me was crystallize the functionalist assumption—namely, that the brain was at the implementation level and cognition was at the program level. By putting his views so clearly, Fodor allowed me to begin to feel my way toward a different approach. I could understand the basic point that the same program could be run on different machines, but I suppose I was always dubious that cognition could be understood as analogous to a program. You must understand that in the beginning most of my ideas were pretty hazy, and I guess lots still are; but I just could not see how knowing about the brain could fail to shed light on cognitive processes.

Paul and I discussed such questions endlessly. In the summer of 1971 we were crossing the prairies in an old car to reach British Columbia, a trip that takes about four days. Shortly after we entered the province of Saskatchewan, Paul began to try out a new idea of his, namely that representations in general might not be at *all* like sentences. As Paul sized it up, we use a theory to explain behavior in terms of beliefs *that p* and desires *that p* and so on, but this may be as mistaken as "commonsense" ideas about the earth being stationary and spirits causing the tides. He argued that sensory representations and maybe even most cognitive representations were not best understood on the model of sentences such as *p and q*. Hence content, intentionality, meaning, and the whole pudding had to be rethought from scratch. As a matter of principle I took the opposite view, and at first I could not really understand what on earth he was getting at. We argued and argued about it during the entire four days, and by the time we reached Vancouver, I thought, I bet he's right.

Davidson is wrong, Fodor is wrong, Putnam is wrong. Most people are looking at the problem of representation in the wrong way because they are looking at it as tied to language, as languagelike. Language is probably not necessary for representing the world, and probably lots of what we call reasoning does not involve anything languagelike either.

Since we seemed to agree on what was wrong, the next question was, What is right? How *does* the mind-brain represent the world? How can we go about finding out how it represents the world? What will such a theory look like? We both knew that a priori reflection would not tell us much. And we agreed with Quine that there is no first philosophy, no fundamental difference between philosophy and science. So against the backdrop of these rather radical ideas, studying neuroscience was almost inevitable. In many ways it was Paul's willingness to pick something up, turn it around, and look at it from an entirely different perspective that was absolutely crucial in encouraging and making respectable my naturalistic impulses. We also just thought it was enormous fun.

Which sciences do you think play an important role in cognitive science?

Experimental psychology, linguistics, and psychophysics, because they help specify what exactly are the capacities of humans and other animals, and they can also say something about the interconnections between those capacities. Neuropsychology, because by studying patients with brain damage, we get a far deeper perspective on the scope and limits of psychological capacities. Neuroanatomy, because you have to understand the basic components of nervous systems (neurons) and their specific patterns of connectivity. Neurophysiology, because you have to know how neurons work, and how neurons respond as a function of what other neurons are saying to them, and what transmitters cause them to do, what hormones and modulators cause them to do. Computational neuroscience, because a network model constrained by neuroscience is an important tool for seeing what interactive and dynamic effects can be generated by a circuit. Modeling is an important source of ideas and hypotheses, but it is in no sense a replacement for experimental neuroscience or experimental psychology. Modeling is important because representations generally appear to be highly distributed across many cells, and to understand how properties emerge from networks, computer models can be useful.[5] Developmental psychology, because it is critical to understand how capacities come into existence and change over time and on what their development depends. Developmental neurobiology, because we need to understand how the timetable for capacity development links up to the timetable for brain development. Molecular biology, because we need a perspective on genetic specification and plasticity. Philosophy, because we badly need to synthesize and theorize and ask the questions everyone else is either too embarrassed or too focused or too busy to ask.

Are there neuroscientists currently involved in cognitive science?

Absolutely. In many domains, particularly memory, sensory systems, and attention, there are projects that involve many of the aforementioned fields. For example, I can think of at least five vision labs that do both the psychophysics of vision as well as visual system physiology and anatomy. Patricia Goldman-Rakic[6] is a striking example of a neuroanatomist whose physiological research into working memory has uncovered particular cellular responses during a task requiring holding of spatial information for a few seconds, and these responses appear likely to be an important piece in the puzzle of working (short-term) memory. Fuster does related work, but with working memory for visual patterns rather than spatial location.[7] Mark Konishi's work on the auditory capacity and auditory system of the barn owl is surely one of the stunning success stories of cognitive neuroscience.[8] Several graduate students in the Philosophy Department are doing joint work in neuroscience, and at least six others are doing joint work with psychologists and some with computer scientists. It seems to be just part of the ethos of UCSD that cross-disciplinary projects are encouraged.

Was it coincidence that you are here, where neuroscience is so strong?

The cognitive neuroscience community here was an important attraction as we were considering moving to San Diego. It is large, diverse, and lively. At some point or other, everybody passes through, so there is lots of contact, and that means there is the opportunity to learn a lot and keep abreast of new developments. There is also considerable interest and activity in computational neuroscience, so I feel very much at home. I am an adjunct at the Salk Institute, and I currently work in Terry Sejnowski's laboratory. The lab has tea every afternoon, and various people drop by—often Francis Crick, and some of the visual psychologists, such as Ramachandran—and we discuss everything from consciousness and free will to apparent motion to the NMDA receptor.

What, in your opinion, is the role of Artificial Intelligence in cognitive science now?

I like to know whether a computational hypothesis is probably true or not. Consequently, to the degree that a model is constrained by the facts, I am more interested. Speculative models can be fun, and they may have theoretical significance. If, however, a model aims at describing how some mind-brain performance is accomplished and it is constrained only at the behavioral level, then given the vastness of computational space, we will not know whether the model is even close to being right. The point is that there are many, many ways of performing a task such as motion perception, but my interest lies in how brains really do it. Consequently, my work tends to focus on models of fairly low-level phenomena, such as vestibulo-ocular movement, motion perception, perception of 3-D depth, and so on. And I think what we understand at these levels will help us grasp the principles underlying what is going on at higher levels, such as reasoning and language.

In some respects, network simulations are all too easy. Anybody can go to the psychological section of the library, pull out a journal, build a network model to mimic the curve in some paper or other, and imagine that a psychological property has been explained. But the network model is not very interesting unless we have some idea about whether it tells us something true about how the brain really computes or represents. For example, a model of language comprehension might characterize semantic memory simply as a list. Now, from a neurobiological point of view, we know that memory in brains cannot be accurately characterized as a list; so although the list model might be behaviorally adequate, we are pretty certain it is wrong. And it becomes very interesting to ask how a more neurobiologically plausible memory model will constrain a theory of language learning and language use. Several postdoctoral fellows at UCSD have been tracking down that question, and, from their perspective, what it is to learn and understand a language begins to look very different from how the conventional wisdom lays it out.

What is your relationship with Artificial Intelligence? Or, to put the question differently, are Artificial Intelligence workers really interested in how the brain works? Has there been a change over the last years in the sense that they are now more interested in philosophy and neuroscience?

It depends on what you mean by "Artificial Intelligence." Since about 1986, when McClelland and Rumelhart published the two-volume set on PDP (parallel distributed processing), lots of people have become keenly interested in connectionist models of information processing. They are exploring such matters as the properties of feedback models of different configurations, models that develop modulelike specializations, models that can handle compositionality, deduction, and so on. There is also a major effort underway to address the constructionist problem—that is, the problem of how to make devices that interact with the real world in real time, as opposed to simulating a function on a computer. Carver Mead, for example, builds artificial retinas and artificial cochleas using analog VLSI (very large scale integration) technology, by building chips with hundreds of thousands of transistors.[9] Others are interested in technological applications of neural nets, some in the mathematics and statistics of neural nets, some in the range of ways of training neural nets. But conventional Artificial Intelligence, of the kind exemplified by the work of Roger Schank or Allen Newell, for example, is less interesting from my point of view than neural net research because—so far, at least—it does not appear to be very revealing about how the brain really represents and computes. There is a role for simple models and a role for realistic models, and researchers can learn from one another. It looks very likely that many kinds of representations are patterns of activation across a set of neurons (activation vectors) and that a good bit of computation consists, roughly speaking, in pushing vectors through matrices.

How do you see the history of Artificial Intelligence in retrospect? What were the failures, the successes?

Artificial Intelligence in the conventional mode gave the following idea a good run for its money: perception, deciding, figuring out, replying, and so forth is (or is very, very like) running a program on a serial digital computer. It essentially involves symbol manipulation in much the same manner that running software involves symbol manipulation. Here are some disanalogies: the brain is not a serial machine but a parallel machine. The brain is not digital but an analog machine. Memory in nervous systems is not stored in a single location (memory bank), but learning and information processing appear to take place in the very same structures, and there appear to be several different but connected memory systems. The architecture of the nervous system is plastic, and it changes continuously. Neurochemicals can affect the system in many ways, and one way is to effectively rewire circuits. Nervous systems, unlike serial, digital computers, can sustain some damage without grinding to a halt—they are fault-tolerant. Moreover, algorithms are not independent of architecture; a given algorithm may fall gracefully onto one architecture but be ponderous and clumsy on some other architecture. If we knew how the brain really does work, we could probably make an artificial brain out of silicon, for example.

If we do not know how the brain works, then the structure and organization is likely to provide indispensable clues. We are unlikely to find out much about the nature of cognition if we restrict ourselves to behavioral-level data and to writing programs whose goal is to simulate the behavior of humans. By analogy, to really understand how physical traits are transmitted from parents to offspring, one has to understand the basic structure and organization of DNA. We were unlikely to solve this problem simply by writing programs that when executed had as output values standing for offspring with certain proportions of their parents' characteristics. AB (Artificial Biology) cannot dispense with finding out the physical mechanism.

Many people say that the business of cognitive science is with the physical symbol hypothesis. How could one combine it with neural nets?

I do not think that is the business of cognitive science. The business of cognitive science is to understand perception, emotion, memory, learning, and so forth. It is an empirical question whether very much of this involves symbol manipulation or whether it involves processes of a very different sort. One major effort in neuroscience is to untangle the problem of sensory coding. Konishi's work on how the barn owl's brain uses the delay between the time a signal reaches the ear as a basis for computing location in the horizontal plane and interaural amplitude differences to compute location in the vertical plane is a high-water mark. On the basis of the anatomy and physiology, he has a detailed hypothesis about how the coding for elevation and azimuth is done, and how the two are integrated, how noise is reduced, and so forth. And Mead is modeling the Konishi data in analog VLSI. So there is an example where we understand pretty well what it is for a pattern of brain activity to represent a certain location in space, and how to represent "to the left of" and "to the right of" and do other relational problems. Along similar lines, aspects of representation of motor plans in terms of patterns of neuronal activity are being figured out. My hunch is that quite a lot might come of sneaking up on language via nonlinguistic forms of representation. Consequently I am especially interested in the child development and language acquisition research, and I am also keenly interested in the work of the Rumbaughs on the bonobo chimpanzees. These animals are very different from the standard chimpanzee, and they appear to be picking up and using language in much the same way that human children do. Because we understand very little about how language is accomplished by the brain, it is probably a good idea if research is very diverse rather than all in one basket.

What do you think about the Turing Test?

I don't think about it very much. I assume that if someone made a robot with sensory-motor control much like my own, including language, then that would be pretty interesting, and I would probably have as good reasons for saying it is conscious as for saying you are. Indeed, if it is to be behaviorally comparable to a human, it will probably have to have a similar basic computational configuration (parallel architecture, for example) to perform in real time, and it will likely have to have a similar functional organization—our kinds of attentional mechanisms (however many there are), short-term memory (however many kinds of these there are), working memory, long-term memory for episodes, semantic memory, and so on. For something to have these capacities might also require, if it is to function efficiently and in real time, the capacity for awareness—sensory awareness, at least. Notice, however, that my comments are full of "probably's," "might-be's," and assorted other qualifiers. The difficulty is that we do not really know very much about what it takes for sensory awareness, attention, and so forth, so the Turing experiment is desperately underdescribed. Consequently, it is easy for people to have unshakable but opposite convictions and to waste a huge amount of time with countless imagined scenarios. The scenarios are all hopelessly underdefined, so no one can make significant progress. I find that rather unrewarding. I would prefer to try to make a robot that really can do something, such as, for example, Rod Brooks's insect robots, which can really walk on uneven terrain, follow warm things, avoid obstacles, and so on. With a simple beginning, we can learn some basic principles and then build on those. I am fascinated by the idea of building more humanlike things, but I do not have the faintest idea how to do that—there are too many things we have yet to learn from the brain about how to make a visual system, how to make a system adaptive, and so on. We are learning, but evolution has had many millions of years to luck onto workable solutions. It may take our brains a bit of time to sort it all out.

For you the most important question would be the question of consciousness?

What I think about most are the problems of sensory coding and of sensory-motor control, and the related questions concerning adaptation, decision making, the role of the appetite, and the emotions. I sometimes wonder if the mechanisms of consciousness will sort of fall out of the wider theory of neural representation and computation, once it is developed a bit further, in the way that various things began to fall into place once the structure of DNA was understood. You have to be careful, of course, because even after the double helix, an absolutely staggering amount had to be—and is still being—figured out. But the discovery provided the sort of organizing point to unscramble how proteins got made, what exactly enzymes did, what RNA was for, and so on. Now, with the brain, I rather doubt that there will be a comparable piece of the puzzle such that once we have that piece, what to do next will be clearer. The brain is really very complex, and I think we shall have to understand a very great deal at the level of the single cell before we can really make the cards tumble. Nevertheless, the way I tend to envisage it, the nature and mechanism of sensory awareness will not be such a big mystery once we get a better fix on the computational hierarchy, the role of back projections, and how time is managed by nervous systems.

What are the most important current problems that Artificial Intelligence has to overcome?

For neural network research, there are a number of puzzles. One concerns scaling: how can we make nets that are as large as a rat's brain; how does specialization of brain regions work; how is integration of information from specialized pathways achieved? We also need to understand a lot more about net-to-net interactions and about oscillators and coupled oscillators and what they can do for a nervous system; and we especially need to construct neural nets with dynamics and with sensitivity to real time, not just to order of events.

Recommended Works by Patricia Smith Churchland

Neurophilosophy: Toward a Unified Science of the Mind-Brain. Cambridge, Mass.: MIT Press, 1986.

(With T. J. Sejnowski.) *The Computational Brain.* Cambridge, Mass.: MIT Press, 1992.

(With T. J. Sejnowski.) "Neural Representation and Neural Computation." In *Philosophical Perspectives*, vol. 4: *Philosophy of Mind and Action Theory*, ed. J. E. Tomberlin. Altascadero, Calif.: Ridgeview Publishing, 1990. Reprinted in *Mind and Cognition*, ed. W. G. Lycan, pp. 224–51. Oxford: Blackwell, 1990.

(With T. J. Sejnowski and C. Koch.) "Computational Neuroscience." *Science* 241 (1988): 1299–1306. Reprinted in *Encyclopedia of Neuroscience.*

Notes

1. W.V.O. Quine, *Word and Object* (Cambridge, Mass.: MIT Press, 1960).

2. J. A. Fodor, *The Language of Thought* (Cambridge, Mass.: Harvard University Press, 1975).

3. E.g., A. Pellionisz and R. Llinas, "Space-Time Representation in the Brain: The Cerebellum as a Predictive Space-Time Metric Tensor," *Neuroscience* 7 (1982): 2249–2970; F. Crick and C. Asanuma, "Certain Aspects of the Anatomy and Physiology of the Cerebral Cortex," in D. Rumelhart, J. McClelland, and the PDP Research Group, *Parallel Distributed Processing: Explorations in the Microstructure of Cognition* (Cambridge, Mass.: MIT Press, 1986), vol. 2, pp. 333–71.

4. R. Gregory, *Eye and Brain* (London: McGraw-Hill, 1966).

5. P. S. Churchland and T. Sejnowski, *The Computational Brain* (Cambridge, Mass.: MIT Press, 1992).

6. P. S. Goldman-Rakic, "Circuitry of Primate Prefrontal Cortex and Regulation of Behavior by Representational Memory" in *Handbook of Physiology: The Nervous System*, vol. 5, ed. V. B. Mountcastle, F. Bloom, and S. R. Geiger (Bethesda, Md., 1987).

7. J. M. Fuster, *The Prefrontal Cortex: Anatomy, Physiology, and Neuropsychology of the Frontal Lobe*, 2d ed. (New York: Raven Press, 1989).

8. E.g., M. Konishi, "Centrally Synthesized Maps of Sensory Space," *Trends in Neuroscience* 9 (1986): 163–68.

9. C. Mead, *Analog VLSI and Neural Systems* (Reading, Mass.: Addison-Wesley, 1988).

PAUL M. CHURCHLAND

Neural Networks and Commonsense

I would say that I have been in cognitive science my entire career, because I think a philosopher of science, someone who works on epistemology, is a cognitive scientist. For a long time, philosophers were the only cognitive scientists. Many philosophers pursued their interests in a nonempirical way, and I think that was unfortunate. But I have always been inclined to do my philosophy in a way that responds to empirical data, so I think I have been doing cognitive science for twenty years. But it is more rewarding now than it was twenty years ago, because there are so many interesting things coming from neuroscience and from Artificial Intelligence.

A good and honest answer to the question of how I came to cognitive science would be the attraction of the cognitive science group over in the psychology department (at the University of California at San Diego), centered around David Rumelhart, Francis Crick, and David Zipser,[1] as well as other people. When Pat (Churchland) and I arrived here seven years ago, it was an interdisciplinary group that we joined, and it has had a major influence on us since then. But we were already interested in those things when we arrived, and this is part of the reason why we came.

You said that it is now more interesting because of input from other sciences. Could you explain what cognitive science is in your opinion? What are the links to other sciences?

Many and various. It is difficult to summarize them. I can give you the main ones that are salient here. One of the major connections is with neuroscience. We have a large population of neuroscientists and neurologists and neuropsychologists here, and so we have access to interesting information about memory, clinical cases, and perceptual losses from the hospital. There are people working on ERP research (evoked response potential research, using EEG technology). We have people

Paul M. Churchland studied philosophy at the Universities of British Columbia (Canada) and Pittsburgh, specializing in the philosophy of mind, perception, and neuroscience as well as in epistemology and history of science. He earned his Ph.D. in 1969 from the University of Pittsburgh. From 1966 onward he taught at different universities in the U.S. and Canada and became full professor at the University of Manitoba in 1979. In 1984, he moved to the University of California at San Diego, where he has been Professor of Philosophy since.

working on vision, on learning—the neuroscientists have quite an impact on what goes on here. We have the group in cognitive science itself. They are doing simulations, exploring neural nets and their properties in learning and behavior. Psychology contributes some because we have a wonderful group that works on vision. The cognitive science group itself came off as a part of the old psychology department. We have mathematicians, computer scientists, anthropologists, philosophers. The lines of interaction are not so well defined right now because they are changing so quickly. Computer science and neuroscience are going to have the biggest impact over the next five to ten years, I think, because we are doing such wonderful things—we can do artificial models. And neuroscience is making interesting advances in both the theoretical and the practical part.

Why has San Diego become this center of cognitive science?

Partly it was coincidence, and partly it was design. It was coincidence that we had a very strong neuroscience community. It was coincidence that we had a psychology department with people who were doing network modeling. That had been an underground research program for a long time. People did not think it was worthwhile as a kind of Artificial Intelligence or psychology. Very few places did that at all. We had David Rumelhart here, and earlier Geoffrey Hinton and James McClelland and David Zipser[2]—it was an accident that they were here rather than somewhere else. It was an accident that philosophers interested in learning theory and in neuroscience—Patricia Churchland and myself—came here: partly for the weather, partly because of the neuroscientists, partly because we knew there was a cognitive science group. We had visited them and knew they were good. We did not know *how* good they were— they turned out to be wonderful. It was also a coincidence that back propagation was invented, or reinvented, here and that some of the earliest research was done here.

It is design, of course, that we brought in certain people as soon as there was a nucleus. We brought in Terry Sejnowski, who has contributed enormously and will contribute more to the development of this area. To summarize: it is partly luck and partly design. There were good people here before they realized they had something in common. And as soon as they realized it, they brought in more good people.

What is your opinion about the contribution of Artificial Intelligence to cognitive science? How should it contribute, and what is its importance in cognitive science?

The question should be divided into two parts: the first is what Artificial Intelligence *has* contributed to cognitive science over the last twenty or thirty years. The answer is: a nontrivial amount, but not near as much as was hoped or expected. It was one of the respects in which good old-fashioned Artificial Intelligence was a failure. The contribution of standard architectures and standard programming Artificial Intelligence was a disappointment.

The answer to the second part of the question is: Artificial Intelligence will contribute enormously to cognitive science in the future, now that the architecture of the machines we are exploring is more brainlike, given that the hardware is more similar to the brain and that the behavior is more similar to behavior in living creatures. Then the discoveries we can make in Artificial Intelligence will contribute enormously to cognitive science. Artificial Intelligence has a vital, central role to play in cognitive science, because we can experiment and explore the properties of neural networks more easily in artificial systems than we can in natural ones. We can never give up exploring the natural ones—that is the empirical data—but as an adjunct, Artificial Intelligence is going to be very important.

What is the very idea of Artificial Intelligence? Would you include connectionism in Artificial Intelligence?

Indeed I would. Connectionist architectures are going to assume the central stage in Artificial Intelligence research in the next five or ten years, if they have not already. In general, I think of Artificial Intelligence as the discipline that attempts to re-create all the cognitive talents that we find in living creatures. I don't confine this to humans. The animal kingdom holds hundreds of thousands of models and examples of cognitive achievements. We have not been paying enough attention to natural creatures in our attempt to create artificial versions of them. With the advent of network research we are back to paying a more respectable amount of attention to the real phenomena, and we are profiting from it.

How could one define the central theoretical assumptions of Artificial Intelligence?

Its aim is to re-create and understand the cognitive abilities of living creatures generally. We have an old approach that has shown itself a failure, and a new approach that has shown itself promising in many respects. The old approach used the standard architectures and wrote programs. It was a wonderfully easy way to do interesting things very quickly, but it ran into trouble. While many of the standard architectures were very large in their memories and very fast in their processing, they could not do the number of computations that the biological brain does in the same amount of time, because they were serial. We need massively parallel architectures if we want to re-create what biological creatures do simply and easily. Now that we are attempting to do that, things are going to move very much faster.

However—to anticipate your next question—it is true that most of the simulation of neural networks has taken place on standard architecture machines, which are slow. There is a virtue to this: it allows us to explore different architectures, because you can simulate one on a computer more easily than you can build a parallel architecture. But this has produced a problem: we are unable to simulate networks with more than about 10^3 units, because the computational problem becomes too hard. Training takes too long on a standard machine. An artificial network with 10^3 units is very, very tiny, because you have at least 10^{11} units in your brain, which is an enormous difference. We've got to break through this wall, and in order to do so, we have to make artificial neural networks that are real networks, not simulated. We have to make VLSI chips that re-create parallel networks, rather than try to simulate them on standard machines. That is a breakthrough that needs to be made in order to continue to explore neural nets and their promise, because at the moment there is a size limit on how big a network can be simulated.

There is a view that cognitive science is mostly concerned with the manipulation of physical symbols—the physical symbol systems hypothesis. According to this view, how can neural network modeling be combined with cognitive science?

One could, but what one has to do first is give up the old idea in its old form. It is an idea that is tied to the standard architectures and the standard programming techniques with which we are familiar. We need to get away from the idea that we are going to achieve Artificial Intelligence by writing clever programs. There is a sense in which neural networks manipulate symbols, but it is not the familiar sense. You can think of a neural network as manipulating its input vector. That is a symbol, if you like, which represents the input, and that symbol is processed; it goes through a matrix of synaptic weights, it produces another vector that you can regard as another symbol. You can see the neural networks as symbol processors, but they are processing a very different kind of symbol from what we are familiar with, and they are processing them in very different ways from what we are familiar with. I think the honest thing is to admit that a great change has taken place here. The old symbol processing paradigm will never go away entirely, and it should not, for it is useful for many things. But it is not going to be useful for understanding how natural intelligence does its job or for re-creating high-powered Artificial Intelligence.

Could you outline the contributions of philosophy and Artificial Intelligence to this common undertaking? What is the task of philosophy in cognitive science, and what is the task of Artificial Intelligence?

I would guess that philosophers are going to play a relatively minor role, as philosophers play a relatively minor role in any scientific endeavor that has gotten on its feet. Philosophers help to get endeavors on their feet, but once they are on their feet—that is, once they are interacting with the experimental data—philosophers recede into the background. In this case, it is so entertaining that it is going to keep the attention of philosophers, and there is so much methodological argument going back and forth that philosophers will play a useful role, but the interesting science will come from the theorists and the experimentalists.

As to where theory and experiment are going to come from, there are going to be two or three major sources of experiments: the first main source of interesting experiments is the neurosciences. We can now learn things with our experimental techniques that we could not learn five or ten years ago. We are going to learn a great deal about how natural creatures do it, and we can import that into the artificial area.

Another area of experiments will be to explore what neural networks can do by training them up on various kinds of problems and by using different kinds of learning algorithms. Experiments of this kind have caused this explosion in the last few years within cognitive science.

A third area is going to be engineering exploration, as we explore alternative ways of realizing these parallel processing concepts. We can do it electronically, we can do it optically, we can implement learning algorithms in novel and interesting ways that engineers will think of. Hardware is one of the areas of exploration, because the new style of Artificial Intelligence looks for new architectures. So there is room for exploration of new technologies to implement these kinds of architectures.

So there are three areas of experimentation. So far as theory goes, we will get some theorizing from philosophers, because we are still addressing things psychologists have not done very well: problems of consciousness and rationality and the nature of science itself. But theorizing will come from all over. Mathematicians will contribute some interesting theories about the general properties of neural networks. Psychologists will contribute interesting theories about how these networks relate to what creatures do, how they drive sensory-motor coordination, how they provide motor control. This is a very complex matter that we are only beginning to address. It was a failure in the old-fashioned Artificial

Intelligence, because motor systems are too complex to deal with in real time. Neural networks can deal with them in real time very easily, so I think we will have some advances there, too.

In your experience, were there difficulties to overcome for a real interdisciplinary approach?

Philosophers are supposed to learn the business of other disciplines. Our job is a synoptic job, a unifying job. Some philosophers saw fundamental problems in the way old-fashioned Artificial Intelligence was working. I don't want to blame only Artificial Intelligence workers— many philosophers were working with exactly the same assumptions, people doing inductive logic and epistemology and philosophy of science. They gave the traditional Artificial Intelligence lots of support. But just as traditional Artificial Intelligence has failed, much of traditional philosophy has failed, too. We are forced to new theories of epistemology, new theories of perception, new theories of the nature of mind.

There are areas where we could seem mistaken in fundamental assumptions, and there is one I would really like to emphasize, to go back to good old-fashioned Artificial Intelligence: the assumption that the architecture did not matter. People knew that Turing machines could, in principle, compute any function. And those premises were true then, and they are true now, but they are deeply misleading. Because even though in principle a Turing machine might be able to do what you or I or a squirrel or a mouse could do, it might be that when you are limited to that kind of architecture, it would take a Turing machine 10^{89} years to do it. In principle it could, but it cannot do it in real time, it cannot come even remotely close to doing it in real time.

The claim that architecture did not matter was a justified claim, because in principle it did not matter. You could compute any function with an architecture like that or with a von Neumann architecture, to come a little closer to reality, but you could not do it in real time. If you want to do it in real time as mice and birds and creatures who have to live in the real world do it, then you have to go to massively parallel architectures. The architecture does matter profoundly, even to do the job at all.

And once you get to the new architectures, you see that they have many other properties that you did not suspect, properties that you had trouble re-creating in the old architecture: generation of similarity spaces, sensitivity to high-dimensional similarities, categorization, facial recognition, motor control. Discovering that the architecture matters was a major breakthrough. That is one of the reasons why the neurosciences are going to have a bigger impact over the next ten years than they have had over the last thirty.

Why is neural network modeling becoming so important now? *You mentioned that there had been a project going on for years. And the idea is not new; it started with Rosenblatt[3] way back.*

Yes, it started with Rosenblatt. The answer is a little bit technical, but I will give you the fast answer. Part of the reason why the neural network paradigm went into eclipse is that it lost the early battle against standard program-writing Artificial Intelligence. The early networks were limited; they had only two layers; they typically had linear units. Such networks are indeed very limited in what they can perform, as Minsky and Papert quite correctly pointed out at that time.[4] People generalized from that conclusion, and that was a mistake. But it was not just *that* mistake.

While everybody could see these limitations, you could write programs on the bigger and faster machines—there was a new bigger and faster machine every month, it seemed—and you could get much more interesting results much more quickly without building hardware, without studying the brain. It was easier and, in the early stages, more rewarding. So it is no surprise that the community of Artificial Intelligence researchers all went over to that paradigm. They were not being wildly irrational; they were rational.

Traditions acquire a momentum, however, and it is unfortunate that opinions swung so hard. There were only a few people pursuing their neural network research, and it took a long time for people to realize that if you made a three-level network—or four or five, but in any case more than two—and if you gave the network nonlinear response properties, you could get networks like that to do things that were not limited in the way Minsky and Papert said. The real breakthrough came when back propagation was invented, not because that is a particularly wonderful learning algorithm from either a biological or a technical point of view but because it was an easily applicable and ruthlessly efficient one. It allowed people to train up these networks—which can get quite complicated pretty quickly—fairly easily and then see what they could do.

Once we had found an easy way to train neural networks, we soon discovered that they could do an amazing variety of things. It also allowed us to examine the networks once they were trained and to see how they were doing it. That gave us interesting insights almost immediately. I am thinking here of the analysis of the hidden units, when we discovered the categorical structures that were spontaneously generated inside the network. It was very exciting, and back propagation made it possible.

In the long run, back propagation will fade into the background and will be replaced by better learning algorithms, even in Artificial Intelligence. We will discover that the brain uses quite different learning algorithms. When we do "natural" cognitive science, back propagation will not be too important, but it was very important to make the break-through happen. It gave us a way to manipulate the networks to a degree we had never been able before, and we learned exciting things very quickly.

The claim that the architecture does not matter was connected with the claim that this was the way to overcome Cartesian dualism. But in a certain way, it is another kind of dualism.

That is right. We have a name for it, my wife and I: we call it "theory dualism." They claim that there is a special level of conceptualization or understanding. You could understand how a cognitive system worked at a level that abstracted from the architecture and the physical details. And since you could realize the same abstract thing in different ways, it did not matter what the lower-level stuff was. Well, it does matter—it matters very much in *fact*. It may not matter in *principle*, but the principle has turned out to be so unrealistic that we have to push it aside. It is amazing how much insight we have gained in such a short time once we started paying attention to alternative architectures. It is a very hard lesson here that we are nevertheless going to learn.

In your book Matter and Consciousness[5] *you propose eliminative materialism as a new theory of mind that is able to overcome the weaknesses of functionalism. Are neural networks the practice that goes with eliminative materialism?*

I think so, but I don't want to go too far. The first claim that elimina-tive materialism wants to make is that the important kinematical and dynamical factors inside the brain—the things that really explain human behavior and human development over time—may not be desires, be-liefs, fears, hopes, and wishes, all the familiar things we ascribed to the mind. Those may not be a useful way to conceive what is kinematically and dynamically important. They may be a completely false theory, or a superficial one. It might turn out that they are not important. On this score, we already have dramatic evidence.

On the neural network models, the important representations are high-dimensioned vectors, vectors in the sense of a pattern of stimula-tion or activation across a large population of neurons. Those are the representations—not sentences, not propositions, not beliefs, and so on. The important computation is transformation of one pattern into an-other pattern. The parallel would be that the important computations for beliefs, desires, and so on are inferences from one to the other. That is what commonsense thinks, what traditional Artificial Intelligence thinks. The new view says that activation patterns are transformed by banks of synaptic connections. That is a dramatic new conception of the kinematics and dynamics of what is going on inside the brain. It owes nothing to the old views.

We are now in a position where maybe we can understand how the brain works, how cognition works, without ever using the old notions. It does not mean we have to give up the old notions; they may be explainable in terms of the new. Maybe we can explain what beliefs are, and they may turn out to be a small part of our cognitive activity, but a real part. And maybe we won't explain them, and in this case, we misconceived ourselves. It is too early to say yet. But the case for eliminative materialism continues to grow.

One argument against eliminative materialism is that it eliminates consciousness, too. What about qualia?

In the case of qualia, I am disinclined to be an eliminativist. Our current neural theory explains qualia surprisingly well. Qualia, namely the features of our sensations, can be explained in terms of these multidimensional spaces, and a particular quale is going to be identified with a particular vector. We will be able to explain qualia very nicely. I don't think we are going to eliminate qualia. We keep qualia, but we have a deeper understanding now than before. This is an example of how something need not be eliminated but can be explained and therefore kept.

In the case of beliefs and desires, this may happen too, and I will be happy with it. I don't believe in eliminative materialism because I want it to be true. If beliefs and desires do get explained, this will show that the identity theory is true, that reductive materialism is true; and this will make me just as happy, because what is interesting here is getting the truth, getting a unified theory of how the brain works. But the story is not over yet. We don't know how and whether we can explain beliefs and desires in terms of transformations among high-dimensional vectors. Maybe we will learn how; in that case they will stay. Maybe they won't; maybe there is no way to explain them, in which case—at least in the long run, in science, in the laboratories—we will stop using that vocabulary. We may go on using it in the marketplace and over the dinner table.

In your book you mention that the mainstream is functionalism and the case for eliminative materialism is not so strong right now. Has there been a change in the meantime?

What is happening is that resistance to the idea is fading. People now see it as a possibility. Before, people were inclined to say it is not possible and too radical an idea. People are not frightened any more because they can see it would not be so bad to eliminate our old vocabulary, if the reason for eliminating it is the discovery of a better one. When eliminative materialism came up first, we did not have a very clear conception of what an alternative framework would be. But neuroscience and connectionist Artificial Intelligence and cognitive psychology are starting to produce an alternative conceptual framework, and we can see what it might look like. So the idea is no longer so frightening. Many people now regard it as a possibility, at least. But you are not going to find people who are proud eliminative materialists—not even me—because it is an empirical question, and it will only be decided by more research, research that is going on right now.

You mentioned earlier that the first networks were beaten by traditional Artificial Intelligence. What would be a measure of success? Do you think the Turing Test would be one?

I don't think the Turing Test is a very deep or useful idea. Of course we want to simulate what cognitive creatures do, and one can ask what a test for a successful simulation would be like. The Turing Test invites us to look at external things, to look at behavior. Actually, it is quite ruthless in that regard: it asks us to think of an interaction as being carried out on a typewriter and a printout. We should instead be looking at a test for adequate simulation that works inside the cognitive creature. We don't want to re-create only the external behavior; we also want to re-create the causal processes that produce it. And we are going to succeed, because now we are looking inside the box to see what architecture is there, how it functions.

The Turing Test remained popular just as long as good old-fashioned Artificial Intelligence could convince us of the idea that architecture does not matter, that it was external functional form that mattered. Of course the external functional form matters, but if you want to reproduce it effectively you have to look inside the machinery that is sustaining the cognitive structure. I am not impressed by the Turing Test. I think it will be replaced by more penetrating criteria for successful simulation.

What is the reason for the popularity of the Turing Test? On the one hand, cognitive science tries to look inside the brain; and on the other, a behavioristic test is still popular, especially with Artificial Intelligence people themselves.

There are a number of reasons. One is that it is easy to understand. Another reason is that it does push aside an area of potential disagreement between people. The question "Could a machine be conscious?" immediately raises the question "Does it have thoughts inside?" People have religious views, metaphysical views, all sorts of views that make it impossible to discuss things effectively. The Turing Test has the advantage of locating the argument in an area where people can agree. Certainly behavior is relevant to the question of intelligence, and the test caught on because Turing posed the challenge: "Look, it can produce all the behavior that is relevant to intelligence—how could you withhold the agreement that it must have intelligence inside? What could be more relevant or important than the external behavior?" That is a forceful idea, and I am not surprised that the Artificial Intelligence tradition seized on it.

However, since it turned out that the Artificial Intelligence tradition has been remiss in trying to understand what intelligence is—their models have not explored the brain and how the brain works to the degree that they should—I think it is an irony that the breakthrough, neural networks, should come at a point that shows that the Turing Test is not a very deep idea.

Let me mention another famous Gedankenexperiment *in that area: what do you think about Searle's Chinese Room Argument?*

I am of two minds about Searle's thought experiment. I think he is right in criticizing the view of traditional Artificial Intelligence, the idea that we can re-create intelligence by an abstract symbol manipulating machine. It is not a promising research strategy. However, I don't think his particular argument is effective. If I were a good old-fashioned Artificial Intelligence researcher, I would not be much moved by Searle's argument. I might be moved by other arguments, but not by the Chinese Room Argument. It is a peculiarly philosophical argument that derives its force from its appeal to our preanalytic intuitions, to our prejudices, if you like. And I think our prejudices need to be reeducated. But I agree with Searle that traditional Artificial Intelligence has to be replaced by something that is more faithful to the brain.

What do you think about the so-called commonsense problem?

I think it is a real problem. It is an area where traditional Artificial Intelligence failed. It was not able to deal with the large databases that any normal human has, it was not able to deal with the problem of fast and relevant retrieval. It is dramatic that neural networks solve this problem at a blow. All the knowledge is embodied in the enormous configuration of synaptic weights, and it is accessed by the incoming stimulus itself, automatically. The incoming stimulus has in it the information that allows it to tap into the relevant knowledge. It is deeply dramatic how this chronic problem of the old-fashioned view, called the commonsense knowledge problem, should be solved so swiftly and elegantly by the new neural networks. Because what you have in your head is a neural network, and what you draw on as a person is commonsense knowledge, and it is no accident that the two go together.

What is the future prospect of Artificial Intelligence? Will there be some particular practical outcome? Will there be problems with the social impact of Artificial Intelligence?

I think the projections that good old-fashioned Artificial Intelligence workers made are, in fact, going to be realized. They made them twenty years ago, and the discipline is famous for not having realized many of them, but the reason they were not realized was that they were tied to a hopeless architecture. Now that they have a new architecture, more like systems that we know can solve these problems—that is, living creatures—they will succeed in their rather glorious aims. It is hard to predict, but I think the impact will be at least as great as that of the steam engine and the Industrial Revolution that followed. It is going to be extraordinary, but I hesitate to predict specific claims.

You see how the technology can become popular immediately. Expert systems of the old-fashioned kind are going to be replaced by neural network expert systems. One example that amuses me every time I think of it is the system that evaluates loan applications that are used by banks. Banks have many filing cabinets full of old files containing someone's initial loan application. They tell what their salary is, what houses they have, lots of information. On the basis of that the loan officer makes a judgment: shall I give him a loan, shall I not? And then the file contains records of how the person repays the loan.

Now, all that data just begs to be put into a neural network, so that the neural network can be trained up on all these examples. The input is a loan application, the output is a prediction about repayment behavior. Banks want to be able to predict repayment behavior. So somebody trained up a neural network on a large sample of past loan applications so that it would correctly predict repayment behavior. Once trained, they make better predictions on new applications than loan officers; they already work better than the people do. That is a humdrum example, but it is not humdrum in financial terms. Because even if they predict only 5 percent better than humans—and I think they are rather better than that—it can mean hundreds of millions of dollars for a large bank over the course of a year.

The pull to bring the technology to work will be enormous. There is going to be an enormous market for neural networks. This, however, is not very philosophical, and I am no better than anybody else in predicting these things. Let me think of something philosophical. . . . When we learn to build big neural networks, maybe as big as you and me, they will help us to automate the process of scientific exploration itself. I think that Artificial Intelligence will only become good enough in perhaps fifteen or twenty years to assist in thinking up new scientific theories, in helping us to do conceptual things. Computers have always helped us to process experimental data, but they did not help us with anything conceptual and theoretical. The neural networks, if we can make them big enough, will be able to help us do that. They will be able to do it 10^7 times faster than we will, because if we make neural networks that work on electricity and not on slow-moving electrochemical impulses on neurons, then we will have a brain that thinks as we do, but 10^7 times faster than we do. That is useful, but it is also a little frightening.

Recommended Works by Paul M. Churchland

"Eliminative Materialism." In *Matter and Consciousness*, rev. and expanded ed. Cambridge, Mass.: MIT Press, 1988. Reprinted in *Introducing Philosophy*, 4th ed., ed. R. C. Solomon, pp. 449–53. Chicago: Harcourt Brace Jovanovich, 1989.

"Eliminative Materialism and the Propositional Attitudes." 1981. Reprinted in *Mind and Cognition*, ed. W. Lycan, pp. 206–23. Oxford: Blackwell, 1990.

"Folk Psychology and the Explanation of Human Behavior." In *Philosophical Perspectives*, vol. 4: *Philosophy of Mind and Action Theory*, ed. J. E. Tomberlin. Altascadero, Calif.: Ridgeview Publishing, 1990.

Matter and Consciousness. Rev. and expanded ed. Cambridge, Mass.: MIT Press, 1988.

A Neurocomputational Perspective: The Nature of Mind and the Structure of Science. Cambridge, Mass.: MIT Press, 1989. Paperback ed., 1992.

Notes

1. F.H.C. Crick and C. Asanuma, "Certain Aspects of the Anatomy and Physiology of the Cerebral Cortex," chap. 20 of *Parallel Distributed Processing: Explorations in the Microstructure of Cognition*, vol. 2: *Psychological and Biological Models*, by D. E. Rumelhart, J. L. McClelland, and the PDP Research Group (Cambridge, Mass.: MIT Press, 1986); D. Zipser, "Biologically Plausible Models of Place Recognition and Goal Location," chap. 23 of ibid.

2. All contributed to the important book *Parallel Distributed Processing*, considered the "bible" of connectionism; see also n. 1.

3. Frank Rosenblatt is considered the pioneer of neural network modeling. In the early fifties, he developed the perceptron, a network model for pattern recognition. F. Rosenblatt, *Principles of Neurodynamics* (New York: Spartan, 1962).

4. M. L. Minsky and S. A. Papert, *Perceptrons: An Introduction to Computational Geometry* (Cambridge, Mass.: MIT Press, 1969), reprinted in an expanded ed. as *Perceptrons* (Cambridge, Mass.: MIT Press, 1988).

5. P. M. Churchland, *Matter and Consciousness: A Contemporary Introduction to the Philosophy of Mind* (Cambridge, Mass.: MIT Press, 1984).

AARON V. CICOUREL

Cognition and Cultural Belief

I first got interested in language and thought when I took my master's degree in 1955 at UCLA, where I was introduced to the work of Edward Sapir[1] by Harry Hoijer, an anthropological linguist. In summer of 1955, I met Alfred Schutz[2] and extended my familiarity with his work. These contacts stimulated me to go outside of sociology.

In the spring of 1954, I began a series of small research projects with Harold Garfinkel[3] that led to a serious interest in everyday reasoning and social interaction. These activities influenced my dissertation research. My dissertation got a lot of people in sociology upset because I talked about a number of issues that were about language, social interaction, and reasoning among the aged. Without using the word *cognition*, I was concerned with the phenomenal world of the aged and at the time was stimulated by phenomenological interests in philosophy. . . . I corresponded briefly with Alfred Schutz at that time. I found phenomenological ideas very useful, but I was bothered eventually by the fact that there was not much empirical research attached to that tradition. My preoccupation with the local ethnographically situated use of language and thought in natural settings was not always consistent with this tradition.

By 1958, at Northwestern University, I had already read Chomsky's book *Syntactic Structures*[4] and George Miller's[5] works on memory and language. I was also strongly interested in work by Roger Brown[6] and his students on language acquisition, but I could not find anyone in sociology who cared about or gave any importance to these areas. So I had to do it sort of on my own, as a kind of underground enterprise. I saw Schutz's work and my contributions to what was to be called ethnomethodology as a way of framing some of these ideas on language and cognition. But eventually, I felt that ethnomethodology was too inbred and not open to outside ideas and needed to be integrated with other

Aaron V. Cicourel earned a B.A. degree in psychology from the University of California at Los Angeles in 1951. He pursued his studies at UCLA (M.A.) and at Cornell University, where he earned his Ph.D. in anthropology and sociology in 1957. His professional career was initiated with a postdoctoral fellowship at the UCLA Medical Center and then led him to Northwestern University, the University of California at Riverside, Berkeley, and Santa Barbara, and finally UC San Diego, where he has been Professor of Sociology in the School of Medicine and the Department of Sociology since 1970 and Professor of Cognitive Science, Pediatrics, and Sociology since 1989. He has been a visiting lecturer, scholar, and professor in numerous countries all over the world, including Argentina, Mexico, France, Germany, Spain, Australia, Japan, and China.

areas such as work by Austin, Grice, and Strawson,[7] cognitive and linguistic anthropology, developmental psycholinguistics, and general cognitive issues.

In 1969–70 I was attracted to work in linguistics and psychology in San Diego and was invited to give a lecture in linguistics on my research in sign language. In 1970–71, I received a NSF (National Science Foundation) Senior Postdoctoral Fellowship to work on British Sign Language in England, where I replicated, with deaf subjects, work on hearing children's language competence and repeated some of this research in California.

I moved to San Diego in 1971 and got in touch with the cognitive group in psychology and also with people in linguistics. At UCSD, an informal group interested in cognitive science began meeting, and this led to the development of a cognitive science doctoral program and the foundation of the first Department of Cognitive Science.

I think of myself, however, as being somewhat more marginal to cognitive science than people in Artificial Intelligence or linguistics or psychology, because coming from sociology and anthropology has not been the usual route to this new area of research. It has been a bit strange for people in cognitive science to understand my social science background, but I sometimes remind them that my first degree was in psychology from UCLA, and this background made it easier to talk to colleagues in cognitive science than to many sociologists.

These remarks provide a very brief history of my initial involvement in cognitive science: I think it started very early and maybe had to do with the fact that I was exposed, as both an undergraduate and a graduate student, to experimental psychology, a lot of philosophy, and aspects of linguistics, statistics, and mathematics. As a graduate student in sociology, however, I got into trouble because I felt that the use of mathematics in sociological research was often misleading. The sociological data seldom seem to meet the assumptions of the models used.

Following my early use of the work of Alfred Schutz on socially distributed knowledge, I became interested in Roy D'Andrade's and Edwin Hutchins's recent work in the eighties on culture and cognition and in socially distributed cognition.[8] D'Andrade's work on the cultural basis of cognition was of special interest, and I found it useful while pursuing my interests in the way locally organized interaction framed work in sociolinguistics.

The two notions of socially distributed cognition and socially distributed knowledge overlapped considerably because cognition is always embedded in cultural beliefs about the world and in local social practices. These notions have led me recently to the link between neural child development, human information processing and the way socially organized ecologies influence the brain's internal organization and the child's capacity for normal problem solving, language use, and emotional behavior. These conditions tend to be ignored by social scientists.

What could be the reasons why social scientists ignore this fact and why they do not play an important part in cognitive science now?

It is the other side of the coin of why cognitive scientists ignore the environment and social structure or interaction. Cognitive scientists and social scientists each tend to take as self-evident each other's concepts.

Social scientists treat cognition as if it were self-evident, like in the classical rational man model. In sociological social psychology, that is, primarily symbolic interactionism, a lot of emphasis is placed on what is situational without examining language use and information processing constraints during social interaction. It seems to me that this view also ignores the constraints that the brain has placed on the way you can use conscious and unconscious thought processes, emotions, and language, and the way that language places further restrictions on what you can or want to communicate about what you think you know or don't know.

A central issue of human social interaction are constraints that stem from cognitive processing. But information processing is also constrained by culture as constitutive of and emergent in local environments. Social scientists tend to ignore the role of human information processing for the production of social interaction and more complex forms of social organization. Cognitive scientists, however, tend to take for granted the influence and constraints that complex forms of social organization and locally organized social interaction can have on information processing. Cognitive scientists, however, have, in recent years, started to recognize the important role that external displays (or external representations) play in problem solving.

Behavioral ecologists in biology working with nonhuman animals recognize the role that the physical environment and social ecology have in influencing behavior. But we have seldom looked at human cognition in the same way. Work on the neurobiology of human cognitive processes tends to pay only lip service to "experience" or the role of the local environment or social ecology on brain maturation. Looking at the role of social ecology and interaction requires a lot of longitudinal, labor-intensive research.

My point is that when you take for granted culture and the way it is reflected in a local social ecology you eliminate the contexts within which the development of human reasoning occurs. When culture, language use, and local interaction are part of cognitive studies, subjects are asked to imagine a sequence of events that have not been studied independently for their cultural basis or variation except for the self-evident variation assumed by the experimenter. But in everyday life, cognition and culture include locally contingent, emergent properties while exhibiting invariant patterns or regularities whose systematic study remains uneven.

But what is complicated about cognitive science's focus on individual human information processing is what happens when two minds interact. There are many constraints on how interaction can occur. For example, during exchanges in experimental settings and everyday life, problem solving can be influenced by how well subjects know each other as part of the local social ecology, how long they have known each other, the nature of the task, individual variation, and the socially distributed nature of cognitive tasks. These conditions are integral aspects of a surgical operating room, a courtroom, formal and informal exchanges in a public or private organization, and when people meet in noninstitutionalized open spaces. Environmental constraints are never topics of relevance when we focus on individual minds except when the experimenter builds them into an experiment in a self-evident way. The conditions for setting up an experiment or engaging in an exchange is never self-evident. For example, in experimental settings, the investigators must have polite exchanges, and they try to establish a mini–social world or ecology that includes paying attention to their appearance, who they are, what they are about, and the conditions needed for further interaction and solving a problem.

Do you think that the last ten years have brought a change in the sense that people try to introduce the social environment in cognition—for instance Habermas with his Theory of Communicative Action,[9] *or Lucy Suchman's* Plans and Situated Action?[10]

Lucy Suchman is an anthropologist who has been interested in conversational analysis and ethnomethodology. She is obviously aware of environmental conditions and applies her knowledge of ethnomethodology and conversation analysis to issues in Artificial Intelligence and the situated nature of local task accomplishments. When it comes to Habermas, I think that is a different story. The linguistic basis of Habermas's model of communication derives mainly from speech act theory and is limited vis-à-vis the linguistic practices of people in the everyday world. And although his model of socialization is broad in nature, it is limited in terms of how it could be tested empirically. It needs to be linked to empirical research that people have done on children's cognitive and linguistic development. Even though I think it is an intellectually challenging and highly detailed theory, it is empirically difficult to assess despite its strong appeal.

The most interesting parts for me are his intermediate reflections, where he brings in the cultural background as a horizon. Perhaps Searle and Dreyfus had some influence on him when he stayed in Berkeley? What do you think about this notion of "background"?

For me, the issue is: what is "background"? Searle sustains primarily a formal theory of language. Despite his important work in speech act theory he works within the analytic tradition. That tradition limits the extent to which the environment comes in. Dreyfus has been interested in the background notion because of his phenomenological perspective, but the relevance of a locally organized social ecology is not addressed as directly as can be seen in work by Alfred Schutz. Habermas, however, has been influenced by a broad range of theorists, including those interested in everyday interaction, such as G. H. Mead[11] and Schutz, and in his other works he recognized that the environment plays a key role. Nevertheless, I think that his conception of environment and background remains rather abstract. His former student, Oevermann, however, tries to address socialization issues directly.

For Habermas, this is like a universal aspect.

But to say it is universal is almost to stop you from demonstrating the limits of this empirically. What we have to ask is: what is the empirical significance of such views? Pierre Bourdieu's work addresses some of the empirical issues more directly even though it is a strongly structural point of view.[12] He also tries to take into account processes, and his ideal type notion of habitus is an important aspect of socialization. Habermas also talks about something similar to Bourdieu's notion of habitus or lifespace, where the child learns about personal power and to cope with the everyday world.

I think Bourdieu's notion of habitus is an important structural idea because he senses the relevance of process. But neither he nor Habermas has pursued the socialization process deeply enough.

Do you have a guess why it is that sociologists could not introduce some notion of cognition into their work—why they are not interested?

What I said earlier applies: sociologists avoid cognition because from Durkheim on they worry that such a view would reduce society to psychology. When I was a graduate student, people used to criticize me when I used any psychological notions. They always warned me that I was going to psychologize society in a reductionist sense.

I think this is mistaken; you can benefit from the inclusion of different levels of analysis without being reductionist. These levels do not reduce but they interact, and you have to understand their mutual influence.

What I have been trying to say is that the structure of social organization is important as a constraint, as well as how interaction is going to guide its reproduction. Social ecology puts limits on the kind of human cognition and information processing that can occur in a given physical space, at a given time for particular kinds of problem solving or decisions. On certain occasions, you can use only certain types of language, you can use only certain topics, you have to be seated in a certain place, and you have certain regularities about who gets to talk and when. This is the case in all social ecologies just as it applies to you and me sitting in this room. You do not expect me to get up and to lie down on the table. If I did, you would get worried about what I am doing and if I am serious. . . . Everything we do involves these constraints as well as the fact that social ecology or organization facilitates certain kinds of interaction and inhibits other kinds. Well, I see all of this as the interaction of different levels of analysis.

Do you think that the new direction of connectionism is not so important? The connectionists claim that the hardware really matters, that you cannot say anything about the computational level. Their view is that you cannot study—as functionalism did—the mind without the brain but that you have to study the brain, too.

I do not see why studying the brain is a problem. The question is: is studying the brain the answer to what the mind is? It seems to me that you have to be careful about how you characterize that. The people I know in connectionism are saying that often you want to be aware of neural structures and functions that are relevant for understanding cognition. But they do not always study cognition in order to understand how the brain works but ask how knowing the ways in which the brain works can help us understand how the mind works. Researchers would like to know what neural substrates are relevant for cognitive functions while also examining and understanding these functions at their own level of analysis, despite not always having a clear picture of the neurobiology.

With what the connectionists claim—do you think that it is an advance from Newell's and Simon's physical symbol system hypothesis?

I am not an expert in this area and cannot give you a clear response. There are notable differences of opinion here. Recent work by Newell and his colleagues involves a powerful symbol-processing simulation of human intelligence (called Soar). Connectionist views address a more microlevel analysis and time frame.

I am interested in a more macrolevel bandwidth, such as decision making in such areas as caretaker-child interaction during the acquisition of cognitive, cultural, and linguistic knowledge and practice, medical diagnostic reasoning, and the role of expert systems that employ rules within a knowledge base but where the time frame can be highly variable. The problem of modeling an individual mind is so complicated right now that most modelers are not prepared to do more than that. But Hutchins and others are concerned with applying connectionist models to the cognitive interaction of two or more individuals. I have been working on how to model medical doctors' diagnostic reasoning—doctors talking to the patient and to other doctors—and in collaboration with colleagues hope to use both production and connectionist architectures.

Many Artificial Intelligence workers have made great claims. They have said, for instance, that in programming they were building the human mind.

Many people make claims about Artificial Intelligence. But there are a lot of people who recognize that exaggerations exist, and are sensitive to the limitations of the field. The mass media often are guilty of promoting such claims.

The key is that Artificial Intelligence modeling can help clarify basic mechanisms involved in cognition. The value of simulation, in my opinion, depends on the way it makes use of empirical research in a given domain.

What do you, as a sociologist, think about the Turing Test as a test to see if your modeling is successful?

I think it is difficult to apply a Turing Test to sociology because the field is too diffuse and lacks adequate articulation between general theory and the concepts that research analysts explore empirically. Few restricted test conditions exist for systematic study of basic concepts and a Turing Test because the notion of environment remains ambiguous and unexplored.

You mentioned the restricted test conditions and the more complicated environment. What do you think about the so-called commonsense problem?

The notion of commonsense suggests a contrasting term like logical thinking or rational choice. Specifying behavioral tests for these notions implies descriptive accuracy about when such reasoning can be observed. For example: what happens in actual decision making that can be observed and tape-recorded? In some of my current research, I am trying to challenge the clinical basis of the knowledge base employed in the use of a medical system. My colleague in this research is a clinician, and we look at the question of how you actually get the knowledge from the patient that is required to use an expert system. The people who have written the Artificial Intelligence programs have ignored that part, but I think they are beginning to recognize its necessary inclusion.

What do you think about the discipline of Artificial Intelligence called "knowledge engineering"? Don't you think their concept of "knowledge" is totally formalized?

My colleague Don Norman is interested in everyday cognition and the use of artifacts that could be seen as a part of the knowledge base of persons who design artifacts and test their utility.[13] Norman's knowledge engineering is highly practical yet retains many theoretical issues relevant to Artificial Intelligence. He addresses the cognitive processes of users of artifacts that are not addressed adequately by their producers. The notion of knowledge can be examined as a process, a set of practices and mental models about the processes and what they are to accomplish. Let me assume that the physician is a knowledge engineer. How should this "engineering" be addressed? I go to the actual situation in which a physician talks to a patient about symptoms. I tape-record each session and look at the material to see how the physician makes an inference based on her or his knowledge base and from what the patient says. The general point is that knowledge should be conceived as a process both internally and when instantiated in experimental or practical settings.

Is it right that you claim not only that cognition is underdeveloped in social theory but also that social scientists do not have experiments or ideas for experiments?

Let's take the classical theorists—Weber, Durkheim, Parsons—or recent theorists like Habermas or Bourdieu. Weber was supposed to have done a survey in a factory, and Bourdieu has done some fieldwork and a number of surveys. To my knowledge, none of them have examined everyday social interaction experimentally. The experimental work done by sociologists has been confined to what is called "small group research" by a subset of social psychologists at a few universities. In the United States, such work has been conducted at the University of Michigan, the University of Washington, Iowa University, and Stanford University. . . . Let me elaborate.

The study of human affect, language, learning, and decision making has progressed primarily by its reliance on laboratory research and the use of sophisticated modeling programs. The laboratory and modeling research, however, seldom includes information about the behavioral ecology of humans in natural settings or the social ecology of the laboratory.

My recent concern with the acquisition of emotions and social inter-action uses experimental data obtained by Judith Reilly, Doris Trauner, and Joan Stiles on children with and without focal brain lesions. One focus of this research is the notion that processing and expressing emo-tions requires an intact right hemisphere and that this hemispheric spe-cialization may occur quite early in human development.

I pursue a perspective on the development of affect and facial expres-sions that builds on work in behavioral ecology—for example, the idea that primates and humans express emotion when learning to create and sustain coherent and appropriate social interaction as part of their ge-netic and behavioral adaptation to local social environment.

In general terms, we seek a developmental explanation of human af-fect and facial expression and suggest that this development is a function of the role of neural substrates and the learning of pragmatic elements of verbal, paralinguistic, and nonverbal behavior. This interpenetration of levels of analysis is necessary if the infant is to produce behavior that adults will perceive and comprehend as appropriate emotional commu-nication. This view is consistent with the notion that language and affect seem to be independent communicative systems and, according to Reilly and Trauner, suggest different time trajectories in the brain for the later-alization of individual cognitive functions.

Briefly, Reilly and Trauner suggest that infants with focal brain le-sions might benefit from intensive stimulation from their social ecolo-gies. Yet, some caretakers seem to become sufficiently depressed in socializing such infants that this can eliminate the possibility of enhanc-ing the infant's or child's ability to adapt to typical behavioral routines that occur in local social encounters.

My goal is to examine the extent to which experimental work on cog-nition presumes stable forms of social interaction the infant and child must acquire to be seen as "normal." This social and communicative competence is thus necessary to be able to conduct experiments about the infant's or child's cognitive development.

Do you see any social problems coming out of Artificial Intelligence work, for example, with expert systems?

It seems to me that any problems that emerge can be useful if they force us to question the way we use existing teaching systems, medical systems, legal systems, and so on. We may be forced to recognize issues we don't want to see. That is where a problem could come up because you are now tampering with cultural beliefs that some people don't want to change. So if you try to institute a model that you worked out by some simulation, you will have to point out things that might happen, things that make the system go wrong, that people don't want to hear about.

Let me give you a sense of my own views about how current Artificial Intelligence programs or expert systems can be supplemented.

My concern with medical diagnostic reasoning revolves around problems of actually implementing existing systems. I think traditional production systems and the use of neural network modeling capture the essential formal criteria used to arrive at an adequate diagnosis. But a kind of "front end" is needed for implementing a system and using it as a teaching device for novice health care professionals. By "front end," I mean the following:

First, what minimal knowledge should the novice have in order to access the expert system and to understand the kinds of requests it will make of her or him?

And second, what kinds of instructions should the novice have in order to orient her or his medical history and physical examinations to the kinds of questions that would be necessary to use an expert system?

We are concerned with the language frames used in eliciting or accessing information from patients that facilitate and mediate responses to the queries of an expert system.

Don't you think that there is a structural similarity between the digital computer on the one hand and the society on the other hand? What is the reason for the huge social impact of the computer?

I have difficulty making such a link because our knowledge about "society" is incredibly impoverished compared to what we know about the digital computer. I would prefer to say that digital computers are analogous to our ideal conceptions of what bureaucracies should be like—for example, the existence of hierarchy and general and specific rules governing decisions.

In more general terms, digital computer capabilities seem to resemble a notion of ideal problem solving, and this leads to the definition of problems to be solved in terms of the hardware and software that are currently available. But it does not necessarily mean that you are going to solve problems adequately. It can also mean that you may oversimplify the world so that they can be addressed efficiently by your models.

Recommended Works by Aaron V. Cicourel

Cognitive Sociology: Language and Meaning in Social Interaction. New York: Free Press, 1974.

"The Integration of Distributed Knowledge in Collaborative Medical Diagnosis." In *Intellectual Teamwork: The Social and Technological Foundation of Cooperative Work*, ed. J. Galagher, R. Kraut, and C. Egido, p. 221–42. Hillsdale, N.J.: Lawrence Erlbaum, 1990.

"The Interpenetration of Communicative Contexts: Examples from Medical Encounters."
Reprinted in *Rethinking Context: Language as an Interactive Phenomenon*, ed.
A. Duranti and C. Goodwin, pp. 291–310. Cambridge, Mass.: Cambridge University
Press, 1992.

Method and Measurement in Sociology. New York: Free Press, 1964.

"Review Article on John R. Searle's Intentionality: An Essay in the Philosophy of Mind."
Journal of Pragmatics 11 (1987): 641–60.

"Semantics, Pragmatics, and Situated Meaning." In *Pragmatics at Issue* 6, no. 1, ed.
J. Verschueren, pp. 37–66. Amsterdam: John Benjamins, 1991.

Notes

1. E. Sapir, *Language* (San Diego: Harcourt Brace Jovanovich, 1921).

2. A. Schutz, *The Phenomenology of the Social World* (Evanston, Ill.: Northwestern
University Press, 1967); see also *Alfred Schutz on Phenomenology and Social Relations:
Selected Writings*, ed. H. R. Wagner (Chicago: University of Chicago Press, 1973), and
The Collected Papers of Alfred Schutz, 3 vols. (The Hague: Nijhoff, 1962–66).

3. H. Garfinkel, *Studies in Ethnomethodology* (Englewood Cliffs, N.J.: Prentice-Hall,
1967).

4. N. Chomsky, *Syntactic Structures* (The Hague: Mouton, 1957).

5. G. Miller, "The Magical Number Seven, Plus or Minus Two: Some Limits on Our
Capacity for Processing Information," *Psychological Review* 63 (1956): 81–97.

6. R. Brown, *A First Language: The Early Stages* (Cambridge, Mass.: Harvard University Press, 1973); R. Brown and U. Bellugi, "Three Processes in the Child's Acquisition of
Syntax," *Harvard Educational Review* 34 (1964): 133–51.

7. J. L. Austin, *How to Do Things with Words* (Oxford: Oxford University Press,
1965); H. P. Grice, "Logic and Conversation," in *Syntax and Semantics*, vol. 3: *Speech
Acts*, ed. P. Cole and J. Morgan (New York: Academic Press, 1975); and P. F. Strawson,
"On Referring" (1950). Both "Logic and Conversation" and "On Referring" are re-
printed in the anthology *The Philosophy of Language*, ed. A. P. Martinich (New York:
Holt, Rinehart, and Winston, 1985).

8. See, e.g., R. D'Andrade, "A Folk Model of the Mind," in *Cultural Models in Lan-
guage and Thought*, ed. D. Holland and N. Quinn (Cambridge: Cambridge University
Press, 1987), pp. 112–50; E. Hutchins, *Culture and Inference: A Trobriand Case Study*
(Cambridge, Mass.: Harvard University Press, 1980), and "Myth and Experience in the
Trobriand Islands," in *Cultural Models in Language and Thought*, pp. 269–89.

9. J. Habermas, *The Theory of Communicative Action (Theorie des kommunikativen
Handelns)*, 2 vols., trans. Thomas McCarthy (Boston: Beacon Press, 1984).

10. L. Suchman, *Plans and Situated Action* (Cambridge: Cambridge University Press,
1987).

11. G. H. Mead, *Mind, Self, and Society* (Chicago: University of Chicago Press, 1934).

12. E.g., P. Bourdieu, *La distinction: Critique sociale du jugement* (Paris: Editions de
minuit, 1979); and *Le sens pratique* (Paris: Editions de minuit, 1980).

13. D. Norman, *The Psychology of Everyday Things* (New York: Basic Books, 1988).

DANIEL C. DENNETT

In Defense of AI

Before there was a field called cognitive science, I was involved in it when I was a graduate student at Oxford. At that time, I knew no science at all. I had had a purely humanistic education as an undergraduate. But I was very interested in the mind and in the philosophy of mind. I was completely frustrated by the work that was being done by philosophers, because they did not know anything about the brain, and they did not seem to be interested. So I decided that I had to start to learn about the brain to see what relevance it had. I became an autodidact neuroscientist, with the help of a few professors.

What I found out was that people who knew about brains did not have a lot to say about the mind, either. In those days, unlike today, it was very hard to come across much of anything in neuroscience that had the ambition of addressing any of the philosophical questions about mind.

Just about that time, I learned about Artificial Intelligence. This was in 1963–64. I got quite interested. There was a nice little book edited by Alan Anderson, *Minds and Machines*,[1] which was philosophical but raised some issues. I talked with Anderson who was in England that year. When I got my first job at the University of California at Irvine in 1965, there was a small Artificial Intelligence group there. One day one of them, Julian Feldman,[2] came into my office and threw a paper down on my desk, saying: "You're a philosopher; what do you make of this?" The paper he threw on my desk was Bert Dreyfus's first attack on Artificial Intelligence. It was his Rand memo, called "Alchemy and Artificial Intelligence."[3] I read it and thought it was wrong, that Dreyful made mistakes. He said he wanted me to write an answer, showing where the mistakes are.

Since it fit with my interests anyway, I wrote an article responding simultaneously to Dreyfus and to Allen Newell's view at that time. Newell had come to Irvine to give a talk, and I thought he was wrong, too, in a slightly different way. So I wrote a piece that responded to

Daniel C. Dennett was born in 1942. He earned his B.A. degree from Harvard University in 1963 and his D.Phil. from Oxford in 1965. From 1965 to 1971, he taught at the University of California at Irvine. He moved to Tufts University in Medford, Massachusetts, in 1971, where he has been since 1975. Since 1985, he has been Distinguished Professor of Arts and Sciences and Director of the Center for Cognitive Studies at Tufts University.

Dreyfus and Newell. It was actually my first publication, in 1968. It was an article published in the journal *Behavioral Science*, called "Machine Traces and Protocol Statements."[4]

That hooked me. I was very interested in the debate, and I was sure that Dreyfus was wrong. The people in Artificial Intelligence were glad to have a philosopher on their side. An interesting relationship began to develop, and it has continued over the years.

From where did you get your knowledge about Artificial Intelligence and computer science?

I got it out of interactions of this sort with people, growing gradually over the years, by talking with people and reading accessible literature. But I was not really computer-literate. In 1978–79, John McCarthy set up a group at the Center for Advanced Studies in Behavioral Science in Palo Alto on philosophy and Artificial Intelligence. That was a wonderful year. There were six of us there, all year long: John McCarthy, Zenon Pylyshyn, Patrick Hayes, Bob Moore, John Haugeland, and myself. Two philosophers and four Artificial Intelligence people.[5] In the course of the year, a lot of other people came around for short periods. I learned a tremendous amount; we talked a lot and debated a lot. I still did not come out of that year very computer-literate, but I made some considerable progress. In the years following, I developed that much further, even to the point where, a few years ago, I taught a computer science course here. I am not a serious programmer, but I know how to program, and I know some languages.

Was it difficult to discuss things with people in computer science, with your totally different background and training?

It has been difficult, but the surprising thing is that the real difficulty arises from the fact that although they are trained as computer scientists they use a lot of terms that philosophers use, and it takes a long time to discover that they don't mean the same things by them. Their terms are "false friends." You can have a conversation going on for quite a long time before you realize that you are talking past each other because you don't use the words in the same way. This was particularly apparent during the year at Stanford. It would be hard to find four more philosophical people in Artificial Intelligence than McCarthy, Pylyshyn, Hayes, and Moore. And it would be impossible to find two more Artificial Intelligence–minded philosophers than Haugeland and me. So you would think we could talk together very well.

You have the feeling that in this debate between Artificial Intelligence and its critics, you are on the side of Artificial Intelligence?

I am happy to identify myself as the defender of Artificial Intelligence, of strong Artificial Intelligence. I have been a critic of numerous errors I found in Artificial Intelligence, but on the overall issue, I think a very good case can be made for strong Artificial Intelligence, and a very good case can be made for the importance of details in the research.

You mentioned your criticism of Dreyfus. What are your main points in this issue?

When Dreyfus first started criticizing Artificial Intelligence, he did not know very much about the field, any more than I did. He had some good instincts about where the difficult problems for Artificial Intelligence lay, and he pointed them out. But he exaggerated. He claimed that things were impossible in principle, that these were obstacles that no sophistication in Artificial Intelligence could ever deal with. I thought his arguments were not plausible at all.

If you soften his claims and read him as saying that Artificial Intelligence has underestimated the difficulties of certain issues, that these are the outstandingly difficult problems for Artificial Intelligence, and that Artificial Intelligence would have to revolutionize some of its thinking in order to account for them—if that had been what he said, he would have been absolutely right.

I think there is still a great difference between your view and his. He is criticizing the whole rationalist tradition, whereas you come from this tradition and defend it.

Actually, I am not defending the particular line he is criticizing. Like most philosophers, Dreyfus tries to find something radical and absolute rather than a more moderate statement to defend. I think his current line about the bankruptcy of rationalism, if you like, is just overstated. Again, if you moderate it, I agree with it. I think that people in Artificial Intelligence, particularly people like John McCarthy, have preposterously overestimated what could be done with proof procedures, with explicit theorem-proving style inference. If that is what Dreyfus is saying, he is right.

But then, a lot of people in Artificial Intelligence are saying that, too, and have been saying it for years and years. Marvin Minsky, one of the founders of Artificial Intelligence, has always been a stern critic of that hyperrationalism of, say, McCarthy. Dreyfus does not really have a new argument if he is saying something moderate about rationalism. There is only a very limited and radical group of hyperrationalists who fall into this criticism. If, on the other hand, Dreyfus is making a claim that would have as its conclusion that all Artificial Intelligence, including connectionism and production systems, is bankrupt, he is wrong.

He is saying that our thinking is not only representational. There are things you cannot represent, like bodily movements, pattern recognition, and so on. I understand that he says the human mind is not merely a problem solver, as the physical symbol systems hypothesis would imply.

Everything in that statement is true except the claim at the end: that this is what the physical symbol systems hypothesis has to hold. Any sane version of the physical symbol systems hypothesis—that is not my way of thinking, but I defend Allen Newell on this ground—is quite prepared and able to grant that there are transitions in such a system that are not inferences. These transitions are changes from one state to another that are not captured by any informational term—that is, from the standpoint of the symbols, they are inexplicable. They are not the adding of a premise, they are not the adding of a hypothesis, they are not a revision of a representation—they are a change of state, but they cannot be captured in the language of representation. It is possible to ignore that possibility and think that you can have a theory of the mind that does not permit such changes. But this is a very strong and gratuitous claim, and there is no particular reason to hold this view.

In my own work, *The Intentional Stance*,[6] I give a number of examples of ways in which items, states, and events can have a determinate effect on a cognitive system without themselves being representations. In putting forward these examples, I was not arguing against any entrenched view of Artificial Intelligence.

So you think that one can combine the physical symbol systems hypothesis with the intuition that you cannot explain everything with a symbol?

Yes, I don't see why it could not be done. It is an artificial watershed that has been created as if you could not step across that line and be true to your view.

What about connectionism? It criticizes the physical symbol systems hypothesis and seeks alternative explanations for the human mind.

Indeed, connectionism is a perfect case of what we were talking about. It is the clearest vision we yet have of that mixture. If you look at the nodes in a connectionist network, some of them appear to be symbols. Some of them seem to have careers, suggesting that they are particular symbols. This is the symbol for "cat," this is the symbol for "dog." It seems likely to say that whenever cats are the topic, that symbol is active; otherwise it is not.

Nevertheless, if you make the mistake of a simple identification of these nodes—as cat symbol and dog symbol—this does not work. Because it turns out that you can disable this node, and the system can go right on thinking about cats. Moreover, if you keep the cat and dog nodes going and disable some of the other nodes that seem to be just noisy, the system will not work. The competence of the whole system depends on the cooperation of all its elements, some of which are very much like symbols. There is no neat way of demarcating the events that are symbolic from the events that are not. At the same time, one can recognize that some of the things that happen to those symbols cannot be correctly and adequately described or predicted at the symbol level.

Even in this case, you would not say that there is a watershed between connectionism and the physical symbol systems hypothesis?
There is a big difference in theoretical outlook, but it is important to recognize that both outlooks count as Artificial Intelligence. They count as strong Artificial Intelligence. In John Searle's terms, connectionism is just as much an instance of strong Artificial Intelligence as McCarthy's or Schank's[7] or Newell's work. The last time I talked with him he was very clear about this—that you could be a strong Artificial Intelligence connectionist. He was just as skeptical about it as of any other kind of Artificial Intelligence.

Both Searle and Dreyfus generally think that it could be a more promising direction.
Sure, it is that. There is a bit of a dilemma for both of them. If they want to say: "Hurrah for connectionism, this is what we meant, this is what the alternative was supposed to be," then they are embarrassed to admit that it is strong Artificial Intelligence, just a different brand. Then their criticism was directed not at Artificial Intelligence in general but at a particular brand of Artificial Intelligence. On the other hand, if they say: "You are right, it is Artificial Intelligence, and Artificial Intelligence is impossible," then how can it be such a promising avenue? Searle could say that it is a promising avenue for weak Artificial Intelligence, but then we have to look and see whether he has any independent arguments against strong Artificial Intelligence that don't depend on his focus on the brand of Artificial Intelligence that is not connectionist.

Let's follow Searle's criticism of strong Artificial Intelligence. He has refuted strong Artificial Intelligence with his Chinese Room Argument. What do you think about that?
I think that is completely wrong. First of all, the Chinese Room is not an argument, it is a thought experiment. When I first pointed this out, Searle agreed that it is not an argument but a parable, a story. So it's neither sound nor an argument; it is merely a demonstration. I have pointed out in different places that it is misleading and an abuse of the

thought experiment genre. Searle's most recent reaction to all that is to make explicit what the argument is supposed to be. I have written a piece called "Fast Thinking," which is in my book *The Intentional Stance*. It examines that argument and shows that it is fallacious.

Briefly, Searle claims that a program is "just syntax," and because you can't derive semantics from mere syntax, strong Artificial Intelligence is impossible. But claiming that a program is just syntax is equivocal. In the sense that it is true, it is too strong for the use he makes of it (in the sense in which a program is just syntax, you can't even get word processing or weather forecasting out of it—and computers manifestly can perform those tasks). Otherwise, the premise is false; but beyond that, the claim that you can't derive semantics from syntax obscures a possibility that Artificial Intelligence has exploited all along: that you can approximate the performance of the ideal semantic engine (as I call it) by a device that is "just syntax."

One of the things that has fascinated me about the debate is that people are much happier with Searle's conclusion than with his path to the conclusion. For years I made the mistake in talking to people about the Chinese Room—it has been around for ten years now—of carefully showing what the fallacy is in the argument, in the sense that I said, "If this is an argument, it has to be *this* argument, which is fallacious, or it has to be *this* argument, which is fallacious," and so on. I would go very carefully through his arguments, but I think people were not interested. They did not care about the details of the argument, they just loved the conclusion. Finally I realized that the way to respond to Searle that met people's beliefs effectively was to look at the conclusion and ask what it actually is, to make sure that we understand that wonderful conclusion. When we do this, we find that it too is actually ambiguous.

There is another statement that is, at first appearance, very close to Searle's conclusion. This statement is no nonsense and might even be true. It is an interesting if not particularly likely empirical hypothesis: the only way to achieve the speed of computation required to re-create artificial intelligence is by using organic computation. The people may be thinking that this is what he is arguing for, but he is not. That is in my chapter "Fast Thinking."

I read your and Hofstadter's comment in The Mind's I.[8] *I think Searle would classify that reply as the "systems reply," because you argue that nobody claims that the microprocessor is intelligent but that the whole system, the room, behaves in an intelligent way. Searle modified the argument in response to this. He says: let's assume that all the system is in my head and that I learn everything by heart; then I would also behave intelligently. He says that the systems reply does not work because it is only one step further. He finds it strange that you would agree with him that the human being manipulating the symbols without knowing what they mean does not have intentionality but that the whole system— paper and pencil and all—does have intentionality.*

This is not recent. If you go back to *Behavioral and Brain Sciences,*[9] you will find my commentary in there, which, at the time I wrote it, was already old. I had already been through that with Searle many times. In the little piece "The Milk of Human Intentionality" I point out that Searle does not address the systems reply and the robot reply correctly. He suggests that if he incorporates the whole system in himself, the argument still goes through. But he never actually demonstrates that by re-telling the story in any detail to see if the argument goes through. I suggest in that piece that if you try it, your intuitions at least waver about whether or not there is anything there that understands Chinese or not.

We have got Searle, a Chinese-speaking robot, who has incorporated the whole system in himself. Suppose I encounter him in the street and say: "Hello, John, how are you?" What does your imagination tell you that happens? First of all, there seem to be two different systems there. One of them does not understand English—it is the Chinese robot. That system should reply to me something like "I am sorry, I don't speak any English!" We can tell the story in different ways. If we tell the story in a way in which he is so engrossed in manipulating his symbols that he is completely oblivious to the external world, then of course I would be certainly astonished to discover that my friend John Searle has disappeared and has been replaced by a Chinese-speaking, Chinese-understanding person. The sense that *somebody* understands Chinese is overpowering—and who else but Searle? We might say: it looks as if the agent Searle has been engulfed by a larger agent, Searle II, who understands Chinese. But that was the systems reply all along.

If Searle had actually gone to the trouble of looking at that version of his thought experiment, it would not have been as obvious to people anyhow.

Sure, you could ask the robot questions, observe its behavior, and ascribe it intentionality, as you wrote in your paper. But to ascribe intentionality does not mean that it really has intentionality.

That is an interesting claim. I have argued, of course, that that, in the limit, is all there is. There is not any original, intrinsic intentionality. The intentionality that gets ascribed to complex intentional systems is all there is.[10] It is an illusion that there is something more intrinsic or real. This is not a radical thesis, but a very straightforward implication of standard biological thinking. We are mechanisms, mechanisms with very elaborate purposes. Ultimately the raison d'être looked like a decisive refutation of any representational theory for a long time, until we had computers.

When computers came along, we began to realize that there could be systems that did not have an infinite regress of homunculi but had a finite regress of homunculi. The homunculi were stupider and stupider and stupider, so finally you can discharge the homunculus, you can break it down into parts that are like homunculi in some regards, but replaceable by machines. That was one of the most liberating conceptual contributions of computers. It showed us how we could break the regress.

Notice that we have many examples of the regress broken, but we could argue, if there were any point to it, about how to describe the result. Should we say that computers don't really represent anything because there is nothing in them if you look at them? Or should we say that they are representations but they don't need complete minds inside to appreciate them? They can get by with systems that are only a little bit like homunculi. In any case, the regress is broken there.

Infinite regress seems to arise in other places. Let's take commonsense. If you think that commonsense is composed of a bunch of rules that are being followed, then you need a commonsense to know which rule to follow when. So you have to have metarules to tell you at which rules to look. But then you will have to have commonsense about how to use the metarules, and so on. If we try to do it with rules forever, we have an infinite regress of rules.

But we should learn from the first regress. This is not an infinite regress. It is simply an empirical question: how many layers of rules do you have to have before you get down to dispositions that can be replaced by a machine? This is an open question. It is strongly analogous to the question of how many layers of language there are between a particular computer application and the machine code. That is an empirical question, too: it may be hard-wired, it may be a virtual machine, it may be a virtual machine running on a virtual machine running on a virtual machine. . . . If it is the latter, there are several layers of directions, of rules, that the whole system is depending on. And indeed, it looks to see what it should do by consulting the next rule on the list. If you want to know how it does that, you may find some other rules. Finally you hit the bottom, and there are no more rules. You have reached the microcode, as they call it; now you are in the hardware.

So your view is, in general, optimistic?

In general, I think that infinite regress arguments are signs that point in the direction of their own solution. If you think about them, you find that it cannot be an infinite regress. So you start tinkering with the assumptions, and then you see how a finite regress is probably the answer.

What do you think are the major problems that cognitive science has to overcome at the moment?

We are just beginning to develop biologically plausible models of consciousness, and they suggest that the computational architecture of a brain capable of sustaining conscious thought is vastly more subtle and ingenious than the architectures we have developed so far. In particular, the standard insulation of function in engineer-designed architectures (each element has one task to perform) is almost certainly a feature that systematically prevents us from exploring the right part of design space. The brain is not magical, but its powers seem magical because they emerge from a tangle of multipurpose, semiautonomous, partially competitive elements. Working out the design principles of such systems is a major task facing Artificial Intelligence and computational neuroscience, but tantalizing insights are arising. Some of these are explored in my new book, *Consciousness Explained.*[11]

Recommended Works by Daniel C. Dennett

Brainstorms: Philosophical Essays on Mind and Psychology. Cambridge, Mass.: MIT Press, 1978.
Consciousness Explained. London: Penguin, 1992.
The Intentional Stance. Cambridge, Mass.: MIT Press, 1987.
(Ed., with D. Hofstadter.) *The Mind's I: Fantasies and Reflections on Self and Soul.* New York: Basic Books, 1981.
"Why the Law of Effect Will Not Go Away," "True Believers: The Intentional Strategy and Why it Works," "Making Sense of Ourselves," and "Quining Qualia." All in *Mind and Cognition: A Reader*, ed. W. G. Lycan. Oxford: Blackwell, 1991.

Notes

1. *Minds and Machines*, ed. A. R. Anderson (Englewood Cliffs, N.J.: Prentice-Hall, 1964).

2. At that time best known for the book co-written with E. A. Feigenbaum, *Computers and Thought* (New York: McGraw-Hill, 1963).

3. H. Dreyfus, "Alchemy and Artificial Intelligence" (Rand Corporation, Santa Monica, Calif., 1965, memo).

4. D. C. Dennett, "Machine Traces and Protocol Statements," *Behavioral Science* 13 (1968): 155–61.

5. John Haugeland is the other philosopher. McCarthy was founder and first director of the Artificial Intelligence labs at MIT and Stanford University. See, e.g., J. McCarthy, "Programs with Common Sense" in *Semantic Information Processing*, ed. M. Minsky (Cambridge, Mass.: MIT Press, 1968). pp. 403–18; and Z. W. Pylyshyn, "What the Mind's Eye Tells the Mind's Brain," *Psychological Bulletin* 8 (1973): 1–14. A major, more recent work by Pylyshyn is *Computation and Cognition: Toward a Foundation for Cogni-*

tive Science (Cambridge, Mass.: MIT Press, 1984); P. Hayes, "The Naive Physics Manifesto," in *Expert Systems in the Electronic Age*, ed. D. Michie (Edinburgh: Edinburgh University Press, 1969), pp. 463–502; "The Second Naive Physics Manifesto," in *Formal Theories of the Commonsense World*, ed. J. R. Hobbs and R. C. Moore (Norwood, N.J.: Ablex, 1985), pp. 1–36. R. C. Moore, "The Role of Logic in Knowledge Representation and Commonsense Reasoning" in *Readings in Knowledge Representation*, ed. R. J. Brachman and H. J. Levesque (San Mateo, Calif.: Morgan Kaufmann, 1985), pp. 336–41.

6. D. C. Dennett, *The Intentional Stance* (Cambridge, Mass.: MIT Press, 1987).

7. R. C. Schank, *Conceptual Information Processing* (Amsterdam: North-Holland, 1975); *Dynamic Memory: A Theory of Learning in Computers and People* (Cambridge: Cambridge University Press, 1982); R. C. Schank and R. Abelson, *Scripts, Plans, Goals, and Understanding* (Hillsdale, N.J.: Lawrence Erlbaum, 1977).

8. D. R. Hofstadter, "Reflections on John R. Searle: Minds, Brains, and Programs" in *The Mind's I* by D. R. Hofstadter and D. C. Dennett (New York: Basic Books, 1981).

9. The journal *Behavioral and Brain Sciences* 3 (1980), where Searle's article containing the Chinese Room Argument "Minds, Brains, and Programs" was first published, together with a series of replies.

10. See the chapter "Intentional Systems" in D. C. Dennett, *Brainstorms* (Cambridge, Mass.: MIT Press, 1978).

11. D. C. Dennett, *Consciousness Explained* (Boston: Little Brown, 1991).

Hubert L. Dreyfus was born in 1929. He studied at Harvard and held fellowships for research in Europe. He taught at the Massachusetts Institute of Technology from 1960 to 1968. Currently he is Professor of Philosophy at the University of California at Berkeley.

HUBERT L. DREYFUS

Cognitivism Abandoned

How does a philosopher get involved in cognitive science? What is the connection between your work and cognitive science?

Well, I got into it even before there was cognitive science, through Artificial Intelligence, because I was teaching at MIT about 1962 or 1963. Students from what was then called "Robot Project"—that is, from Minsky's job—came and said that they had already solved or were solving the problems that philosophy was worried about, like understanding and knowing and so on. It seemed to me at the time that that was unlikely, but if they had, it was something a philosopher had better know about.

So I went to the Rand Corporation in 1965 to see what was going on in Artificial Intelligence. Rand Corporation was, in those days when Stuart, my brother, and I were there in Santa Monica, a think tank paid for by the military, where a bunch of high-quality intellectuals could sit around and do pretty much what they pleased. That was also where Newell, Shaw, and Simon were doing their work.[1]

What I discovered was that far from being antiphilosophy—and far from what I had suspected—these people were indeed working on the very problems that philosophers had worked on, and they were in some way—even if they weren't making much progress—at least turning these problems into a research program in which one could make progress. I also discovered that they represented just that rationalist tradition in philosophy that the people that I taught—people like Heidegger and Wittgenstein and Merleau-Ponty—were calling into question. It struck me as interesting and ironic that the people in Artificial Intelligence, with their view of the mind as fundamentally rational, problem-solving, and rule-governed, had inherited the rationalist or intellectualist tradition that comes from Descartes to Leibniz to Kant to Husserl, whom I also was teaching. It seemed to me at that point, and it still seems to me, clear that if they were right—that is, if Artificial Intelligence succeeded, then the main philosophical tradition had been on the right track; but if they failed, it would show that Wittgenstein and Heidegger and Merleau-

Ponty, who were already criticizing this very tradition for holding that there were representations in the mind that were manipulated according to rules, were right. It would be good evidence for the validity of the phenomenological critique.

So, in your opinion, there was an overlap between the same interesting questions in both Artificial Intelligence and philosophy? Which questions were these?

The really important questions were: does the mind work by having in it symbolic representations of the world? The Cartesian tradition says that the world is copied in the mind, in brain language, as Fodor would say,[2] or in a physical symbol system, as Newell and Simon would say.[3] That's one thesis. And that these representations are manipulated according to strict rules, which is what they somehow, of course, all hold and what Descartes, Kant, and in a way Husserl also held. Those are the two theses.

Which brings me, by the way, to cognitivism, because representationalism is not just a thesis of Artificial Intelligence, it is a thesis of cognitivism as I understand it. It seems to me that Artificial Intelligence and then cognitivism as a generalization of Artificial Intelligence is the heir to traditional, intellectualist, rationalist, idealist philosophy. That is what fascinates me: that they took up the tradition at almost exactly the moment when people in the Anglo-American world stopped believing in it. Wittgenstein's *Philosophical Investigations* were first published in 1953,[4] shortly before Newell and Simon started to believe in symbolic representations. Of course, Heidegger had already criticized this devastatingly in 1927,[5] and Merleau-Ponty had applied Heidegger to criticizing what he called intellectualism in 1945.[6] So, the people in cognitivism have inherited a certain research program—which seems to me a fascinating but wrongheaded and ultimately abandoned direction in philosophy.

Why do you think they took this approach in cognitive science and Artificial Intelligence, just at the time when philosophical criticism was getting stronger?

Well, I think it's really a kind of "trickle-down" effect, in which what was at one time a philosophical position that needed defending got to be taken more and more for granted, and filtered down, so that in a way everybody just believed it. So that, by the time Artificial Intelligence and cognitivists come along, it doesn't even seem to be an assumption anymore. It just seems to be a fact that the mind has in it representations of the world and manipulates those representations according to rules. What I talked about in *What Computers Can't Do*[7] was the amazing way that the early people in Artificial Intelligence—Newell and Simon and Minsky—didn't even try to argue for this notion of representations and rules but just assumed them. That's how philosophy works, I think: what seems to be a difficult philosophical position and needs arguing finally gets down to being accepted by every academic as just self-evident. That's the only reason I can think of why they should believe it. And, I guess we have to add, computers: the idea that computers work just by following programs—namely rules—and can manipulate representations, that was the great idea of Newell and Simon, which was quite brilliant if wrongheaded. Their idea was that you could treat a computer as "a physical symbol system," that you could use the bits in a computer to represent features in the world, organize these features so as to represent situations in the world, and then manipulate all that according to strict rules so as to solve problems, play games, and so forth.

The computer made it possible to make the intellectualist tradition into a research program just as connectionism has now made it possible to make the empiricist, associationist tradition in philosophy, which goes from Hume to the behaviorists, into a research program. In 1957, these two traditions were competing: the empiricist-associationist tradition in Rosenblatt's perceptron and the idealist–rule-following tradition. There are lots of different reasons why the rationalist tradition won out; there is no one simple answer to that.[8]

You mentioned several times the notions of cognitivism and cognitive science. Could you explain what you mean by them?

I want to distinguish cognitive science from cognitivism. I think cognitive science is simply any theory about how the mind does what it does—that is, cognitive science is anything but behaviorism. It is any theory that contends that there is a mental level of processing. Cognitive science is any attempt to explain how the mind and brain produce intelligent behavior. So cognitive science is certainly here to stay. It's just a name for a natural confluence of all the disciplines that study the mind-brain: philosophy, linguistics, computer science, anthropology, neuroscience, psychology. All those disciplines are naturally converging on this question, and what they do is cognitive science.

But cognitivism is a special view. As behaviorism is the view that all you can study is behavior and maybe all that there *is* is behavior, so cognitivism is the special view that all mental phenomena are fundamentally cognitive, that is, involve thinking. That is why it is rationalism. Cognitivism is the peculiar view that even perception is really thinking, is really problem solving, is really manipulation of representations by rules. It seems, when you think of it, counterintuitive that even skills, emotion, perception—those areas that are opposed to cognition and thinking—are just unconscious cognition and thinking. But that is, I think, the thesis of cognitivism. So cognitivism could turn out to be wrong—in fact, I think cognitivism has pretty well already turned out to be wrong. I think that there isn't any philosophical argument or empirical evidence that supports it. But that would simply leave cognitive science channeled in different directions.

What role does Artificial Intelligence play in these notions?

Artificial Intelligence is a good example of cognitivism, and it was the first version of cognitivism. It is an example of trying to make a model of the mind, of making something that does what the mind sometimes does, using rules and representations and cognitive operations such as matching, serially searching, and so on. So here is Artificial Intelligence, trying to do what the mind does, using computers in a special way, namely as physical symbol systems performing cognitive operations. That is a good model for cognitive science: cognitive science just adds one element, saying: "We won't be satisfied if this thing just does what the mind does; we want it to do it *the way* the mind does it." Then you have not Artificial Intelligence but cognitive science as a theory of mind. To make this distinction clear: if you had a machine playing chess by brute force, just simply counting out, it would still be Artificial Intelligence, but it certainly would be no contribution to cognitive science, because everyone knows that chess players could not possibly be processing information fast enough to be consciously or unconsciously counting out all possible positions up to seven-ply, which is something like twelve million moves, as good chess machines now do. So cognitivism is a development out of a certain subpart of Artificial Intelligence.

The relationship between them is like that between theory and practice?

No, it is not like theory and practice, although that's some aspect of it. If anybody had a theory in cognitivism, for example, how one recognizes speech, one could try then to build a computer that used it. That would be theory and practice. And they do have that relation of engineering to science. But the theory and practice distinction is not the right one. You find that distinction right in Artificial Intelligence, between the people who want to solve general problems about cognition—who are like the theorists—and the people who want to build expert systems and sell them using some of these techniques, who think of themselves as knowledge engineers. The distinction is right between the broader field that says: "Let's do the sort of things that the mind enables people to do, using computers." That is Artificial Intelligence. And a subpart of Artificial Intelligence is a contribution to cognitivism: "Let's do that using a computer, and let's do it following the very same steps that experiments show that the mind follows." So a part of Artificial Intelligence is a part of cognitive science, but only a part, because there are other ways to understand what the mind's program is—for example, in linguistics, by looking at people's judgments about grammaticality, or in psychology, using fast reaction times. And there are other ways to build models than to make actually running computer programs. It is a complicated relationship.

How would you outline the history of Artificial Intelligence from the starting point until now?

There are several periods. The first period was the heuristic programming period, about five years, which was when the action was around Newell and Simon, at Rand and at Carnegie-Mellon. Then MIT took over the scene with their notion of semantic information processing,[9] and that is when Minsky and Papert were the leaders. They said that to try to solve problems without knowledge was no use, so they would store some knowledge in their programs. They called this semantic information processing, because they believed that the knowledge would enable them to get the meaning of what they were processing. But that turned out to be a trick, too.

So there are roughly five years of heuristic programming, or "cognitive simulation," from 1957 to 1962, five years of semantic information processing from 1962 to 1967, and then, as I see it, one big failure from, roughly, 1967 onward: fifteen years of frustration and stagnation. Minsky and Papert at MIT remained the leaders with their discovery, around 1967, of "microworlds." Terry Winograd became the star there, and the idea was to try to discover a domain that is sufficiently self-contained and simple, and yet sufficiently complex, too, that you could write a program that works and was interesting without running into any hopeless problems.[10] What microworlds were staving off and hiding becomes obvious around 1972 (of course, this chronology is schematic): it was the commonsense knowledge problem—that you would really have to understand the whole background of commonsense that people understand, and all that is involved, for example, in just having a body that moves forward more easily than backward, that orients itself in space, and so on. Where and how could such knowledge be stored?

That commonsense knowledge problem seems to totally resist Artificial Intelligence's attempts to deal with it: if you got all that knowledge into the computer, you would not know how to retrieve it. But no one even knows what it is or how to put it in. The only idea is Lenat's, who wants to put the Encyclopaedia Britannica in,[11] but nowhere in the Encyclopaedia Britannica does it say that people move forward more easily than they move backward, or that doctors wear underwear, as John

Searle's latest example goes. The background knowledge is precisely what is not in an encyclopedia. It is what every person knows just by growing up in our culture. And a lot of it is not even facts. People do know facts like that doctors wear underwear, but that you move forward more easily than backward—people know that just by doing it. You can turn it into a fact—I just turned it into a fact by spelling it out—but people know a lot without having ever thought of it. For almost twenty years now, this problem has resisted any breakthrough. So I think Artificial Intelligence is stuck in its third phase.

Do you think that this problem with commonsense knowledge has already been realized in the Artificial Intelligence community?

It is realized, not as a failure but as a "much harder problem than we thought." So Roger Schank has been saying for five years that the commonsense knowledge problem is the big problem and writing what seems to me very ad hoc programs trying to deal with it, all of which—because they are ad hoc—are not leading anywhere and not gaining any followers outside of Schank's own group. This is my perception of it. Then there are Newell and Simon. Newell has an architecture called Soar,[12] which is supposedly going to help, but as I see it, it is again too oversimplified. It totally evades the commonsense knowledge problem and does not gain any followers outside the Carnegie group. Then there is Lenat and his wonderful but, I think, hopelessly misguided attempt to program the whole of human knowledge into the computer.

Then there are the people at Stanford who had "commonsense summer." They were trying to formalize commonsense physics. I don't know what *they* concluded, but *I* concluded, reading their report, that their results were ludicrous. They could not formalize even a simple-seeming intuition we have that if you push down on a piece of putty in a cylinder with a hole in it the stuff will extrude out of the hole. The formalization of this takes pages and pages, and it is hard to believe that any child has got this in his consciousness or unconsciousness—or needs it. Behind all this is my assumption that people will eventually see that what children know are a whole lot of special cases that they acquire playing endlessly with blocks, sand, water, and so forth, and what is needed to capture a child's understanding is a lot of pairs of particular situations and results.

Connectionism—or more exactly, the neural networks—stand a much better chance of capturing this learning by examples paired with results. But let's go back to your question: it seems to me that nobody in Artificial Intelligence has said that there has been a failure in symbolic Artificial Intelligence. But there is nothing like a unified research program. Each of these three or four schools—Carnegie-Mellon, Yale, MIT, and Stanford—thinks that it is on a promising track that will eventually solve the commonsense knowledge problem. None of them has made enough progress to convince anybody outside their school or group.

So, in your opinion, the commonsense knowledge problem will cause the breakdown of Artificial Intelligence?

Yes, symbolic processing will fail to solve the commonsense knowledge problem. But there is the theory in Artificial Intelligence that this is a question only of memory. Given enough time and capacity, they say, they will not exactly solve the problem but will store a lot of information, and so, they say, they "will build machines that will behave like human beings in certain respects."

Either you have to go back to microworlds and take a domain like the blocks world or the domains of expert systems, which are so limited that they are cut off from commonsense knowledge and have only a very special domain of knowledge, or else you have to put in a vast amount of commonsense knowledge. In that case, I don't see—and I think none of them has even faced the problem—how you get out the relevant knowledge.

At the time being, the problem is how you could get this vast amount of knowledge into that superbig memory. I would doubt that you could, because the most important part of commonsense knowledge is not "knowledge" at all but a skill for coping with things and people, which human beings get as they grow up. But let's suppose they do put in the Encyclopaedia Britannica, or something else ten times bigger than the Encyclopaedia Britannica that has all the background stuff that is not in the Encyclopaedia Britannica, then I do not see how they could possibly access the part that they needed in any given situation of the sort these programs try to deal with—let's say, language understanding, motion of autonomous vehicles, or whatever it is they are trying to deal with in the everyday world. Nobody has ever made a program that deals with anything in the everyday world successfully.

So, commonsense knowledge does not consist of propositions, and therefore you cannot represent it in a computer? And, besides, if it consisted of propositions, you would have the problem of retrieval?

Right.

You mentioned that connectionism perhaps has a better chance of solving these problems.

Connectionism, or what is apparently more and more called neural network simulation, was always in the background, since 1957, with Rosenblatt. It becomes important as a really serious research program only around 1985 or 1986. Certainly since 1986, when *Parallel Distributed Processing* was published,[13] it has been a big deal. Since that point, cognitivism and Artificial Intelligence look new, because now there are two competing research programs within cognitivism and Artificial Intelligence. I find this very interesting. In Artificial Intelligence, there is the attempt to make machines intelligent using rules and representations, which has been stuck for fifteen years over the commonsense knowledge problem and which could never deal with learning; and then there are those people who discovered that you could train nets so that learning could be basic. You do not have to give the nets any facts at all; you just give them a lot of pairs of inputs and outputs. That becomes another research strategy in Artificial Intelligence.

That means now that cognitive science has two tracks. It has cognitivism, which is the old, failed attempt to use rules and representations, which takes up the rationalist-intellectualist tradition. Now, nets, when generalized into cognitive science and not used just in Artificial Intelligence, take up the philosophical associationist and behaviorist tradition, empiricist as opposed to rationalist—namely that there is nothing in the mind, that the mind is a blank tablet that gradually gets filled with association pairs of inputs and outputs, and that learning consists simply in having associations. This looked like a discredited view while rationalists were running the show, but now it looks to many people like a very promising approach to intelligence.

Already in 1957 Rosenblatt was working on neural nets. What is the reason for the great success of connectionism in recent years?

Mainly because we've got big, big computers—and it takes lots and lots of computing time to do anything. The fact that we now have supercomputers enables people to carry Rosenblatt's intuitions through. And then, there are new algorithms like back propagation, which is also interesting because it makes a clear distinction between Artificial Intelligence and cognitivism. Just as brute force on the symbolic level cannot count as cognitive science or cognitivism, because it is not the way the mind could work, so back propagation as an algorithm for neural network learning cannot count as a theory of neuroscience or psychology, because everybody knows that the brain is not wired this way to do things. So "neural net Artificial Intelligence" might and, I think, will succeed in getting computers to do something intelligent like voice recognition and handwriting recognition, but no one would think of that as a contribution to cognitive science, because it will clearly not be the way people do it. Then there will be another part of neural network simulation that is a contribution to cognitive science and neuroscience, because it will have different algorithms, the sort of algorithms that could be realized in the brain. Those working in this area can make progress that Rosenblatt could not make partly because he did not have computers big enough, but partly because he lost all his support for his projects after the symbolic people convinced everybody in our rationalist world that obviously thinking was the way it would go; whereas there is no thinking in a neural net, it is just associative.

But behaviorism did not work very well, either. My guess is that neural nets were just coming in at the time when behaviorism was going out. The end for behaviorism began around 1967, that is, just about the time when support for neural nets was completely lost. I would say something like: in our rationalist world, with computers that were not very powerful, it just seemed obvious that symbolic information processing was the way to go. Only now, when symbolic information processing has been given a chance and shows that it is not the way to go, when we have computers big enough to try to do neural nets again, will all sorts of associationist models come back into cognitive science.

What do you think about the Turing Test? Is it a useful idea?

Lots of philosophers have pointed out—and I am sure that they are right—that the Turing Test is not a good test of anything. Something can pass the Turing Test and still not have consciousness or intentionality. But I have always liked the Turing Test, because unlike philosophers like Searle and others I want to pose my questions in ways that are most favorable for the people in Artificial Intelligence and cognitivism and see if they can succeed by their own standards. I want to make it as easy as possible for them and show that they still cannot pass the test. So I would be happy with the Turing Test, and I would say that they cannot even pass the Turing Test—that is to say: they cannot even pass their own test for success. I would not impose any higher standard on them by saying, "Look, your machine has got no consciousness." I accept their ground rules and want to show that they are failing on their own terms.

Will there be any important development in some fields of applications in the next years, or is Artificial Intelligence just a complete failure? What about expert systems, for example?

I think Artificial Intelligence is a great failure, but I think also there are a few useful applications. So it is not a total failure. But I think expert systems are almost dead already. It is turning out to be obvious, I think, that expert systems are terribly expensive to build, terribly expensive to keep up-to-date, and never as good as the experts whose rules they supposedly are running.

And this last part is a philosophically interesting one: it seems to me that expertise is not based on rules. When you ask an expert for his rules, he regresses to what he remembers from when he was not an expert and still had to use rules. That is why expert systems are never as good as intuitive experts.[14] That leaves only two areas where expert systems will be interesting, and these are relatively restricted areas.

In some domain cut off from commonsense and where it is not important to be expert—maybe moneylending or something like that, where it is more important to be routine and cheap, and not go on strike or get tired—there could be useful expert systems, I imagine. But where it is a matter of life and death, like in medicine, I do not think expert systems will actually be used. There are systems, some of them in use, like the PUFF system, but it turns out you can algorithmatize the whole thing, and you do not need any rules from experts. There will certainly be places for algorithmic systems like MACSYMA, which does high-level math, but that is not an expert system, by my definition—it is not based on rules elicited from experts.[15] Where you can find a domain in which there are algorithms for solving problems, you can certainly use computers, but I would not call this Artificial Intelligence anymore, because it is obvious to everybody that human beings do not use these algorithms; they are doing it some other way.

The only two areas where expert systems are useful are where you can have a system that is just competent in a domain that is isolated from the rest of the world, and—here is the most interesting use, but I have heard of only two examples—find an area of expertise where the experts are not intuitive and have to make calculations, because the domain is so complex and changing, for instance the famous R1 or XCON system by DEC, which configures VAXes.[16] It turns out that the people doing the job of customizing mainframe computers had to make a lot of calculations about the capacities of components on the basis of the latest manuals, and so, naturally, machines can do that better. There were no intuitive experts who had a holistic view of configuration. Another example is the system that the Defense Department uses, called ALLPS, which was developed by SRI (Stanford Research Institute) and was not meant to be an expert system. It was not developed by Artificial Intelligence people. It is an expert system, however, in that it uses rules obtained from people who load transport planes. It can figure out much faster than they can how to arrange cargo in a transport plane, so as to get the most inside and keep the center of gravity in balance and so that you can get everything out in the right order. It is a big, big complicated problem. Two experts had to calculate for many hours to do it. Now, one program can do it in a few minutes. That's a real success, and it is real Artificial Intelligence, a real expert system. But one can expect such success only in isolated domains, where calculation is required for expertise.

Feigenbaum once said that expert systems would be the second and the real computer revolution.[17] But in your opinion the social changes caused by Artificial Intelligence and expert systems won't be so drastic?

Feigenbaum has a complete misunderstanding of expertise, again a rationalist misunderstanding, as if experts used rules to operate on a vast body of facts that they acquire. Then expert systems would eventually take over expertise and wisdom, vastly changing our culture. But because this is not the case, you already get remarks in the *New York Times* and the *Wall Street Journal*. Let me give you some examples.

In 1986, in the *San Francisco Examiner*, an Artificial Intelligence researcher from the Rand Corporation says: "There has been great progress in other areas of computer science while Artificial Intelligence is still saying 'Give us more time!' How long should we wait? They are not going to make any breakthroughs, their stuff is not working, we are investing lots of money, but Wall Street is not going to throw away money forever!" The best one is an article from the *New York Times*, I think, that says that you don't have to worry anymore about the expert systems taking away your job if you are an expert, because it has now turned out that they are not good enough to take over anybody's jobs. So it is becoming clear already. And I gather, though I do not know all the details, that the whole Fifth Generation program in Japan has pretty well failed. It seems to have failed, surprisingly enough, even to produce the promised spin-off. Lots of people said, "Well, it will fail to produce intelligent systems in ten years, but it will produce lots of other interesting things." I haven't heard any interesting things that have come out of the Fifth Generation, and it's been around for five or six years, I think.

So the *New York Times*, which was all for expert systems in 1984, saying that they opened up a new prospect of computer-aided decision making and more wisdom than any one person can contain, said later, in 1987: "True expertise, it turns out, is a subtle phenomenon and one that can rarely be replicated in the pre-programmed rules that enable a software to simulate the thinking of its creators." This is part of the article that says that experts do not have to worry about expert systems.

So what are the future prospects of Artificial Intelligence, in your opinion?

I will make a prediction. Since I am always beating on everybody else for making predictions, it seems only fair that I expose myself: my guess is that symbolic information processing and cognitivism will disappear within ten years, because, on the one hand, now that people see another possibility, they will begin to see how strange it was to think it was obvious that you could produce intelligence using rules and representations—and that together with the fact that the commonsense knowledge problem remains unsolved, and nobody has even a clue about how to solve it. On the other hand, attacks from people like Searle on the philosophical foundations of cognitivism have convinced people like Robert Cummins, for instance, to give up cognitivism. And Fodor is very much on the defense now, with his notion of brain writing and so on. Fewer and fewer people find cognitivism philosophically plausible. I predict that within ten years there won't be any cognitivism or symbolic Artificial Intelligence around and that researchers will have turned to neural network simulation in Artificial Intelligence, and to a Gestalt/associationist/prototype-oriented work in cognitive science.

Recommended Works by Hubert L. Dreyfus

Being-in-the-World: A Commentary on Heidegger's Being and Time, Division I. Cambridge, Mass.: MIT Press, 1991.
Mind over Machine. New York: Free Press, 1986.
What Computers Still Can't Do: A Critique of Artificial Reason. Cambridge, Mass.: MIT Press, 1992. 3d ed. of his 1972 book criticizing Artificial Intelligence, *What Computers Can't Do: The Limits of Artificial Intelligence.* New York: Harper and Row, 1972.
(With S. E. Dreyfus.) "Making a Mind vs. Modeling the Brain: AI Back at a Branchpoint." *Daedalus* (1988). Reprinted in *The Artificial Intelligence Debate,* ed. S. Graubard. Cambridge, Mass.: MIT Press, 1988.
(Ed., with H. Hall.) *Husserl, Intentionality, and Cognitive Science.* Cambridge, Mass.: MIT Press, 1982.

Notes

1. See the interviews of A. Newell and H. A. Simon in this volume.
2. J. Fodor, The Language of Thought (Cambridge, Mass.: MIT Press, 1975).
3. A. Newell, "Physical Symbol Systems," *Cognitive Science* 4 (1980): 135–83.
4. L. Wittgenstein, *Philosophical Investigations* (Oxford: Blackwell, 1953).
5. M. Heidegger, *Sein und Zeit,* vol. 8, first published in *Jahrbuch für Philosophie und phänomenologische Forschung,* ed. E. Husserl (Halle, 1927), and in English as *Being and Time* (New York: Harper and Row, 1962).
6. M. Merleau-Ponty, *Phénomenologie de la perception* (Paris: Gallimard, 1945), published in English as *Phenomenology of Perception* (London: Routledge and Kegan Paul, 1962).
7. H. L. Dreyfus, *What Computers Can't Do: The Limits of Artificial Intelligence* (New York: Harper and Row, 1972).
8. See H. L. Dreyfus and S. E. Dreyfus, "Making a Mind versus Modeling the Brain: Artificial Intelligence Back at a Branchpoint," in *The Artificial Intelligence Debate,* ed. S. Graubard (Cambridge, Mass.: MIT Press, 1988), pp. 15–44.
9. M. Minsky, ed., *Semantic Information Processing* (Cambridge, Mass.: MIT Press, 1968).

10. T. Winograd, "Understanding Natural Language," *Cognitive Psychology* 3 (1972): 1–191; the program became famous as "SHRDLU."

11. D. B. Lenat and R. V. Guha, *Building Large Knowledge-Based Systems: Representations and Inference in the CYC Project* (Reading, Mass.: Addison-Wesley, 1990).

12. See J. Michon and A. Anureyk, eds., *Perspectives on Soar* (Norwell, Mass.: Kluwer, 1991).

13. D. Rumelhart, J. McClelland, and the PDP Research Group, *Parallel Distributed Processing: Explorations in the Microstructure of Cognition*, 2 vols. (Cambridge, Mass.: MIT Press, 1986).

14. See H. L. Dreyfus and S. E. Dreyfus, *Minds over Machine* (New York: Free Press, 1986).

15. PUFF is an expert system developed by the Heuristic Programming Project (Stanford) that analyzes patients' data in order to diagnose pulmonary infections; see E. A. Feigenbaum and P. McCorduck, *The Fifth Generation: Artificial Intelligence and Japan's Computer Challenge to the World* (Reading, Mass.: Addison-Wesley, 1983), p. 96. MACSYMA is a program for mathematical theorem proving, using heuristic pattern-matching techniques. It was developed at MIT in the late sixties; see H. C. Mishkoff, *Understanding Artificial Intelligence*, 2d ed. (Indianapolis: Howard W. Sams, 1988), p. 94.

16. XCON (or R1) was first built by Digital Equipment Corporation and Carnegie-Mellon University in 1980. It is a production-rule system that assists engineers in configuring VAX minicomputers to customer specifications; see H. C. Mishkoff, *Understanding Artificial Intelligence*, pp. 85–86.

17. In E. A. Feigenbaum and P. McCorduck, *The Fifth Generation*.

Jerry A. Fodor was born in 1935 and attended Columbia College, Princeton University, and Oxford University. He holds a Ph.D. in philosophy from Princeton University (1960). From 1959 to 1986 he taught at MIT, first as Instructor and then as Assistant Professor (1961–63) in the Department of Humanities and as Associate Professor in the Departments of Philosophy and Psychology (1963–69). From 1969 onward he has been Professor in the Departments of Philosophy and Psychology. In 1986, he became Distinguished Professor and in 1988 Adjunct Professor at CUNY Graduate Center. Since 1988, he has been State of New Jersey Professor of Philosophy at Rutgers University.

JERRY A. FODOR

The Folly of Simulation

I was trained as a philosopher at Princeton. My first academic job was at MIT, about 1960, when all the Chomsky[1] stuff was starting. There was a lot of linguistics around there, which I picked up a little. Then I spent a year visiting the University of Illinois, where I was involved in a program of Charles Osgood's.[2] I had a position in a research program he was running. I talked a lot with graduate students about psychology, and so I got involved in doing some experiments. For fairly fortuitous reasons, I picked up some information about philosophy, linguistics, and psychology. Since a lot of it seemed to bear on the same set of problems, I have worked back and forth between those fields ever since. It turned out that my own career had elements of the various fields, which converged to become cognitive science, except computer science, about which I know very little.

In your opinion, which sciences are part of the enterprise of cognitive science?

The standard consensus is that the core disciplines are computer science, linguistics, philosophy, probably mostly cognitive psychology. In my own view, cognitive psychology is the central discipline. Then neuroscience, and maybe parts of anthropology and fields like that. My own view is that cognitive science is basically just cognitive psychology, only done with more methodological and theoretical sophistication than cognitive psychologists have been traditionally trained to do it. If you are a cognitive psychologist and know a little bit about philosophy of mind, linguistics, and computer theory, that makes you a cognitive scientist. If you are a cognitive psychologist but know nothing about these fields, you are a cognitive psychologist but not a cognitive scientist.

So cognitive science is dominated by cognitive psychology?

I am not sure that this is right as a sociological claim, but it is as far as the ideology of the field goes, as far as its intellectual structure is concerned. I think it is, basically, just sophisticated cognitive psychology. Unfortunately, cognitive psychology as people are trained to practice it, at least in this country, has been traditionally committed to methodological empiricism and to disciplinary isolationism, in which it was, for example, perfectly possible to study language without knowing anything about linguistics. Traditionally, cognitive psychologists have had the view of the world that experimental psychologists are inclined to. Cognitive science is cognitive psychology done the way that cognitive psychology naturally would have been done except for the history of academic psychology in America, with its dominant behaviorist and empiricist background.

Which role does Artificial Intelligence play in cognitive science? Do you think Artificial Intelligence is a science on its own, or is it part of computer science?

I think what I guess a lot of people in Artificial Intelligence think: that it is basically engineering. The relation to cognitive science is actually fairly marginal, except insofar as what you learn from trying to build intelligent mechanisms, to program systems to do things, may bear on the analysis of human intelligence. You have to distinguish between cognitive science as an attempt to understand—roughly speaking—thought and any attempt to build intelligent machines, which is an engineering problem and has no intrinsic scientific interest as such.

And again, you have to distinguish between a certain model of what thought processes are like, namely that thought processes involve transformations of representations, that thinking is essentially a matter of operating on mental symbols—an idea that, basically, starts with Turing and has been elaborated by a lot of people since. You have to distinguish between that very profound idea, the central idea of modern psychology, which is endorsed by most cognitive science people and by most Artificial Intelligence people, and any particular purposes to which that idea might be put. Artificial Intelligence attempts to exploit that picture of what mental processes are like for the purpose of building intelligent artifacts. It shares the theory with cognitive science viewed as an attempt to understand human thought processes, but they are really different undertakings. The project of simulating intelligence as such seems to me to be of very little scientific interest.

To build intelligent machines, without making claims about the mind? In your opinion, it is more interesting to make the claims and then build the machines?

I don't see the point of building machines. You don't do physics by trying to build a simulation of the universe. You don't try to build a device that would pass the Turing Test for God; that would be a crazy way of doing physics. And there is a reason why it is a crazy way: the observational phenomena, the things you can see when you look at the

world naively, pretheoretically, are the effects of horrendously complicated interactions of the underlying mechanisms. It may be in principle
possible, but in practice it's out of the question to actually try to reconstruct those interactions. What you do when you do science is not to
simulate observational variance but to build experimental environments
in which you can strip off as many of the interacting factors as possible
and study them one by one. Simulation is not a goal of physics.

It seems to me that exactly the same principles would apply in psychology. What you want to know is what the variables are that interact
in, say, the production of intelligent behavior, and what the law of their
interaction is. Once you know that, the question of actually simulating
a piece of behavior passing the Turing Test is of marginal interest, perhaps of no interest at all if the interactions are complicated enough. The
old story that you don't try to simulate falling leaves in physics because
you just could not know enough, and anyway the project would not be
of any interest, applies similarly to psychology.

There are two paradigms of exploration in the field. One says: the way
to understand intelligent phenomena is to simulate them. The other says:
the way to understand intelligent phenomena is to view them as the interaction of underlying mental processes, and then to build experimental
environments in which you can study the processes one by one. The
former strategy is the characteristic of Artificial Intelligence, insofar as
Artificial Intelligence is thought to be relevant to human behavior at all.
The latter is the strategy of cognitive psychology. It seems to me that all
the scientific precedent favors the latter way.

This is a long answer to a short question. The short answer would
be: aside from engineering, aside from the fact that it would be nice to
make robots and so on, I don't see the scientific interest of simulating
behavior.

*You mentioned the Turing Test. What do you think about it? Is it a
useful idea in the enterprise of cognitive science?*

If you don't think that simulation is the goal of cognitive science, then
you are not likely to be impressed by the Turing Test. But anyway,
though I am an enormous admirer of Turing—I think he had the only
serious idea, since Descartes anyway, in the history of the study of the
mind—I don't think the Turing Test is of any interest. What the Turing
Test asks you to do is accept a hypothesis—that the machine is thinking
or so—on the basis of finding no difference.

Try to get an experimental paper accepted in a journal on the following ground: I have got an argument that says that Pluto does not have
any rings, namely I looked at it with this telescope and I could not see
them. What the editor of the journal will tell you is that that shows one
of two things. Either Pluto does not have any rings, or your telescope is

not sensitive enough to see them. Similarly for the Turing Test: if a machine can pass the Turing Test, that shows either that the machine is
intelligent or that the judge did not ask the right questions.

It would be interesting if the test were given by a judge who knows
what questions to ask. But that raises the whole question over again,
namely what the questions are that define intelligence. As a proposal
for verification of theories in cognitive science, the Turing Test is a bad
idea.

*What would be the research strategy to follow to study the mind, in your
opinion?*

What you do in any other field. It is what you would do to study
automobiles if automobiles turned out not to be artifacts. You would
develop theories that explain the observable behavior. Then you construct artificial environments in which details of the theory can be tested.
I don't think there is any interesting answer to the question of what we
should do in cognitive science that would be different from the answer to
the question of what we should do in geology.

*What do you think is the central point or the very idea of cognitive
science?*

The idea that mental processes are transformations of mental representations. The basic question in cognitive science is, How could a mechanism be rational? The serious answer to that question is owing to
Turing, namely: it could be rational by being a sort of proof-theoretic
device, that is, by being a mechanism that has representational capacities—mental states that represent states of the world—and that can operate on these mental states by virtue of its syntactical properties. The
basic idea in cognitive science is the idea of proof theory, that is, that you
can simulate semantic relations—in particular, semantic relations
among thoughts—by syntactical processes. That is what Turing suggested, and that is what we have all been doing in one or the other area
of mental processing.

*In the last few years, this view came under challenge—for example, the
thought experiments of Searle, who says with his Chinese Room parable
that you cannot understand with only syntactical processing and formal
representation.*

There is a lot to be said about that. The first thing is that Turing's idea
was not that you can answer the question syntactically: what is it for a
mental state to have content? It is an answer to the question, How do
you construct a device whose state transitions or processes will respect
or preserve the content of mental states? You have a device whose repre-

sentations of the world have both semantic and syntactical properties. The question the Turing machine is an answer to is, How could the state transitions of such a device be respected in their intentional, semantic properties?

So you have a state that constitutes the belief *that p and q*. How can you guarantee that this state will cause a state that constitutes the belief *that q*? Answer: by exploiting parallelisms between semantic and syntactical properties. That is not the answer to the question, What is it for a mental state to have truth conditions or intentional properties? Given you have an answer to the first question, whatever it is, and given a mechanism whose states have intentional properties, it is an answer to the question, How could you imagine the causal processes of that mechanism preserving these properties? This question is quite different from Searle's.

Do you think the question of intentionality does not matter at all?

There are two questions that you want to answer. The answer to one is Turing's; the other is still open. They come in reverse order. The first question is, What is it for a physical state to have semantic properties? That is the question Searle is worrying about. I entirely agree that the Turing idea does not answer that question and is not supposed to.

The other question is, Given that a state of a physical system has semantic properties, how could the state transitions, the causal processes of that system, preserve those properties? How is it that we go from one true thought to another? That does not answer the question of what gives our thoughts truth conditions, and that is the problem of intentionality, basically.

Turing's answer seems to me to be the right one, namely that the states that have truth conditions must have syntactical properties, and preservation of their semantic properties is determined by the character of the operations on the syntactical properties.

To resume: Searle is not talking about intelligence, he is talking about intentionality. Turing talks about intelligence, not about intentionality. That is quite different. Artificial Intelligence is about intelligence, not about intentionality. I think people in Artificial Intelligence are very confused about this.

But Searle claims that you cannot say anything about intelligence without intentionality.

I know, but there is the following division of the question. Given that you have a device whose states have intentional properties, what kind of device would be such that its state transitions are intelligent? These are two different questions. The whole cognitive science enterprise does not bear on the question of what it is for the state of a mechanism to have intentional properties. But it does bear on the question of what it is for the state of a mechanism to be intelligent, namely what it is to preserve truth.

Even if you knew what it is for a thought to be true—that is the intentionality problem, basically—you would be faced with the question, How is it that our thought processes take us from one true thought to another? How come our thought processes are not incoherent, in the sense that we don't go from the thought "It is raining outside" to the thought "Pluto has rings." The question of what makes our thought processes coherent remains even after you know what it is for them to have intentional properties. It is the former of these questions that the cognitive science enterprise bears on.

The main idea is to exploit the fact that syntactical properties can preserve semantic properties. That is the basic idea of proof theory. Searle is perfectly right when he says that the Turing enterprise does not tell you what it is for a state to have semantic properties. Whatever the right theory about intentionality is, it is not what cognitive science is going on about. I think you can have a theory of intentionality, but this is not what cognitive science is trying to give. If we assume intentionality, the question becomes, Can we have a syntactical theory of intelligence?

If I understand you correctly, then all his criticism does not have any force? What about Dreyfus's commonsense argument?

What Dreyfus thinks is a much deeper issue. He thinks that it is impossible to mechanize intelligence. Dreyfus thinks that no syntactical theory will account for the coherence of mental processes, whatever it is that accounts for their intentionality. Now, I don't see any reason why he should be right about that, and there have been no serious arguments given. There is an enormous amount that we don't understand in syntactical terms. You can clamp all these problems together under the problem of—roughly speaking—relevance. I see perfectly well that we don't have a reasonable syntactical account of that.

But there is something that Bert (Dreyfus) and John (Searle) don't take sufficiently seriously: there has been an enormous amount of progress in our understanding of all sorts of cognitive processes, particularly in areas like visual and language perception. We just know an enormous amount that we did not know twenty years ago, both in terms of what the experimental data are and in terms of what kinds of models are available to capture them. There have been extraordinary successes in this field, and they are all syntactical theories.

But both Searle and Dreyfus claim that the traditional approach of Artificial Intelligence with the physical symbol systems hypothesis has failed.

What is the evidence for that? The evidence is that we don't have simulations of intelligent processes. We don't have simulations—but it was crazy to try to get simulations in the first place. One of the things that Searle and Dreyfus don't do is distinguish between Artificial Intelligence and cognitive science. I absolutely agree that it is crazy to try seriously to simulate intelligent behavior, but that is not surprising.

I think it is crazy to try to simulate the behavior of the Great Lakes—I mean, except for very schematic purposes. Nobody who is interested in hydrodynamics would be interested in simulating the behavior of the Great Lakes. And if somebody said, "Look, we haven't got a running simulation of the behavior of the Great Lakes," that would not be an argument that hydrodynamics has failed. The argument that hydrodynamics has succeeded is that 1) we can state a lot of interesting laws; and 2) in all sorts of experimental environments, where we can test things under highly controlled, artificial conditions, we can predict the outcomes. That is also true in cognitive science.

What Searle and particularly Dreyfus do is identify cognitive science and, in particular, cognitive psychology with Artificial Intelligence. Then they argue that Artificial Intelligence has failed, which indeed it ought to. Simulation is not a reasonable goal for a scientific enterprise. But when they go on to say that the syntactical theory of mind is no good, that is a mistake.

In their criticism, they mainly referred to people like Minsky and McCarthy, who made those claims.

Minsky and McCarthy have gotten what they deserved from Searle and Dreyfus. I don't think that, at this stage, talk about success or failure is even relevant. We are at the very margins of exploring a very complicated idea about how the mind might work. A lot of people said a lot of irresponsible things in Artificial Intelligence—about how we would have a simulation of God within fifteen minutes or so. All this is crazy; it was never sensible. Nobody in his or her right mind ever thought that it would happen.

But on the other hand, the fact that those predictions cannot be carried out just shows that they should never have been made. The fact that the Artificial Intelligence project has failed does not show anything about the syntactical account of cognitive processes. What Bert and John have to explain away are all the predictions, all the detailed experimental results that we have been able to account for in the last twenty years. Simulation failed, but simulation was never an appropriate goal.

Do all the philosophers share your view that simulation failed? People like Dennett or Hofstadter[3] believe in simulation, and they have a very strong view of it. They say that a good simulation is like reality.

The question does not arise, because there are not any good simulations. That is the first thing. The second thing is what I have already said before: there is no science in which simulation is the goal of the enterprise. The goal of the enterprise is to understand. When you deal with cases where the surface phenomena are plausibly viewed as interactions, you don't try to simulate them.

Let's change our direction. What do you think about connectionism?

As I said, Turing had the one idea in the field, namely that the way to account for the coherence of mental processes is proof-theoretic. Connectionism consists of the thought that the next thing to do is to give up the one good idea in the field. I don't think that can be taken seriously. Connectionism fails for all the reasons associationism fails—that is, it cannot account for productivity, it cannot account for systematicity, it cannot account for rationality of thought. All the reasons we had to think that associationism was not a serious approach to the mind are the reasons we have for thinking that connectionism is not a serious approach to the mind.[4]

Why it has become so fashionable is an interesting question to which I don't know the answer. It is very hard to stop people from being associationists. One of the things connectionism really has shown is that a lot of people who were converted to the Turing model did not understand the arguments that converted them because the same arguments apply in the case of connectionism. I suppose this is another one of those winds of fashion that will blow itself out in the next four or five years.

It seemed to me that they have a series of powerful arguments, for instance, that the human brain, on the basic level, is a connectionist network.

Nobody knows what the mind is like on the basic level. What is true is that it is a lot of things connected to one another. There is this recurrent problem: what psychology is about is the causal structure of the mind at the intentional level of description. It may very well be, for all I know, that at some neurological level it is a system that obeys connectionist postulates. That is of no interest to psychology. And the reason why it is of no interest to psychology is that nobody doubts either that you can instantiate a connectionist network in a Turing machine or that you can instantiate a Turing machine in a connectionist network. The question of what the causal structure of the brain is like at the neurological level might be settled in a connectionist direction and leave entirely open what the causal structure of the brain is like at the intentional level.

Nobody disputes that you could instantiate a neural network on a Turing machine, but the difficulty is—and the connectionists point this out proudly—doing it in real time.

You have to be very fast. This probably shows that you have to be parallel; it does not show that you have to be a connectionist system. You might be a community of Turing machines, for all we know.

The connectionist model is one among infinitely many possible parallel models. There are parallel computers available now. They are not connectionist machines. They are, roughly speaking, von Neumann machines running in parallel and exchanging information among one another. One issue you could seriously raise when thinking about cognitive architecture is, How serial is the mind? The answer I don't know. It may be that the speed argument suggests that there is a lot of parallelism. If you are a modularity theorist as I am[5]—if you are a faculty theorist, as it were—then what you think is that it must be very parallel, because of all these modular systems running at the same time and almost independently from one another. They are not even parallel, they are autonomous: they don't even exchange information.

For all sorts of reasons it is plausible to view the mind as, let's say, a community of interactive von Neumann machines or some other classical machines. But that of course leaves entirely open the question whether they are networks. The states of networks have no internal syntactic structure. The states of von Neumann machines do, and that is independent of whether the von Neumann machines are running in parallel.

Let me stress this: here are two orthogonal questions. You can have any combination of answers you like. One question is, Are the internal states of this machine syntactically structured? Second question: does the machine run in serial or in parallel? Everybody I know who holds a version of the classical theory holds the following view: The answer to the first question is yes; the states are syntactically structured, and their semantic coherence has to be explained proof-theoretically. And the answer to the second question is also yes; the syntactically driven devices run in parallel.

What about the property of graceful degradation that connectionists underline? Can it be achieved in a serial mechanism with symbol manipulation?

Sure it can.[6] The character of degradation has to do with the relation between the virtual machine and its physical implementation. The intentional machine is projected onto its physical realization. A classical machine has discrete computational states—discrete states of the machine and discrete states of the tape if it is a Turing machine. There is no inference from that to any claim about the implementation, the hardware that implements that. If the hardware that implements these discrete states is diffuse, then to that extent the device will exhibit graceful degradation.

Nobody, in fact, has a machine, either the classical or the network kind, that degrades in anything like the way people do. So in a sense, it is all loose talk. But insofar as there is anything to it, it seems to rest on the standard confusion between the virtual and the implemented machines. In the sense in which graceful degradation is built into a network machine, it has to do with the way the network is implemented. It is not going to be built into a network machine that is implemented in a Turing machine, for example.

What do you think about the future prospects of cognitive science? Do you think that connectionism will become more powerful?

I think connectionism is, just as I said, the latest form of associationism. People will come to see that and give it up for the same reason they gave up associationism, that is, associated connections between mental states will not account for the rationality of mental processes and for learning—the hypothesis that learning is a kind of statistical inference has been tried over and over again and has failed, always for the same reasons. In the only interesting cases, learning is not statistical inference but a kind of theory formation. And that is as mysterious from the point of view of network theory as it is from the point of view of the classical theory. I think connectionist networks will just disappear the way expert systems have disappeared.

As for the future of cognitive science, we are about two thousand years from having a serious theory of how the mind works.

Did you say that expert systems will disappear?

They have disappeared. No one takes them seriously in psychology.

People in computer science have a different view.

I said in psychology. I don't think anybody seriously agrees that what the mind is is a series of catalogues, scripts, and expert systems. The capacity of that kind of picture to explain psychological phenomena has been asymptotically approaching zero. In psychology, this program has never been very lively and is now conclusively dead. And for the usual reasons: you have problems of novelty, of systematicity, of productivity—all that stuff that we don't understand but that makes it implausible that thought is just the union of a bunch of mental habits, which is basically the expert system idea. My guess is that connectionist models will have the same fate.

Connectionist models do have some success with pattern recognition.
    This is a very interesting issue. The question is not whether you can get some success with this machine. The question is whether you can get some success that you could not get by just doing statistics. Up to now, we have been thinking about these devices as inference devices, devices that try to simulate the coherence of thought. Thinking of them as learning devices, I think they are analog statistics packages. To get some success with these devices, to show that it is the architecture itself that is leading to new discoveries and a new understanding, is to do something that you cannot do with statistical analysis. I don't know of any demonstration where this is the case. In all the cases I have heard of, people say that their networks learn something in *n* trials. Now suppose we take the training trials as a data matrix and do statistics on just the data matrix: all the cases I have heard of come out exactly the same way. What you have is not a new theory of learning but a standard statistical theory of learning, only it is embedded in an analog machine rather than done on a classical machine.

So, in your opinion, that cannot explain anything.
    It will explain learning only insofar as learning is a process of statistical inference. The problem is that learning is no process of statistical inference. Here is a rough and crude way to put it. The thing we don't understand about learning is not how, given a hypothesis and a body of data, the hypothesis is statistically tested against this body of data— statistics can tell us that. The thing we don't understand is how the hypothesis is generated. How does the child figure out a picture of the world to test against the data he is given? The networks do not throw any light on that question at all.

You said before that one interesting thing is to understand the rational coherence of the human mind. Dreyfus, for instance, challenges this view. He claims that the human mind is not only rational.
    I think he is wrong about that. There are a number of questions here. One is the question of idealization. In any reasonable science you construct models for idealized conditions, and you try to understand the failures of the models in terms of interactions. Roughly speaking, you construct a theory of gases for gases that don't have all the properties that real gases have, and insofar as the model fails, you say it is because

of interactions between the properties the theory talks about and the properties that the theory treats as noise. Now, that has always been the rational strategy in science, because the interesting question is not whether to do it but which idealization to work with.

I assume that cognitive science is going to work in exactly the same way. We idealize to conceptual processes that you don't actually encounter but that you find degraded and distorted and noisy in actual behavior. That is another reason—in fact, it is the same reason—why you don't want to do simulation. You want to do psychology for ideal conditions and then explain actual behavior as ideal conditions plus noise. The question is whether or not the idealization to rational systems is justified—and, as in any other science, justified in terms of whether or not good science comes from it.

In the case of perceptual psychology, the advance has been extraordinarily rich. The idealization to hypothesis forming and testing devices seems to work extremely well in visual perception, in language perception. In logic, we idealize again to the theory of an ideally rational agent and explain errors as interactions. We may be interested in the interacting variables; we may be interested, for example, in how the internalized theory of validity interacts with memory variables or attention variables. That is something you can study. But you study it by first idealizing and then studying the interactions.

These idealizations have to be tested against actual practice, and so far I don't know any reason at all to suppose that the idealizations that have been attempted in cognitive theory are not appropriate or not producing empirical success. I would have thought that the empirical results suggest exactly the opposite inference.

Let me just say a word about background information. To begin with, I think that Bert is absolutely right that we have no general theory of relevance, that we don't know how problems are solved when their solution depends on the intelligent organism's ability to access relevant information from the vast amount we know about the world. That is correct, and it is one of the problems that will eventually test the plausibility of the whole syntactical picture.

On the other hand, it is an empirical question how much of that kind of capacity is actually brought to bear in any given cognitive task. There is a lot of reason—and this is what the modularity story is about—especially in areas of perceptual integration and possibly in areas of integration of action, that these processes are generally isolated from the relevance problem, that you have computational processes that are quite delimited in their ability to access background information. If that is correct, one might have a theory of those processes even if the general problem of relevance were not solved.

Bert is inclined to take it for granted that perception is arbitrarily saturated by background information. That is a standard position in phenomenological psychology. But I think there is evidence that it is not true. The evidence is that there is a lot of perceptual integration that is highly modularized. That means that you can construct computational models for those processes without encountering the general question of relevance, of how background information is exploited. If you look at the successes and failures of cognitive science over the last twenty or thirty years, they really have been in those areas where this isolation probably applies.

The critiques say that if you want to account for commonsense knowledge, then you have to explore this huge amount of background knowledge, and there is no end in view. If you have answered one question, then you have to assume another one. Searle uses the example "Do doctors wear underwear?" He is claiming that the answer is not found by an inference process.

This is an interesting claim. The question is: are there any arguments for it? It is a serious question, and I don't know the answer to it. To go from the fact that it is a serious question to the conclusion that there is no inferential answer to it seems a little quick. If there really is an argument that shows that no inferential process will work, then the computational theory is wrong.

The argument goes that the background of knowledge consists of skills and practices. If you start to think about them, you cannot do them any more, like typing or skiing. In learning these things, you have to follow rules. But once you have achieved expert level, you don't follow rules.

There are two points. It is not clear whether the analogy between physical skills like skiing and intellectual competence holds. It is not at all obvious. I don't know what the right story is about skiing. But it is not at all obvious, in fact, there are a lot of reasons not to believe that a process like language comprehension, which is highly stereotyped, should nevertheless be noninferential. I could tell you experimental data for a week that look like it required an inferential solution. If you think that the fact that it is stereotyped shows that it is noninferential, you have to explain away the experimental data.

Second, Bert's argument is: it starts out like a highly conscious process, a rule-governed activity. That is not true in the case of language; maybe it is in the case of driving. Even Bert thinks that the fact that after you have become skilled you have no phenomenological, introspective access to an inferential process establishes only a prima facie case that it is not inferential. The trouble is, there are all these data that suggest that it is, that a process like speech recognition, which is phenomenologically instantaneous, is nevertheless spread out over very short periods of time, goes through stages, involves mental operations, and so on. Even Bert thinks that the introspective argument does not determine facts. There are a lot of data that suggest that introspection is not accurate, that people are wrong about whether or not these tasks are inferential.

I always think of the following problem: I am a skilled typist on an English keyboard. Then I have to change to a German keyboard. I know that it is different, but I still confuse the keys.

You have to answer the following questions: is the task inferential? That is one; the other is: what premises are available to the inference? The point of all the modularity stuff is that you have to have tasks that can be independently and plausibly shown to be inferential but where the information to which they have access is not identical to the information to which you have access. That is the sense in which they are encapsulated.

There are all sorts of cases where the things that you know don't affect the things you can do or how things look, depending on whether you are talking about perception or action. That does not show that perception or action are not inferential. What it shows is that some of the things that you know are not accessible to the computational processes that underlie perception and action. That shows that if the processes are inferential, they are also encapsulated. And that is perfectly consistent.

If I understand correctly, I know rationally that it is a German keyboard, but there are unconscious inferences . . .

. . . that don't have access to your rational knowledge. There are plenty of cases that work that way. Take visual illusions. There is a pretty good argument that inferential processes are involved in generating the illusions. But knowing about the illusions does not make them go away. These seem to be cases where on the one hand you have computational processes and on the other hand they don't have access to information that is available to you. What that shows is, among other things, that what is consciously available to you and what is inferential don't turn out to be the same thing. That seems to me quite plausible. In fact, it is what everyone from Freud on has thought, or maybe from Leibniz on. If these processes are inferential, then 1) the inferences are not conscious; and 2) they are encapsulated.

My last question: what are the most serious problems that cognitive science has to overcome at the moment?

There are two problems. One I agree with Bert and John about, namely that there is this enormous problem of relevance. It comes up in all sorts of ways, but, generally speaking, we are confronted with it certainly in problem solving, maybe even in perception, analogical information, imprecise information, information that seems, originally, remote from the problem we are working on—all that stuff that comes up in Artificial Intelligence as the frame problem. And we don't know how to solve it. It really is not clear that you can have a syntactical theory of relevance.

It does not seem to me foolish to argue that the problems for a syntactical theory of relevance are not much different from the problems for a syntactical theory of induction. And the failure of the project of syntactical inductive logic does not bode well for the second. It is probably a galaxy of problems that we don't know how to taxonomize and we certainly don't see how to solve. It may turn out that the syntactical theory does not work for exactly what Turing had in mind, namely thought. The syntactical theory may work best for perception, integration of action, and so on—that is, for the encapsulated phenomena. I agree with John and Bert that we have to make some progress on the problem of relevance or find a different model.

The other issue is: if you divide things up in the way I suggested, then you want to distinguish very carefully the problems of intentionality from the problem of rationality—as I said, there has been one idea in cognitive science, and it was an idea about the second problem. But somebody better had an idea about the first one, because the whole cognitive science view is predicated on intentional realism. It is predicated on the idea—here, too, I agree with John and Bert—that people really do have beliefs and desires and that their mental processes really are causal processes and involve causal interaction among these belief and desire states.

Now, a lot of philosophers think that the whole belief-desire picture is corrupt, that you cannot give a coherent account of what it is for a mental state to have truth conditions or satisfaction conditions or content or whatever. In philosophy, that is the standard view. Quine, Stich, Putnam, the Churchlands—all hold that the view that there are firm facts of the matter about content cannot be sustained. So somebody has to give an empirically respectable account of intentionality. And as I said, the Turing proposal does not touch on that but just takes it for granted. But the problem that has been assumed had better be solved. There, the question is not so much an empirical one as a foundational one. We need a philosophically respectable account. I have great hopes for the resuscitation of informational semantics along the lines of Dretske's work.[7] But who knows whether this will work out? That is a serious foundational problem.

So there is plenty to do.

Recommended Works by Jerry A. Fodor

The Language of Thought. Cambridge, Mass.: Harvard University Press, 1975.
The Modularity of Mind: An Essay on Faculty Psychology. Cambridge, Mass.: MIT Press, 1983.
Psychosemantics: The Problem of Meaning in the Philosophy of Mind. Cambridge, Mass.: MIT Press, 1987.
A Theory of Content and Other Essays. Cambridge, Mass.: MIT Press, 1990.
(With E. LePore.) *Holism: A Shopper's Guide*. Oxford: Blackwell, 1992.

Notes

1. N. Chomsky, Syntactic Structures (The Hague: Mouton, 1957), "Review of Skinner's 'Verbal Behavior,'" *Language* 35 (1959): 26–58. See also J. Katz and J. Fodor, "The Structure of a Semantic Theory," *Language* 39 (1963): 170–210; as well as J. Fodor and J. Katz, eds., *The Structure of Language: Readings in the Philosophy of Language* (Englewood Cliffs, N.J.: Prentice-Hall, 1964).

2. See, e.g., C. Osgood (one of the pioneers of psycholinguistics), *Method and Theory in Experimental Psychology* (New York: Oxford University Press, 1953); and "Psycholinguistics," in *Psychology: A Study of a Science*, vol. 6, ed. S. Koch (New York: McGraw-Hill, 1963), pp. 244–316.

3. D. R. Hofstadter and D. C. Dennett, *The Mind's I* (New York: Basic Books, 1981).

4. The criticism of connectionism is pursued more in depth in J. A. Fodor and Z. W. Pylyshyn, "Connectionism and Cognitive Architecture: A Critical Analysis," *Cognition* 28 (1988): 3–72.

5. J. A. Fodor, *The Modularity of Mind* (Cambridge, Mass.: MIT Press, 1983).

6. For the following arguments cf. also Fodor and Pylyshyn, "Connectionism and Cognitive Architecture."

7. F. Dretske, *Knowledge and the Flow of Information* (Cambridge, Mass.: MIT Press, 1981).

JOHN HAUGELAND

Farewell to GOFAI?

When I was in graduate school, Bert Dreyfus was one of my teachers. I first took a seminar on Artificial Intelligence offered jointly by him and Michael Scriven. They disagreed quite seriously, and I was impressed by Dreyfus's side and got involved in helping to formulate his argument. I did not actually write on Artificial Intelligence for my dissertation, but I became interested in the topic and continued to work on it when I came to Pittsburgh.

Then, I was member of a group at CASBS—Center for Advanced Study in the Behavioral Sciences—at Stanford in 1979 that was devoted to philosophy and Artificial Intelligence. In the course of that year, my work concentrated further.

What are the main ideas of cognitive science, in contradistinction to Artificial Intelligence? Is Artificial Intelligence a part of cognitive science?

I certainly think that Artificial Intelligence is a part of cognitive science. What part it is depends on your outlook, on what you think about cognitive science. There is one way of thinking about cognitive science that takes the word *cognitive* narrowly and takes Artificial Intelligence as the essence of it. That is the physical symbol systems hypothesis, or what I call "Good Old-Fashioned Artificial Intelligence." The basic approach is to investigate how intellectual achievements are possible via rational manipulation of symbolic representations, which is the essence of cognition according to this narrow conception.

There are other conceptions of cognition in which ideas from Artificial Intelligence play a smaller role or none at all.

Which domains should be included in cognitive science? Is it an interdisciplinary field—and if so, which disciplines should take part in this enterprise to study the mind?

Psychology, of course, and Artificial Intelligence—those are the central ones. Linguistics would be the next most important. Sometimes

John Haugeland attended Harvey Mudd College from 1962 to 1966 and studied at the University of California at Berkeley from 1966 to 1967 and again, after working as a teacher with the Peace Corps in Tonga, from 1970 to 1973. He earned his Ph.D. degree from Berkeley in 1976 and began his career at the University of Pittsburgh in 1974 as an instructor. He became Professor of Philosophy at Pittsburgh in 1986.

people talk about cognitive linguistics,[1] although this is again a disputed characterization. We still don't know just how cognition is relevant to linguistics. One importance of both linguistics and psychology for Artificial Intelligence is setting the problems, determining what an Artificial Intelligence system must be able to achieve. Some people also include cognitive anthropology, but that is secondary. And, of course, philosophy is often included. I believe that philosophy belongs in cognitive science only because the "cognitive sciences" have not got their act together yet. If and when the cognitive sciences succeed in scientifically understanding cognition and the mind, then philosophy's role will recede.

Some people, especially in the computer science community, don't understand what role philosophy has to play in this enterprise. Most of the time they don't care for it, because they are more pragmatic. Do you think the distinction is one between theory and experimentation? Traditional psychology also arose from experimental psychology. Philosophy, at the moment, seems to have the theoretical part to play in cognitive science, and this is also one criticism coming from the Artificial Intelligence community: that philosophers don't understand anything about computers.

There are a lot of questions there. I am not surprised that partisans of Artificial Intelligence and cognitive science are perplexed as to the role of philosophy. That fits exactly with what I said: it is because they believe that the science will succeed, that the fundamental problems have been figured out (i.e., what to do, what to investigate, how to go about it). Once you have that view, once you have what Kuhn calls a paradigm,[2] then all that remains is normal science.

I don't want to denigrate that; it is extremely important and difficult and intellectually challenging. But it would not require philosophy. I think that philosophy is still pertinent and relevant, however, because I don't think that the field has arrived at that point yet. Indeed, we are quite possibly in a revolutionary period in this decade.

Do you think that there are different periods in the development of cognitive science? You coined the term Old-Fashioned Artificial Intelligence, *so you seem to think of connectionism as the new direction.*
"Good Old-Fashioned Artificial Intelligence" I call it—GOFAI.

Which periods would you distinguish?
There are probably four main periods. First, there is the period of pre–Artificial Intelligence, before it really got going, from the middle forties to the middle fifties, which was characterized by machine transla-

tion, cybernetics, and self-organizing systems. There was a lot of enthusiasm about how a mechanism—something that we understand—could exhibit goal-directed, sensible behavior and, in grand extrapolation of that, intellect. That extrapolation, alas, did not have much substance, and not much was achieved. All of those efforts were basically failures.

Starting in the middle to late fifties, the physical symbol systems hypothesis began to take over. Minsky's and Papert's book[3] did have an important effect on network research, but I think the symbol processing approach took over because it was, on the face of it, more powerful and more flexible. You had quite impressive early successes, quite restricted successes, of course—and that turned out to be more important than it seemed at the time.

Do you think that without the book Perceptrons *and Minsky's and Papert's critique of Rosenblatt, the development would have been different—that connectionism would have been stronger at that time?*

No, I don't. Maybe a little bit. There is the historical issue of how it affected funding. I really am not in a position to say anything about that. But I think the symbolic processing hypothesis was, at that time, the stronger and more promising hypothesis, and that was the main reason.

Then, by the early seventies, it became clear that intelligence involved the management and access to massive amounts of knowledge, that problem solving—heuristic search in a problem space—and reasoning were in fact not the essence of the issue. They were important, of course, but you could not focus on these and leave knowledge access or commonsense aside. So that led to the third period, in which the emphasis of the work shifted toward the management of massive amounts of knowledge—that is, storing it in a consistent way, finding it when needed, organizing as the underlying theme. And that, in Good Old-Fashioned Artificial Intelligence, remains the last period. It is a bit stale by now—fifteen or twenty years old.

But, by the early eighties, there arose connectionism. That is not so much an improved development of what preceded as it is an offshoot, an alternative. It has dominated the excitement in this decade. All the flash has been there. This is not to say that it has been stunningly successful in what it has actually been able to achieve. But it is the part that still seems most promising.

What are the most difficult problems to overcome, at the moment, for cognitive science and Artificial Intelligence?

It is not a simple question, because there are different ways in which to understand what is meant by "for" cognitive science and Artificial Intelligence. If you understand cognitive science and Artificial Intelligence to mean the classical, "Good Old-Fashioned" approach based on the physical symbol systems hypothesis, then there is a group of closely related problems: the problem of managing massive amounts of commonsense knowledge; the frame problem; and the problem of achieving skillful behavior in real time, of which expertise is a special case. By "in real time" I mean "quickly." Those are all closely related problems, at least within that paradigm, and that is the present roadblock.

Let's explore these difficulties in more detail. Do you think that the so-called commonsense problem is proof that the physical symbol systems hypothesis failed? Simon tells me that it is not, that they are trying to incorporate it in their models. For philosophers like Dreyfus, the idea of the physical symbol systems hypothesis—that is, that the behavior of the human mind is only rational—does not fit reality. In his opinion, the commonsense problem is one proof that Artificial Intelligence with the physical symbol systems hypothesis will never succeed.

I think "proof" is too strong a word. Nobody is in a position to claim that anything has been proved, neither Dreyfus nor Simon. But I do believe that, in the first place, the problem of building machines that have commonsense is nowhere near solved. And I believe that it is absolutely central, in order to have intelligence at all, for a system to have flexible, nonbrittle commonsense that can handle unexpected situations in a fluid, natural way, bringing to bear whatever it happens to know.

Of course, Simon would say—and so would McCarthy and Newell and some others—that by working on that they are going to incorporate it in their models. As I said before, they have been doing that for at least fifteen years. That is the focal problem of Good Old-Fashioned Artificial Intelligence. I am not optimistic about its being solved in those terms. I think that is quite unlikely. The rate of progress in these fifteen years is quite unimpressive.

They think it is only a problem of time and better machines—a problem of scale. For instance, they told me about the approach of Douglas Lenat[4] in Texas, to put the encyclopedia into the computer.

I think Lenat's project is hopeless.

Could you explain that? In your opinion, it is not a problem of scale?

It is conceivable that it is a problem of scale eventually. It is conceivable that the problem of commonsense will be solved in terms of physical symbol processing once the problem of scale has been solved. But it is absolutely not a simple question of having enough memory to store all that knowledge and enough data entry personnel to enter it. It is not just that the knowledge is not yet stored in the machines, it is that it is not known how to store it in the machines. The problem of how to organize, manage, and keep knowledge up-to-date on an immense scale has not been solved in principle. That is why Lenat's project is hopeless, in my view.

There are two ways to say it is hopeless. My personal hunch—in fact, conviction—is that this problem will not be solved in physical symbol systems terms, ever. But I don't think that this is proven or obvious.

Why is that so? People believe that they can represent everything—why not commonsense knowledge?

People believe that because it is the only thing they understand how to do. It is not that they are pursuing research in that direction because they believe that it is the way the mind works and that it will succeed, but the other way around: they believe it is the way the mind works and that they will succeed because that is the way in which they know how to pursue research. This is Kuhn's point about paradigms. That is why the connectionist alternative can come as such an illumination to people: oh my God, there is another way to go! And, in fact, there are other ways as well—none, however, so well developed.

My conviction that the physical symbol systems hypothesis cannot work is mostly an extrapolation from the history of it. I don't think that you could produce an in-principle argument that it could not work, either like Searle or like Dreyfus (who still believes he has an in-principle argument).

I am mostly sympathetic to Dreyfus's outlook. I think it is right to focus on background, skills and practice, intuitive knowledge, and so on as the area where the symbol processing hypothesis fails. But to point this out as the area where it fails is not to explain why it fails or to argue that it has to fail there. And so someone like Newell or Simon is quite right to say, "Well, we will include that in our system." Of course, for them to say so does not mean they can. I believe that Dreyfus is probably right that they cannot.

What about the problem of skills that you mentioned?

It is closely related. In my view, skills—the ability to get along in the world, manage things, deal with things, including perceptual skills and of course linguistic skills, but that is much more complicated to say precisely—are embodied in people, in the brain, in some way quite different from symbol structures (possibly in connectionist patterns in the network—that is a very attractive possibility, but far from established). It is because skills are embodied in this other way that the background implicit in the collection of skills and practices will not succumb to a symbol processing treatment.

You mean that, say, moving around or typing is not a cognitive skill at all? Dreyfus says that "the body just does it," whereas Searle speaks of cognitive skills and that "the brain just does it." But they stress "just does it"—not by symbol manipulation.

That is a terminological dispute. What you call cognitive or not depends on how broadly or how narrowly you use the term *cognitive*. This is the difference in the conception of cognition I mentioned at the beginning of the interview. Whether you say it is the brain or the body—nobody knows. I meant "embodied" to include the brain as part of the body. Whether it is in the midbrain or the cortex or the spinal cord, I don't know—and I don't care. Neither does either of them. To say that it—the brain or the body—"just does it" is vacuous. What does "just" mean? Of course it does it. Even Newell and Simon would agree with that: a body—including the brain as part of the body—does it.

To say "just does it" sounds like saying ". . . and there is nothing more to explain": It is utterly unintelligible and inaccessible to theory, it is something we should never ask about, we could never understand, that there can be no science of it. Nobody is in a position to make those negative claims. I think a psychologist or an Artificial Intelligence professor is quite right to be exasperated by any suggestion that "it just does it" meaning "so don't try to figure it out." That is crazy.

However, none of that means that the way it does it will ultimately be intelligible as symbol processing. That is a much stronger thesis, which Newell and Simon would accept and Rumelhart and McClelland would not, nor would I, nor would Dreyfus.

In one of your articles,[5] you mentioned a third obstacle: moods.

And I would generalize that to include affects, and even affectedness in perception. But I started with mood as the most difficult and perplexing. That is another area in which it seems to me that symbol processing is unprepared and unequipped to deal with manifest phenomena.

There are two basic possibilities. Either affective phenomena, including moods, can be partitioned off from something that should be called more properly "cognitive," so that separate theory approaches would be required for each and there would be no conflict between them. Or we cannot partition off moods and affect (and I would now add self-understanding—Ego), in which case symbol processing psychology cannot cope with it.

Do you think that physical symbol systems hypothesis could cope with the problem of qualia?

I don't know what to think about the problem of qualia. I am inclined to suspect that that problem can be partitioned off. That is different from what I think about moods and self-understanding. Those are integral to intelligence and skills. They are necessary parts and presuppositions of an intellect at all. And so an account that cannot incorporate them cannot succeed. I don't think that of qualia.

Let's turn to connectionism. What are its advantages, and what are the current obstacles to overcome? In one of your articles,[6] ten years ago I think, you already speculated about holograms—is not that related to the idea of connectionism? But when you wrote this article, connectionism was not around yet.

The rise of connectionism has become a power in the field since then. But the idea that intelligence might be implemented holistically in networks, in ways that do not involve the processing of symbols, has been around, in a less specific form, for a long time. In that old article, I was just saying that the basic idea was not trivial or easy to dismiss.

There are three kinds of attraction for connectionism. One, very important, is that the brain looks like a network. People always mention that and then try to say that it is not the central thing. I think what Feldman and Ballard call the "one-hundred-steps constraint"[7] is quite important. The hardware response time of neural systems is in the order of a few milliseconds. That is how long it takes for a neuron to respond to a stimulus and come up with a response. So nothing can be done faster than that in the brain. And the minimum psychological response time for people for anything—recognizing a face and so on, difficult things that no machine can yet do—is of the order of a few tenths of a second, which is about a hundred times the hardware response time. So these most basic, quickest mental achievements must be achieved in about a hundred machine cycles.

That is a very serious constraint on the architecture. It is an especially serious constraint in that the physical structure of the brain is known to be a massive web of simple, generally similar units, interconnected with incredible complexity. If you had all kinds of time, you could implement any kind of system in that hardware. But you don't have all the time you want. So the one-hundred-steps constraint ties you more closely to the hardware. It is one attraction of connectionism that it is an approach fitting this hardware apparently much more naturally.

Couldn't one argue that on the underlying level, the microprocessor and the circuits also form a pattern that is like the brain?

You have to be more specific in what you mean by "pattern." Of course there are patterns in any computer. It only can be the sophisticated kind of equipment that it is because it is highly structured and organized.

That would be the difference between parallel architecture and von Neumann architecture. But people like Newell and Simon say that they now use parallel processing. They claim that on a certain level, connectionism, too, has to introduce symbols and to explain the relation to the real world. In their view, there is not so much difference at this level. At first I thought the main difference was between serial and parallel architecture.

That is not the most penetrating way to put it. It is true that in most of the history of classical Artificial Intelligence, it has been based on serial processing. Not a von Neumann machine, but a LISP machine. But a LISP machine is still usually a serial processor. Yet it is also correct to say that the symbolic paradigm can accommodate a great deal of parallelism—parallel LISP machines, and especially production systems, are amenable to a fair amount of parallelism.

The deeper way to put the difference is that in a connectionist network, you have very simple processors that, indeed, work in parallel. They work in ignorance of what the others are doing. But, more important, they don't process symbols—that is, there is no symbolic interpretation of their input and output states.

Moreover, they are massively interconnected, and those connections have properties. The connections are not homogeneous; they have different strengths. Because the number of connections runs on the order of the square of the number of units, you have a huge number of connections. The way in which those strengths vary on all those connections encodes the knowledge, the skill that the system has.

So that, if one component in a symbol processing machine goes bad, everything crashes, whereas one wrong connection in a network can be compensated, which is similar to the brain. But one question: what is the difference between a von Neumann machine and a LISP machine? I thought that the von Neumann machine is the general machine with one processor, one quartz, determining the machine cycle.

The architecture is not at all a hardware issue. The hardware itself will have an architecture, of course, but the notion of architecture is not tied to the hardware. It is true that if you have a hardware von Neumann machine, which is the most common type of hardware for engineering reasons, you can implement on it—in software—a LISP machine. But it is equally true that if you have a hardware LISP machine— and that is possible—you can implement on it a software von Neumann machine.

The question of whether it is hardware or software is not a deep question. The difference between a von Neumann and a LISP machine is that a LISP machine is a function evaluator. It works by passing arguments to functions and getting back values; it is nothing but functions. When a function evaluates, it may pass further arguments to further functions, which then must be evaluated first. So you get a hierarchical tree structure. The only memory in the machine, apart from the long-term memory to remember the function definitions, is the memory needed to keep track of the pending arguments passed and values returned—that is, the stack.

A von Neumann machine is a machine built around operations performed on specified locations in a fixed memory. The important thing that von Neumann added that makes the von Neumann machine distinctive (compared to ENIAC or Babbage's machine[8]) is the ability to operate on values, especially addresses, in the memory where the program is located. Thus, the program memory and the data memory are uniform. The reason why this is so important is that the machine can keep track of where it is in the program, branch off to another part of the program, and then come back to wherever it was and continue processing from there. It makes the construction of subroutines possible. That is the principal innovation that makes the von Neumann machine so important, plus the fact that it is easy to build. A LISP machine or a production system has a different memory structure and a different program structure.

The programming language LISP is a LISP machine, that is the point. When you run the language, what happens is that the machine you are working with comes to behave like a LISP machine. It is a virtual machine implemented in whatever kind of hardware you have.

I thought one had to make a distinction between hard- and software. And that it was one of the attractions of connectionism that the difference was in the underlying hardware, how it is connected, how it is hard-wired.

It has nothing to do with the hardware, except that the one-hundred-steps constraint keeps it close to the brain. In fact, almost all connectionist research is currently done on virtual machines, that is, machines that are software implemented in some other hardware. They do use parallel processors with connections when they are available. The trouble is that machines that are connectionist machines in the hardware are fairly recent and quite expensive.

One of the points of connectionist architectures is that you don't have a central processor—no one is in charge. In a production system, even if it is parallel, you do have somebody in charge, at least at higher levels. Just as you can get your PC to be a LISP machine, you can also get it to be a virtual connectionist machine.

The first attraction of connectionism you mentioned is that the hardware is closer to the human brain.

The second is: it has been discovered that connectionist systems tend, by their very nature, to fill in missing parts. If you have a noisy or defective signal, the system treats that signal as if it were a noisy or deformed variant of the perfect prototype. And this really is the basis of a great deal of what the systems do.

You can apply that to pattern recognition. You take the pattern to be one part of the signal, and the identification of the pattern (perhaps its name) as the other part. You train the system on these "complete" patterns—for example, pattern plus name—and then, later, you give it only one half of one of them; it will fill in the missing half, thereby "recognizing" the pattern. In pattern association you have the same idea: you give it one part, it will fill in the other. The same with generalization: you give it fifty patterns that have something in common, then one that has the same thing in common but the machine has never seen—it will recognize that, too, by filling in the difference.

And, as you pointed out, if there is a broken connection or node, that turns out to be equivalent to a noisy or deformed signal, and it will make up for the rest. So generalization, pattern recognition, error correction, graceful degradation, and association are, from a higher perspective, all manifestations of a common phenomenon that the parallel distributed processing species of connectionism particularly lends itself to; it emerges from the way it is built.

And that is a broad category of behavior that is quite evident in human behavior, and conspicuously lacking in known symbol systems. They are, as everyone says, brittle. You have those classical examples of failures of Artificial Intelligence systems—you just change the statement of the problem in one trivial word, and the system does not know what to make of it.

So this capability of connectionist systems 1) seems to be evidenced by people who are of course the standard of success for all of Artificial Intelligence; and 2) addresses a lack or apparent failure of classical Artificial Intelligence systems. So this is the second point of attraction, of which hardware is one aspect.

The third attraction is that classical Artificial Intelligence has more or less gone stale for a dozen years. In particular, there have not been major breakthroughs or great ideas. One loses faith as it becomes less promising. It is not over until the fat lady sings, meaning you cannot tell how it works out until the end. And it may in the end turn out that being discouraged at this point by the period of stagnation in GOFAI is wrong.

It could be that when our children look back on this period fifty years from now, they will say that only the faint of heart were discouraged in this period of stagnation, which was a consolidation period with successes continuing afterward. Or it could be that this is the beginning of the end of classical Artificial Intelligence. It went as far as it could; it could not solve all the problems; so it failed and died. Our children and their children will know.

Let me put it in other words: it could be that the traditional view of Artificial Intelligence will fail, but it could also be that it could incorporate the connectionist approach.

The symbol processing hypothesis cannot incorporate connectionism. They are just antithetical.

But it is claimed that they will do it.

One possibility some people are fond of—Fodor and Pylyshyn,[9] for instance, consider that possibility and take it seriously—is that the architecture of the mind, the way to understand how intellect and cognition is possible, must be symbol processing. But the way this symbol processing machine is implemented in people is as a virtual machine in connectionist hardware. That way, they accommodate the fact that the brain looks like a network. And there may even be advantages to doing it that way. But that does not mean to incorporate connectionism into physical symbol system hypothesis but rather to adjust the implementation to the facts about the brain.

I am inclined to think, though it is really hard to tell at this point, that that suggestion is partially right. And everything turns on how much "partially" is. Fodor and Pylyshyn, and probably Newell and Simon, would believe that everything of interest mentally has to be mediated by the symbolic virtual machine. So the "partially" would be "almost completely," except maybe for some peculiarities like reflexes, and physical abilities like walking, and so on.

At the other end, you could say that the "partially" is "very little." We can think symbolically, we can do it consciously: we can reason things out in our heads. And to that extent, symbol processing has to be implemented in whatever the hardware of the brain is. But that is very little of the mind. My own view is that this side is closer to the truth. There is symbol processing, and it has to be implemented in the brain. But it is a rather small facet of mind.

The idea of combining both is that on the lower level—pattern recognition, bodily movement, and so on—connectionist models apply. On the higher level, like language or problem solving, a symbolic model could be right. It is curious that people like Newell and Simon always work on problem solving with tasks that are not everyday tasks.

They conceive of everyday tasks as problem solving—they conceive of speaking a language as problem solving. What you just said about lower level and higher level would be congenial to the outlook of Newell or Pylyshyn. But it is not what I believe. I believe that many "higher level" abilities are not symbolic at all. Abilities like expert intuition, understanding what is going on in a situation, appreciating what is important and unimportant, artistic creativity, and so on are not symbolic, even on their higher level.

I think, in fact, but this is extremely difficult, that most of what has to be understood in understanding how we can speak a language will have to be understood as skills implemented more or less directly in connectionist hardware, but that the linguistic output or input—the words and sentences—are in fact symbol structures. And so it is in the linguistic faculty or module, if you like, that symbol processing first appears, and it is mostly limited to that. Our abilities to think symbolically are themselves developments from the linguistic faculty.

Recommended Works by John Haugeland

Artificial Intelligence: The Very Idea. Cambridge, Mass.: MIT Press, 1985.

"The Intentionality All-Stars." In *Philosophical Perspectives*, vol. 4: *Philosophy of Mind and Action Theory*, ed. J. E. Tomberlin, pp. 383–427. Altascadero, Calif.: Ridgeview Publishing, 1990.

"Representational Genera." In *Philosophy and Connectionist Theory*, ed. W. Ramsey, S. Stich, and D. Rumelhart, pp. 61–89. Hillsdale, N.J.: Lawrence Erlbaum, 1991.

"Understanding Natural Language." In *Mind and Cognition*, ed. W. Lycan, pp. 660–70. Oxford: Blackwell, 1990.

(Ed.) *Mind Design.* Cambridge, Mass.: MIT Press, 1981. (A collection of key articles in cognitive science, among them quite a few by authors who are also interviewed in this book, e.g., Hubert Dreyfus, Daniel Dennett, Jerry Fodor, and, of course, John Haugeland himself.)

Notes

1. See, e.g., G. Lakoff, "Cognitive Semantics," in *Meaning and Mental Representations*, ed. U. Eco, M. Santambrogio, and P. Violi (Bloomington: Indiana University Press, 1988), pp. 119–54; and R. Langacker, *Foundations of Cognitive Grammar*, vol. 1 (Stanford: Stanford University Press, 1987). See also the interview with George Lakoff in this volume.

2. T. S. Kuhn, *The Structure of Scientific Revolutions* (Chicago: University of Chicago Press, 1970).

3. M. L. Minsky and S. Papert, *Perceptrons: An Introduction to Computational Geometry* (Cambridge, Mass.: MIT Press, 1969; reprinted in an expanded ed., 1988).

4. D. B. Lenat, "When Will Machines Learn?" in *Proceedings of the International Conference on Fifth Generation Computer Systems*, vol. 3 (Tokyo: Institute for New Generation Computer Technology, 1988), pp. 1239–45; and D. B. Lenat and R. V. Guha, *Building Large Knowledge-Based Systems: Representations and Inference in the* CYC *Project* (Reading, Mass.: Addison-Wesley,1990).

5. J. Haugeland, "The Nature and Plausibility of Cognitivism," in *Mind Design*, 3d ed., ed. J. Haugeland (Cambridge, Mass.: MIT Press, 1985), pp. 243–81.

6. Ibid.

7. J. A. Feldman and D. H. Ballard, "Connectionist Models and Their Properties," *Cognitive Science* 6 (1982): 205–54.

8. ENIAC (Electronic Numerical Integrator and Computer) was the first electronic computer, built between 1943 and 1946 at the University of Pennsylvania. Charles Babbage (1791–1871) was an English mathematician who built the so-called difference engine (1833), a new kind of calculating machine.

9. J. A. Fodor and Z. W. Pylyshyn, "Connectionism and Cognitive Architecture: A Critical Analysis," *Cognition* 28 (1988): 3–72.

GEORGE LAKOFF

Embodied Minds and Meanings

Probably I should start with my earliest work and explain how I got from there to cognitive science. I was an undergraduate at MIT, where I was a student of both mathematics and English literature. There I had an opportunity to begin to learn linguistics with Roman Jakobson, Morris Halle, and Noam Chomsky.[1] I was primarily interested in working with Jakobson on the relationship between language and literature. And that is an interest that has stayed with me throughout my career. My undergraduate thesis at MIT was a literary criticism thesis, but it contained the first story grammar. It was an attempt to apply Chomskian linguistics to the structure of discourse. I took Propp's *Morphology of a Folk Tale*[2] and recast it in terms of transformational grammar. That led to the study of story grammar in general.

Out of that came the idea of generative semantics. I asked the question of how one could begin with a story grammar and then perhaps generate the actual sentences that express the content of the story. In order to do this, one would have to characterize the output of the story grammar in some semantic terms. This was back in 1963. I noticed that if you took logical forms of the form logicians use to characterize semantics, you could view them as having the same kind of syntactic structure as Chomskian analyses of English sentences have. So I asked myself whether it would be possible for logical forms of a classical sort to be deep structures, actually underlying structures of sentences. This was done before the concept of deep structure was arrived at by Chomsky. "Deep structure" did not really come into being until 1965, with the publication of *Aspects*.[3]

The consequences were the following: if it were the case that logical forms were underlying syntactic representations, then it should be the case that words with complex logical forms should be syntactically decomposable—that is, something like "John heated the milk," which is a causative, should mean something like "John caused the milk to come to

George Lakoff received a Bachelor of Science in mathematics and English literature from MIT in 1962 and a Ph.D. in Linguistics from Indiana University in 1966.
He taught at Harvard and the University of Michigan before his appointment as
Professor of Linguistics at the University of California at Berkeley in 1972. In the
early stages of his career, he was one of the developers of transformational grammar and one of the founders of the Generative Semantics movement of the
1960s. Since 1975, after giving up on formal logic as an adequate way to represent
conceptual systems, he has been one of the major developers of cognitive linguistics, which integrates discoveries about conceptual systems from the cognitive
sciences into the theory of language.

be hot." We eventually found evidence for things of this sort. At the same time, I looked for other kinds of evidence for the existence of logical forms. For example, I sorted the different kinds of properties that logical forms have: they show co-reference between entities, they have to use variables and the binding of variables. They have logical operators that show scope differences. They include notions like propositional functions and predicate-argument structure. I and other generative semanticists like Jim McCawley, John Ross, and Paul Postal[4] found evidence that each of these properties actually did play a role in the grammar of English. That is one of the lasting contributions of generative semantics.

At that time, you were very near the Chomskian view?

It turned out that Chomsky was very much against this view then. Since then, he has come around to agree that logical forms exist in natural language. At that time, he thought it was an outrageously crazy idea. We had to fight considerably to get that idea to be taken seriously, which is nowadays a standard and even conservative view.

At that time, I had certain methodological priorities that governed my work and still do. I had a commitment to the study of linguistic generalizations in all aspects of language. I assumed that there were certain phenomena that defined areas of linguistics; for example, syntactic phenomena include things like grammatical categories, grammatical constructions, agreement, and so forth, and generalizations are generalizations governing the occurrence of grammatical categories or constructions. In semantics, I assumed that there are generalizations governing inferential patterns of the meaning of words, of semantic fields, and so on. In pragmatics, there are generalizations governing speech acts, illocutionary force, presuppositions, and so on. So I took the study of generalizations as a primary commitment.

I also assumed that grammar has to be cognitive and real. I undertook what I called the "cognitive commitment": that everything we discovered about grammar should be consistent with what we know about the mind and the brain. I also had several secondary commitments, one of which was what I called the "Chomskian" or "generative commitment": it was Chomsky's assumption that the correct way to characterize language is in terms of combinatorial systems in mathematics of the sort first discovered by the mathematician E. Post.[5] These are systems in which arbitrary symbols are manipulated without regard to their interpretation. That is what generative—or "formal"—grammars are. So I had a secondary commitment to the study of formal grammar. I also had a "Fregean commitment," that is, that meaning is based on truth and reference. Generative semantics was an attempt to fit all these commitments together—to study the correct linguistic generalizations, have

them be cognitively real, have them fit the Chomskian paradigm and be
describable within symbol manipulation systems, and also have them fit
the Fregean paradigm and have meaning characterized in terms of reference and truth.

The history of my work in generative semantics was, at first, to confirm this, to show that logical form really did show up in grammar. This
was done around 1968. Along with these results, there came to be a
number of results that eventually were contradictions between generalizations and the cognitive commitment on the one hand and the
Chomskian-Fregean commitment on the other hand. The history of my
work as a generative linguist has to do with the discovery of phenomena
where you could not describe language in fully general, cognitively real
terms using the Chomskian-Fregean commitments.

Take, for example, the notion of co-reference, which turns out to be
a very complex notion. Jim McCawley, in 1968, came up with an interesting sentence that could not be analyzed within classical logical forms.
The sentence was: "I dreamt that I was Brigitte Bardot and that I kissed
me." Not "I kissed myself," which would be impossible. But "I kissed
me" should be impossible anyway; it should be reflexive, except that
"kiss" is a nonreflexive predicate. One needs counterpart relations here
between McCawley and Bardot going across what I thought were possible worlds. I suggested, in 1968, a Kripkean possible world semantics to
do this with counterpart relations of the sort characterized by David
Lewis. In 1968, I suggested moving away from the description of logical
form to model theory.

You were still at MIT at that time?

No, no. I was teaching at Harvard after getting my degree from Indiana in 1965, and taught there until 1969. I was working with Ross and
the MIT community. By then, we had begun to disagree with Chomsky.
I was mainly working with Postal, although he left MIT after 1965.

This was about the same time that Richard Montague was suggesting
that possible world semantics ought to be used for natural language semantics, as was Ed Keenan from the University of Pennsylvania. Barbara
Partee sat in on Montague's lectures at UCLA, and the three of us, Partee, Keenan, and myself, were the first ones to suggest that possible
world semantics might be useful for doing natural language semantics.[6]

However, over the years I kept finding more and more cases in which
this was impossible. One could not use possible world semantics so easily for natural language semantics. Between 1968 and 1974–75, many
cases arose where that became impossible. My writings of that time
show what kind of cases they are.[7] What I tried to do during that period
was extend model theoretic semantics to be able to handle more and

more cases in natural language. I also tried to extend the Chomskian theory of grammar to be able to handle the new phenomena that were discovered.

By 1974, it became clear that the Chomskian theory of grammar simply could not do the job. I discovered a class of sentences called "syntactic amalgams" that do not have a single deep structure with derivations. Take a sentence like "John invited you'll never guess how many people to the party." The clause "you'll never guess how many people" takes "John invited such and such many people to the party" as complement. You might think that there is some strange transformation making a complement into the main clause. I noticed, however, that you can iterate these cases: You can say "John invited you'll never guess how many people to you cannot imagine what kind of party for God knows what reason on you'll never guess which day." I worked out the conditions in which this construction was done and showed that you could not keep the standard notion of derivation and deep structure and still account for such sentences.

Around 1975, a great many things came to my attention about semantics that made me give up on model theoretic semantics. One of these was Charles Fillmore's work on frame semantics,[8] in which he showed that you could not get a truth-conditional account of meaning and still account correctly for the distribution of lexical items. Eleanor Rosch's work on prototype theory came to my attention in 1972, and in 1975 I got to know her and Brent Berlin's work on basic level categorization.[9] I attended a lecture at which Rosch revealed fundamental facts about basic level categorization, namely that the psychologically basic categories are in the middle of the category hierarchy, that they depend on things like perception and motor movement and memory. When I saw that it became clear to me that any objectivist account of categorization could not work. Rosch had shown that the human body was involved in determining the nature of categorization. This is very important, because within classical semantics of the sort taken for granted by people both in generative grammar and in Fregean semantics, meaning, logic is disembodied, independent of the peculiarities of the human mind. The mind is viewed as reflecting or capturing some notion of transcendental universal reason. Rosch's work showed that this was impossible. Shortly thereafter, the following year, McDaniel and Kay wrote a paper on color, showing that the Berlin and Kay facts about color categories could be explained in terms of the neurophysiology of color vision.[10] That suggested that neurophysiology entered into semantics. This was completely opposed to the objectivist tradition and Fregean semantics.

The thing that really moved me forever away from doing linguistics of that sort was the discovery of conceptual metaphor in 1978. I would credit it to Michael Reddy, whose paper "The Conduit Metaphor" in Ortonyi's *Metaphor and Thought*[11] showed basically the role of metaphor in everyday language. Reddy looked at the language that characterizes communication and found that it was essentially metaphorical: ideas are objects, linguistic expressions are containers, and communication is sending these objects in containers to someone else who extracts the contents from the words. There are hundreds and hundreds of expressions like "I could not get my meaning across to him," or "the meaning is right there in the words," "I got what you were saying," "the meaning is buried in terribly dense paragraphs," and so on. These examples are fairly convincing. They show a generalization concerning polysemy, that is, where multiple meanings occur for individual words, and the generalization can be characterized in terms of understanding the domain of communication in terms of another, the domain of sending objects.

I noticed then that conceptual metaphor is a natural process. There are hundreds or thousands of generalized mappings in everyday English, and ordinary, everyday semantics is thoroughly metaphorical. This meant that semantics cannot be truth conditional, it could not have to do with the relationship between words and the world, or symbols and the world. It had to do with understanding the world and experiences by human beings and with a kind of metaphorical projection from primary spatial and physical experience to more abstract experience.

So there was this difference: objective semantics with reference and truth conditions on the one hand, and a lot of examples on the other hand suggesting that it could not work like that, that human beings perceiving the world and acting in it give meaning to things.

I began to appreciate this between 1975 and 1979. I was influenced by Rosch's work on prototypes and basic level categories, by the work on the neurophysiology of color vision and its relationship to color categories; there was Fillmore's work on frame semantics, my own work on metaphor. Around the same time, Len Talmy and Ron Langacker began discovering that natural language semantics require mental imagery.[12] Their early works on this are absolutely brilliant. They showed that semantic regularities required an account of image schemata or schematic mental imagery. Reading their work in the late seventies, I found out that it fit in very well with all the other things I had discovered myself or found out through other people.

They all added up to the idea that meaning was embodied, that you could not have disembodied meaning, disembodied reason. That entailed that you could not use the kind of mathematics that Chomsky had used in characterizing grammar in order to characterize semantics. The reason was, as we had first shown in generative semantics, that semantics had an effect on grammar, and we tried to use combinatorial mathematics to characterize logical form. We thought that the use of formal grammars plus model theory would enable us to do syntax and semantics and the model theoretic interpretation.

However, if meaning is embodied, and the mechanisms include not just arbitrary symbols that could be interpreted in terms of the world but things like basic level categories, mental images, image schemas, metaphors, and so on, then there simply would be no way to use this kind of mathematics to explain syntax and semantics. Our work in cognitive linguistics since the late seventies has been an attempt to work out the details of these discoveries, and it changed our idea not only of what semantics is but of what syntax is.

This view is attacked as relativistic by people holding an objective semantics position. What do you think about this?

What is being attacked is a very unsophisticated view of relativism. If you take the idea of total relativism, which is the idea that a concept can be anything at all, that there are no constraints, that "anything goes" in the realm of conceptual structure, I think that is utterly wrong. What is being attacked is that utterly wrong view, which is totally simplistic. It is not even a view that was held by Whorf.[13] The only people who hold such a view these days are deconstructionists—that is, people who think that meaning is purely historically contingent. But I do not know of any serious cognitive scientist who holds that view.

The view that we have suggests that meaning is neither purely objective and fixed nor completely arbitrary and relative. Rather, there are intermediate positions, which say that meaning comes out of the nature of the body and the way we interact with the world as it really is, assuming that there is a reality in the world. We don't just assume that the world comes with objectively given categories. We impose the categories through our interactions, and our conceptual system is not arbitrary at all. It is greatly constrained by the nature of the body, by the nature of our perception, by the nature of the brain, and by the nature of social interaction. Those are very strong constraints, but they do not constrain things completely. They allow for the real cases of relativism that do exist, but they do not permit total relativism.

So, in your view, there is a reality outside the human mind, but this reality is perceived by the human mind. Therefore you cannot establish reality only by logical means.

The conceptual system, the terms on which you understand the world, comes out of interaction with the world. That does not mean that there is a God's-eye view that describes the world in terms of objects, properties, and relations. Objects, properties, and relations are human concepts. We impose them on the world through our interaction with whatever is real.

What do people in the Chomskian tradition say about this?

I have not heard them say anything at all. To my knowledge, they don't even know about it. Perhaps they do, but I have not seen any serious commentary.

What happened when you got these ideas about embodiment?

We began to work it out in detail, trying to show what a positive view of language and thought would be. That is how cognitive linguistics has grown, and a huge amount of research has come out of it. We now have some very solid research on non-Western languages showing that conceptual systems are not universal. Even the concept of space can vary from language to language, although it does not vary without limits but in certain constrained ways.

We have been looking at image schemas. There seems to be a fixed body of image schemas that turns up in language after language. We are trying to figure out what they are and what their properties are. I noticed that they have topological properties and that each image schema carries its own logic as a result of its topological properties, so that one can reason in terms of image schemas. That is a kind of spatial reasoning.

Another thing I discovered was that metaphors seem to preserve image schema structures. In doing so, they seem to preserve spatial logic. The spatial inference patterns that one finds in image schemas when they apply to space are carried over by metaphor to abstract inference patterns. This suggests that abstract reasoning is a metaphorical version of spatial reasoning. We are now trying to work out the details of that.

In your work and that of Talmy, Langacker, and Fauconnier,[14] there appear notions like "mental spaces," "image schemas," and so on. What is the difference between them and the traditional notion of representation in cognitive science?

There have been two generations of cognitive science. In the first generation, it was assumed that mental representation was done in the way suggested by logicians: that there was either a logical form or an image representation made up of symbols and structures of symbols, and that these symbols represented things in the external world or categories in the external world. They were internal representations of some external reality. This was part of the "disembodied mind" view—the view that you could characterize a mind in purely abstract terms using abstract symbols and that this had nothing to do with the body, with perceptual mechanisms, and so on.

The later view, which I call the "embodied mind" vision, showed that categorization had everything to do with the perceptual and motor system, that image schemas have to do with human perception, that metaphors have to do with human understanding, that mental spaces are human constructions. In that view, the mind is embodied. We understand the world not in arbitrary or objectivist fashion but rather through our experience.

That second generation of cognitive science is now being worked out. It fits very well with work done in connectionism. The earlier view fit with traditional Artificial Intelligence, the mathematics of which came out of logic, of course. The physical symbol system hypothesis was basically an adaptation of traditional logic. Connectionism says that the nature of the brain really matters: the fact that it is made up of neural networks, that they are connected in certain ways, that the brain is divided up into a visual system and other parts that characterize its functioning, that the visual system is itself broken up into many subparts. This will all matter for cognition.

We are trying to put the results of cognitive linguistics together with connectionist modeling. One of the things I am now working on is an attempt to show that image schemas that essentially give rise to spatial reasoning and, by metaphor, to abstract reasoning, can be characterized in neural terms. I am trying to build neural models for image schemas. So far, we have been successful for a few basic schemas.

Another kind of work I am involved in is phonology. Traditional generative phonology assumed the symbol manipulation view. It assumed an underlying structure and that you arrive at a surface phonological form by stages. That derivation could only happen in abstract time; it is not something that a real brain could carry out in real time. The technical details just make it impossible. Recently I have been constructing a model of phonology that could be implemented by a real brain, one in which there is basically a three-dimensional structure—a morphological dimension, a phonemic dimension, and a phonetic dimension—with correlations across these dimensions. These correlations can be characterized by neural mappings. John Goldsmith[15] and I are working on a connectionist phonology, which is important to get at a phonology that is cognitively real.

What could be the reason for the rise of connectionist theories?
Already in the sixties, there was an attempt at connectionist theory
by Rosenblatt, which was proved insufficient by Minsky and Papert.[16]

The person who changed all that was David Rumelhart. He showed that the old perceptron theory was too simple, but if you changed it slightly, you could get it to do a lot of interesting things. He proposed various changes that would enable it to compute all sorts of interesting functions. Since then, what is almost an industry has grown out of these results. People have done a lot of engineering work on neural networks, showing what kind of computational capacities you could get from neural networks with what kinds of assumptions. A lot of this is detailed engineering, and there have been some remarkable achievements in neural processing. So much so that there are now thousands of researchers working on neural networks. In fact, between 1986, when Rumelhart's and McClelland's book *Parallel Distributed Processing*[17] came out, and 1989, only three years later, thousands of scientists have gone into the study of neural networks. The progress has been astonishing.

All over the country?
All over the world.

I had the impression that the connectionist view is especially strong here on the West Coast, in San Diego, Berkeley, and so on.
I should distinguish between the engineering version of connectionism and the mathematics of it—the "technical connectionism"—on the one hand and the application of connectionism to cognitive science on the other. Those are two different things. The great advances have come in the engineering aspects and in the mathematics. In terms of cognitive science, the applications have been much more modest, and only a small number of people are working in this field. So, although connectionism has emerged around the world as an engineering device, it has merely begun to be applied to cognitive science. You are right that most of this application has been done on the West Coast.

Do you think that connectionism is the whole story of the human mind? Some people in connectionism claim that it could be used to model the lower level processing of the human mind, like vision, pattern recognition, or perception, but that for more abstract processes like language the symbol manipulation approach could still be useful.
The people who are claiming that are those who come out of Artificial Intelligence, raised in the symbol system tradition. They don't consider connectionism to be biological, they consider it to be another computer science view. According to a biological perspective, the mind is made up of neurons—that is, neural processing must be correct. However, if you are a practicing computer scientist and trying to save the results of earlier computer science by adding a little bit of parallel processing and calling it connectionism, you can do that. Also, if you want to get running systems that do things of the sort that people in Artificial Intelligence could

do in the past, the easiest way to do this is to take your old systems and add a little bit of connectionist networks but not to take seriously the problem of giving neural foundations.

Let me explain: in a physical symbol system, the only way in which a symbol gets meaning is via a connection to the world and to other symbols. But ultimately, there is no grounding in the Artificial Intelligence view; there is no meaning aside from objectivist meaning. What we are trying to do is show how meaning can be grounded in the sensory-motor system, in the body, in the way we interact with the world, in the way human beings actually function. People who try to adapt the physical symbol system view by adding a little bit of connectionism to it are not seriously engaged in that research.

What are you trying to do—what are your epistemological claims when you try to combine linguistics with connectionism? Are you modeling the mind, simulating the mind, or building a mind?

We are trying to explain language and to describe how it works. Let me take an example: why should there be prototype effects or radial categories? One of the things Rumelhart showed in chapter 14 of the PDP book[18] is that radial categories arise naturally from connectionist networks. They arise because once you learn a given pattern of connections, the easiest way to learn related information is to learn it as variations on the existing pattern. Technically this gives rise to radial categories.

We would like to explain why we find what we find: why should language be this way? We are attempting both description and explanation. We are also trying to explain why we have the concepts we have. Why should there be cognitive topology? In the models we have we begin to see that the brain is structured in such a way that modules of retinal maps could naturally give rise to cognitive topology. We have constructed models of such maps, but we do not know whether they actually exist in the brain. At least, now we can make a guess as to why we are finding that cognitive topology exists.

Moreover, one of the things that Rumelhart discovered is that analogical structuring is a natural component of neural networks. It arises naturally out of the mapping from one network to another. That could explain why there should be metaphor, why the abstract reasoning is a metaphorical version of spatial reasoning. That suggests that, given the nature of the brain, it is natural for metaphorical reasoning to arise as soon as you have enough cortex.

Our job as we view it is an explanatory job. Computer models are secondary, but they may allow us to begin to understand why language is the way it is.

You can show that it is at least possible that the mind works like the computer models, and so give an explanation, but you don't claim that the mind is like them?

Right—for that you would need biological evidence. I should say that not all connectionists have this view. Many connectionists are mainly engineers or computer scientists. They are looking for old-fashioned kinds of physical symbol system models inside a connectionist framework. But they are not necessarily concerned with grounding meaning in biology.

What are your predictions for the next five years? What are the most important problems to overcome in linguistics and connectionism?

In connectionism, we need to get good models of how image schemas and metaphors work, and good models of semantics, fitting together the kind of work that has come out of cognitive semantics. We also need to put together a much more serious theory of types of syntax. Langacker's work on cognitive grammar is one kind of a beginning, as is my work in grammatical construction theory and Fillmore's work in construction grammar. But these have not been developed as much as we would like them to be. One reason for this is that we have to do much more work than generative theories of grammar. Generative theories don't have to do all the semantic and cognitive work that our theories are responsible for. It is much easier to make up theories that are limited.

What about the philosophical implications of your work?

There are very great philosophical implications, as Mark Johnson has pointed out. Johnson is probably the principal philosopher who has seen the implications of the second generation of cognitive science research. In his book *The Body in the Mind*[19] and in my book *Women, Fire, and Dangerous Things*,[20] we have both pointed out that the entire notion of epistemology has to be changed within philosophy in such a way that it completely changes most philosophical views. Johnson is now working out the consequences for theories of ethics and morality, and others are working out the consequences for law.

Basically, this prepares a major philosophical shift, because if reason is not transcendental, if categories are not classical categories defined by lists of properties or by Boolean kinds of conditions—rather, if they are bodily defined and if metaphor plays an important part in reason—then all kinds of things change. Not only the analysis of language but also the relationship between ontology and epistemology.

In the traditional view, what is real is independent of one's knowledge of it. If we are correct, then what we take to be real is very much dependent on our conceptual systems. That is, we devise conceptual systems using our bodies and our brains, and then we take this conceptual system as characterizing what is real. These systems fit our experiences of the world in different ways, as there is more than one way to fit our experiences perfectly.

One consequence is that conceptual systems turn out not to be self-consistent. In general, human conceptual systems must be thought of as having inherently inconsistent aspects. No one has worked out the consequences of that yet. In law, there is a commitment to objective categories, to objective reality, and to classical logic as the correct mode of human reasoning. If that turns out not to be correct, laws, for example, should probably be framed differently. In general, the political and social sciences are framed in terms of the old cognitive science, the old objectivist philosophy.

What do you mean exactly by "objectivist philosophy"?

Objectivism makes the following claims: first, it makes a claim about what exists, that is, about ontology. It says that the world is made up of objects, these have objectively given properties, and they stand in objective relations to one another, independent of anybody's knowledge or conceptual system. Categories are collections of objects that share the same properties. Categories not only are objects and properties but they are out there in the world independent of the mind. Therefore, certain kinds of reasoning are out there in the world: if X is in category A, and if category A is in category B, then X is in category B. That means that reason has to do with the structure of the world. If all that is false, if reason is different, if categories are not based on common properties, then our view of reality has to change. Moreover, our view of thought and language has to change. The idea that our mental categories mirror the world, that they fit what is out there, turns out to be wrong.

Correspondingly, our names of mental categories are supposed to be thereby names of categories in the world, and our words can thereby fit the world, and sentences can objectively be true or false. If meaning depends instead on understanding, which in turn is constructed by various cognitive means, then there is no direct way for language to fit the world objectively. It must go through human cognition.

Now, human cognition may be similar enough around the world that in many cases no problems arise. But when you make claims about government or foreign policy, you use metaphors. If you then carry out this foreign policy in terms of those metaphors, then the differences between how people are, how they live in the world, and these metaphorical systems will matter. It will matter that these systems are metaphorical and do not reflect objective reality.

Recommended Works by George Lakoff

"Cognitive Semantics." In *Meaning and Mental Representation*, ed. U. Eco, M. Santambrogio, and P. Violi, pp. 119–54. Bloomington: Indiana University Press, 1988.
Women, Fire, and Dangerous Things: What Categories Reveal about the Mind. Chicago: University of Chicago Press, 1987.
(With M. Johnson.) *Metaphors We Live By*. Chicago: University of Chicago Press, 1980.

Notes

1. Roman Jakobson was a founding member of the Prague School and best known in linguistics for his work in structuralist phonology. See, e.g, R. Jakobson, *Child Language, Aphasia, and Phonological Universals* (The Hague: Mouton, 1968). M. Halle's work was pivotal in generative phonology. See, e.g., *The Sound Pattern of Russian* (The Hague: Mouton, 1959), and R. Jakobson and M. Halle, *Fundamentals of Language* (The Hague: Mouton, 1956), as well as N. Chomsky and M. Halle, *The Sound Pattern of English* (New York: Harper and Row, 1968). N. Chomsky laid out the principles of generative transformation grammar in *Syntactic Structures* (The Hague: Mouton, 1957).
2. V. Propp, *Morfologija Skazki* (Leningrad: Akademija, 1928).
3. N. Chomsky, *Aspects of a Theory of Syntax* (Cambridge, Mass.: MIT Press, 1965).
4. See, e.g., J. D. McCawley, "The Role of Semantics in a Grammar," in *Universals in Linguistic Theory*, ed. E. Bach and R. Harms (New York: Holt, Rinehart, and Winston, 1968), pp. 125–70; J. Ross, "Constraints on Variables in Syntax" (Ph.D. diss., Massachusetts Institute of Technology, 1967); and P. Postal, "Some Further Limitations of Interpretive Theories of Anaphora," *Linguistic Inquiry* 3 (1972): 349–71.
5. See S. C. Kleene, *Mathematical Logic* (New York: Wiley, 1967).
6. On the work of R. Montague, see R. Thomason, ed., *Formal Philosophy: Selected Papers of Richard Montague* (New Haven: Yale University Press, 1974). On Montague grammar, see B. H. Partee, ed., *Montague Grammars* (New York: Academic Press, 1976); and E. L. Keenan, ed., *Formal Semantics of Natural Language* (Cambridge: Cambridge University Press, 1975).
7. E.g., *On Syntactic Irregularity* (New York: Holt, Rinehart, and Winston, 1971), or the article "Hedges," *Journal of Philosophical Logic* 2 (1973): 459–508.
8. C. Fillmore, "Frame Semantics," in *Linguistics in the Morning Calm*, ed. Linguistic Society of Korea (Seoul: Hanshin, 1982), pp. 111–38; "Frames and the Semantics of Understanding," *Quaderni di Semantica* 6, no. 2 (1985): 222–53.
9. On prototype and basic level categories see, e.g., E. Rosch, "Human Categorization," in *Studies in Cross-Cultural Psychology*, ed. N. Warren (London: Academic Press, 1977); "Principles of Categorization," in *Cognition and Categorization*, ed. E. Rosch and B. Lloyd (Hillsdale, N.J.: Lawrence Erlbaum, 1978), pp. 27–48; C. Mervis and E. Rosch, "Categorization of Natural Objects," *Annual Review of Psychology* 32 (1981): 89–115; and B. Berlin et al., *Principles of Tzeltal Plant Classification* (New York: Academic Press, 1974).
10. B. Berlin and P. Kay, *Basic Color Terms* (Berkeley and Los Angeles: University of California Press, 1969).
11. M. Reddy, "The Conduit Metaphor," in *Metaphor and Thought*, ed. A. Ortonyi (Cambridge: Cambridge University Press, 1979), pp. 284–324.

12. L. Talmy, "How Language Structures Space," in *Spatial Orientation: Theory, Typology, and Syntactic Description*, vol. 3, ed. H. Pick and L. Acredolo (Cambridge: Cambridge University Press, 1983), and "Force Dynamics in Language and Thought," in *Papers from the Parasession on Causatives and Agentivity* (Chicago: Chicago Linguistic Society, 1985); R. Langacker, "Space Grammar, Analysability, and the English Passive," *Language* 58, no. 1 (1982): 22–80; and *Foundations of Cognitive Grammar*, vol. 1 (Stanford: Stanford University Press, 1986).

13. B. Whorf, *Language, Thought, and Reality* (Cambridge, Mass.: MIT Press, 1956).

14. G. Fauconnier, *Mental Spaces* (Cambridge, Mass.: MIT Press, 1985).

15. J. Goldsmith, *The Last Phonological Rule* (Chicago: University of Chicago Press, 1990).

16. F. Rosenblatt, *Principles of Neurodynamics* (New York: Spartan, 1962); M. L. Minksy and S. Papert, *Perceptrons* (Cambridge, Mass.: MIT Press, 1988).

17. D. E. Rumelhart, J. L. McClelland, and the PDP Research Group, *Parallel Distributed Processing: Explorations in the Microstructure of Cognition*, 2 vols. (Cambridge, Mass.: MIT Press, 1986).

18. D. E. Rumelhart et al., "Schemata and Sequential Thought Processes in PDP Models," chap. 14 of ibid., vol. 2, pp. 7–57.

19. M. Johnson, *The Body in the Mind* (Chicago: University of Chicago Press, 1987).

20. G. Lakoff, *Women, Fire, and Dangerous Things: What Categories Reveal about the Mind* (Chicago: University of Chicago Press, 1987).

James L. McClelland was born in 1948. He earned his B.A. in psychology from Columbia University in 1970 and his Ph.D. in cognitive psychology from the University of Pennsylvania in 1975. He served as Assistant and then Associate Professor at the University of California in San Diego from 1974 to 1984, until he moved to Carnegie-Mellon University in 1984, where he became Professor of Psychology in 1985.

JAMES L. McCLELLAND

Toward a Pragmatic Connectionism

I started out with a fascination for my own thought processes. I found myself preoccupied with things that were getting in the way of progress. So I started reading on scientific psychology, and I slowly came to realize that I could not come up with an account of what was going on in my head. This was back in high school already.

When I came to college, I had an excellent course from a behaviorist who convinced me that you could actually study what happened, what governed people's behavior. I became extremely involved in this work. I went to a lab and started doing behavioristic experiments, under the influence of the idea that all this stuff I was thinking about my thought processes was irrelevant, that I should study behavior and finally find the laws of behavior. After about a year I started asking questions about what was governing this behavior. I did experiments with animals and wanted to know why they chose this stimulus and not another—because they did not remember, because they saw, because it was like the one before?

When I left college, I knew that I was committed to the scientific study of the mind through looking at behavior. But I was upset with the notion that was prevalent in my university: that you could not study the mind itself but only write down laws of behavior. When I went to graduate school in 1971, I was suddenly made aware of the fact that there was a field of cognitive psychology. In that field, people were actually using the careful study of behavior to make inferences about the nature of mental processes.

The first good example I was exposed to—and I was exposed to it precisely because my adviser wanted me to see the connection with the work of Saul Sternberg[1] on sequential memory search—was this: you run experiments, you measure the time it takes for somebody to do something, and you make an inference about the nature of the mental process that underlies that pattern of reaction time results. That really turned me on: it seemed to me you could actually do experimental research that would give you some insight into the nature of the processes governing people's thoughts.

So although I started graduate school with the idea that I would become a behaviorist, it was not hard to talk me out of it. My adviser, who had become disaffected with behaviorism, pushed me over the edge, and I became a cognitive psychologist. At that time the prevalent theories in cognitive psychology had to do with discrete sequential processing mechanisms. I think it was highly influenced by the conventional von Neumann computer—the early way of thinking about computers as step-by-step procedure-following devices. The idea was to find out what each of the steps was and how long it took. The conception of the steps was that they were things with a definite start and end time and a definite and discrete result—for example, you identified the first letter of a word as a *T*, and that was all you did for that short interval of time.

The problem was that the particular research I got interested in involved looking at the role of context in letter perception. It is very confusing to try to understand that from the point of view of discrete, sequential processing, because you get thrust on the horns of a dilemma very early: either you identify the letters in a sequential way, you get the result of the letter identification process, and then you go to memory to see what word it matches—in which case, memory does not help you to figure out what the letters are. Or else you do some kind of hypothesis testing. But the data did not seem to fit with either of those ways of thinking. It seemed that there were two kinds of influences: context was one, and incoming information was another. I kept wanting to see those influences working together simultaneously. We had thousands of experiments showing that, in fact, we see letters better when they are in contexts that are consistent with them.

When I finished graduate school and went to San Diego, which was in 1974, I immediately started to develop my own theoretical alternatives to this discrete, sequential way of thinking. This was where I became a cognitive scientist as opposed to a cognitive psychologist, I would say. I was no longer doing experiments simply to make inferences from them, but I was also trying to formulate really explicit computational theories. They should allow me to articulate my vague intuitions about what really was going on in a way that was not consistent with existing theories of the time. I had new conceptions, and I needed to formulate a different theoretical framework for them.

How would you outline the difference between cognitive psychology and cognitive science? Is methodology one of the main differences?

I think so. Obviously, the boundary is fuzzy. I like to think of the difference by comparing what happens when I go to two different conferences. One of them is the conference of the Psychonomic Society,[2] a society that was founded out of the same motivations I had to come to psychology in the first place—the idea that to collect data is relevant to finding out what is really going on in people's minds—and which has never lost this very close contact with data. Typical members of a cognitive psychology department stay very close to the data. They might come up with a model explaining the data from this or that experiment, but mostly they stay close to the facts. So when you go to this conference, you hear hundreds of experimental papers on various aspects of cognition.

On the other hand, if you go to a cognitive science meeting, mostly you will hear papers in which people articulate new ideas in the form of an explicit model and then try to relate that idea to data. That "and then" part is actually often lacking in the papers one sees, but the "and then" part is crucial to me. Given where I come from, I don't want to lose track of the data. But the notion that one needs an explicit, computationally adequate framework for thinking about the problem at hand, and that one then tries to use this framework to understand the data, is really the essence.

By computational model, *do you mean that you could model it with a computer?* Computing *traditionally means "calculating"—is that what it means for you?*

Let's see. *Computational model* means that you have to be able to articulate your conception of the nature of the process you are interested in in such a way as to specify a procedure for actually carrying out the process. Now, the word *computational process* gets associated with this, as opposed to just *process*, because it is ordinarily understood, in cognitive circles, that these activities are being done removed from the actual external objects about which they are being done. This is about the whole business of the physical symbol system hypothesis. You are doing some mental activity with a certain result not on physical objects, although there is a physical substrate for the process, but on things that represent those objects, in the same way that we use numbers to calculate bushels of wheat instead of piling up the wheat. In a computational model, we try to figure out what representations we should have, based on what information, and what process applies to these representations.

This is our conception of a computational model, but it also carries with it, for me, the notion of explicitness, of specifying something that allows you to understand in detail how the process actually goes, as opposed to articulating a rather vague, general, overarching conceptualization such as a Piagetian theory, where you have some very interesting but often exceedingly vague concepts that are difficult to bring into contact with facts.

Your definition of a computational model and of the intermediate abstraction seems to contradict the new connectionism, which goes down to the neurophysiological level. One idea of the computational model is that you can abstract from the hardware, that only the intermediate level of symbol manipulation matters.

Let's get this clear. Here is the part where I agree with Newell: the mind is a physical device that performs operations on physical objects, which are not the things themselves but representations of those things. The result of this mental process can be new representations. At some level, this sounds like the physical symbol system hypothesis. We might call it instead the physical instantiation hypothesis, because the process is instantiated in a physical system that is not itself the thing being represented but stands in some relation to it. And the objects in it are traditional symbols.

It is the notion what these objects are, what representations are made out of, how they are, in fact, processed, and what the mechanisms are that lead us from one to the other: there my views may differ from other people's in cognitive science. It is not about the notion that one has to be explicit about these things, that one wants to bring these concepts into contact with empirical data. It is not about the notion that they are, in some sense, symbolic, in some sense not the things themselves. Representationalism is essential in connectionism. Sometimes people misunderstand that.

So the main difference is not about the view that there are representations, that there is symbol manipulation going on.

I would not say "symbol manipulation." I would say: there is some kind of manipulation of representations that stand in a symbolic relation to objects in the world. A symbolic relation to objects in the world simply means that the representation is not the thing itself but something that functions as the object for computational purposes. It can be used as the basis for further processes, which might result in the formation of new representations, just as we have the number 2 to represent the two bushels of wheat from one field and the number 5 to represent the five bushels from another field. We perform an operation on them to come up with the total. We have something—we don't know exactly what it is; we haven't said anything about it yet—that corresponds to something in the world, that we have in our head and that we are manipulating when we think. The things that come out of it are new representations.

But what those representations are, whether they are symbols in the sense of Newell's notion of symbols or not, that is where there is some disagreement.

You told me that you come from the field of perception research. Now, I wonder if it is a coincidence that all people in connectionism worry about perception, hearing, and vision, whereas symbol manipulation people think about the human mind in an abstract way—how it solves problems, and so on. Some people try to combine connectionism and symbol manipulation in this way: on the deeper level, neurophysiology is working in parallel, but on the higher level—for example, in language processing—the symbol manipulation is useful.

Certainly, it is true that the connectionists tend to come from perception, whereas people in symbol manipulation tend to come from high-level domains. But it is by no means a clear-cut division of types of people. I myself certainly do not accept the notion that seems implicit in the way you put the question: that it is right to have adopted a classical symbolic approach if you come from studying thinking. My feeling is that the attempt to understand thinking may progress with the conventional notion of symbol, but that it is limited in how far it can go and that it is only an approximation of the real right way to think about representations. If we are able to understand more clearly how we can keep what the people who think about symbols have been able to achieve, while at the same time incorporating new ideas about what mental representations might be like, we would be able to make more progress in understanding human thought.

A symbol is generally considered to be a token of an abstract type that gains power from pointing to other knowledge. This robs the representation itself of contents. It puts the information about the current object of thought in the background. The symbol itself is not the information but only a pointer to it. When you process the symbol, you have to search down through some complex tree to find that information. That limits the ability of a symbol-oriented system to exploit the contents of the objects of thought. If the objects of thought in the representations we have of them could give us somehow more information about those contents, this would increase the flexibility and the potential context sensitivity of the thought processes.

This is not a check that I have cashed yet, or that anybody really has cashed. But many of us connectionists have the notion that the kind of representations currently used are too static. We want to capture what we can with these representations, but we think that we need to go beyond them to understand how people can be insightful and creative, as opposed to mechanical symbol processors.

So your criticism of symbol manipulation is by far less strong than the criticism coming from some philosophers—for example, Dreyfus or Searle.

One thing I share with the people who formulated the physical symbol system hypothesis is a commitment to sitting down and seeing if something can actually work. I myself am not so expert in the kinds of arguments those people propose, but it is actually my suspicion that many of the things Dreyfus points to are things that result from the limitations of symbols as they are currently conceived of by people like Newell, as opposed to some more general notion of thought as manipulation of representations. Once we achieve a superior notion of what representations are like we will be in a much better position to have a theory that won't founder on that particular kind of criticism.

One point of that criticism is the so-called commonsense problem. Symbol manipulation cannot solve the problem of this huge amount of commonsense knowledge that people have as background. Connectionism is doing better in this respect?

I hope so. The basic notion of connectionism is that a representation is not a static object that needs to be inspected by an executive mechanism. A representation is, rather, a pattern of activity, a dynamic object. If you have a bunch of processing units, and these units have connections to other processing units, and a representation is a pattern of activity over the processing units, then the relations of the current pattern of activity to patterns of activity that might be formed in other sets of units are in the process themselves, in the connections among the units. You are never out of contact with what you know. If what you know is in the connections, the implications of your current representation for the next representation are embedded in the processing machinery itself. If we can get commonsense into the connections, then naturally it is going to influence the thought processes.

A symbolic representation is usually thought of as a list structure, and some mechanism is moving through it and looking at a particular place in the structure at a particular time. When it is looking at that particular place, the context that is in play is only a very local one, as it would be in a LISP subroutine. That makes one think that commonsense can never penetrate the process because this little local process can only use the stuff written in the little subroutine for deciding what to do next. The open-endedness of what people know and how it might be relevant to what we think of doing next—we don't see how to funnel that through the subroutine. The notion that a mental representation is not just a location in the data structure but a pattern of activity that is interacting with the knowledge by virtue of the fact that the knowledge is in the connections between units gives us the hope that we will be able to overcome this particular limitation.

But some researchers—for example, Putnam and Fodor in their interviews here—say that connectionist models cannot do better than statistical methods.

That is a very misleading claim. I would say that we do not yet know the limits of what can be achieved by the kinds of extensions to statistical methods that are embodied in connectionist models. Two specific points are worth making here.

First, Hinton showed that it was possible to learn relations that cannot be expressed in terms of correlations between given variables.[3] What Hinton's network did was discover new variables into which the given variables must be translated. When it had done this, it was able to generate the correct response to inputs that it had never seen. The correlations of which Putnam speaks exist only among the variables that were discovered by the network, not among the given variables.

Second, connectionist models are now being applied extensively to the task of extracting structure from sequences.[4] These networks solve a problem by exploiting a flexible, variable-length temporal window that is not easy to describe in terms of correlations among given variables—except insofar as these descriptions acknowledge the presence of arbitrary embedded structure.

In other words: Fodor and Putnam do not understand the potential of connectionist learning models, and their notion that these models are identical to familiar regression and correlation methods is antediluvian.

One more practical question: the symbol manipulation people invented the programming language for list-processing LISP, mainly to support their point of view and to see how far you could go with it. Are you and your connectionist colleagues working on a programming language to support your own point of view? Are you developing your own tools?

This is an interesting question. . . . When people formulate a theory, they instantiate it in a real computational form. LISP is a real computational form. Of course, it is a very general programming language, as opposed to more specific modeling frameworks in symbol processing. In connectionism, we have a similar kind of situation. We have a basic notion about a set of primitives, a notion of what the general class of primitives might be like. Then we formulate specific frameworks within this general "framework generator," if you like, which can then be used for creating specific models.

At the most abstract level, the idea is that mental processes take place in a system consisting of a large number of simple computational units that are massively interconnected with each other. The knowledge that the system has is in the strength of the connections between the units. The activity of the system occurs through the propagation of signals among these units. There are some constraints on how complicated the computations can be that are performed by each unit.

That is a very general formulation. Within that, people formulate the Boltzmann machine, which is a very specific statement about the properties of the units, the properties of the connections, and maybe about the way knowledge is encoded into the connections. The Boltzmann machine can then be instantiated in a computer program, which can be configured to apply to particular, specific computational problems. What we did in the handbook that Rumelhart and I published[5] was collect together a few of the fairly classical kinds of frameworks for doing connectionist information processing. Then we set up a few example models of how you could apply them to particular small problems. The programs can be generalized by changing the instructions.

So yes, there is a set of tools that we have developed. Other people have other tools. It is not so much a programming language per se, in the classical sense, although some have tried to conceptualize it that way. The tools are more abstract, in general. They are statements about the details of the computational mechanisms, which can then be instantiated in computer programs. Not the computer programs are essential, but the statement of the principles.

The first year here I taught a course without the aid of any software or demos; I just had people read chapters of various books, which was hard. The second year I let them use the simulators that I and my colleagues had developed in our own investigations. Halfway through the course a student came to me and said: "We heard from last year's students that this stuff was totally incomprehensible, but this year we all understand it perfectly." It must be the software . . . maybe it is also my inability to explain things. Personally, I understand basic concepts in linear algebra and differential equations through seeing them in operation, and not from an abstract mathematical point of view. I can see what the abstract mathematics means now that I have seen it in action. We do feel that giving people some firsthand experience maybe gets something into their connections . . . experience is an important part of changing how people think about the nature of processing.

In building computer models, or simulations, what do you claim—that you are building a device like the mind, proposing a hypothesis about how the mind really works or could work? What are the philosophical claims of your constructing a model?

It is really important to understand that people who build connection-ist models have many different kinds of goals. Some people are really trying to model what a particular piece of neural circuitry is actually doing. They know a lot of facts about this neural circuitry and build them into their model. They try to get the model to exhibit behavior that captures other known properties of that physical circuit, perhaps in order to study particular neurons in a particular area with the model incorporating facts about the distribution of connections and so on.

Other people are building connectionist networks because they want to solve a problem in Artificial Intelligence. They wish, for example, to improve machine recognition of speech sounds. They think that the con-nectionist framework will be a successful one for doing this. In that case, there is absolutely no claim made about the status of these things vis-à-vis human cognition. It is just building a better mousetrap—a better de-vice for doing a task. There is often the implicit claim that humans are probably doing something like this if you believe that this is the way to build an artificial system, because humans are so much better than most existing artificial systems in things like speech recognition.

But then there is another goal for this kind of work—actually to model real human cognitive processes. That is the kind of goal I have. Personally, I don't think that any particular piece of work that I do is going to be the final answer to the question, How is it that we are able to perceive, remember contextually appropriate information when we need it, and so on? Rather, I think of the process of model building as being part of an exploration. When I write a paper and propose a partic-ular model, I don't want people to think that this is the final word. I really just want to say, Look, I saw some problems with what had gone on before. It did not seem to capture what I took to be important characteristics. In order to capture these important characteristics, I formulated this new model. It incorporates principles that I happen to think are correct for other reasons as well, and now we will see how well we do with the model. What usually happens is that about 85 per-cent of what I thought I would capture when I started seems to be easily captured.

Of course, when you adopt a particular framework, you make an im-plicit statement that it will be productive to pursue the search for an understanding within the context of this framework. But it is very diffi-cult to evaluate a specific piece of research on this basis: people think one can say, "This or that model accounts for only 85 percent of the data; therefore we can reject the whole framework out of which it was built." Research does not work that way; it is much more complicated. A simi-lar thing happened to symbol processing. They have been able to make a lot of headway in a lot of cases; and the question is not whether they can continue to make headway, because there is always room for some further extension, but whether we might be able to make more of a quantum leap at some point by adopting a new framework.

I am a very pragmatic person, really. I am building models to account for known phenomena. The models certainly are representational. They make claims about what goes on in people's heads. But there is no special ontological claim to these models; they are hypotheses.

So you don't believe in a major breakthrough in the near future but rather in slow and progressive development? What are the biggest problems to overcome currently?

We were talking earlier about the idea that connectionist models could help us understand how commonsense could actually play a role in human cognition. The problem is that this is really an open question at this point. The problems to which we have applied our models up to now have all been problems that Fodor, in his *Modularity of Mind*,[6] would put in one peripheral module or the other. Fodor would like to reserve some central module for "real thought." It is still the case that conventional symbolic models are the way to think about thought. I believe that we will always want a coarse-grain description of thinking processes; that is, we will want to be able to say: "This person has been working on this problem for an hour, and she has suddenly realized that it won't work to follow this path." This sort of high-level description of cognitive activity has to be part of our understanding of thinking. This is the kind of process that symbol processors are always trying more or less directly to encode in their models.

But there is this gap between one step in the symbolic process and the next. Sometimes people have insights that are very difficult to trace: a person may be thinking over the problem, not knowing what else to try, and then—"Oh! Maybe . . ."—and suddenly they have seen the essence of the answer. I don't think symbol manipulation can really deal with this. This is the kind of thing where it seems that connectionist models really ought to be able to say something substantive. But up to now we have not really bridged that gap. That seems to me to be the place where a real breakthrough might be sought.

On the one hand, having a sudden insight in a symbolic model would mean to get a better representation of the problem, and a better match with the data. On the other hand, we have practice and skills. A lot of Artificial Intelligence does not deal with human skills at all, like moving, seeing, typing, playing a musical instrument. How do you think these two could be combined—higher level, rational thinking on the one side, and bodily movement, perception, and so on on the other side?

We already mentioned perception: many people feel that perception is some innate, hard-wired process whereby we see the world. I don't see perception like that at all. In my view, perception is itself an extremely knowledge-rich cognitive skill, which evolves through development and through experience in a domain. I really believe that it is right to say of an expert in a particular domain that "he sees immediately that such and such is the case"—for example, simply by seeing the configuration of pieces on a chessboard. He handles the situation without any deliberate, conscious cogitation. Some less expert person might come to that same realization by some other way, maybe with a lot of slow and laborious checking of possibilities. This slow and laborious checking of possibilities is the kind of stuff that seems to be well captured by conventional Artificial Intelligence approaches. Once we get into this, we get very local. We stop seeing the forest for the trees, we are following a little path through some moves; and while we are doing this we block out the fact that there might be another threat from behind . . .

In essence, experience allows us to encode into the mechanisms of more or less immediate perception an indefinite amount of background, relevant experience, and knowledge, so that we need not engage in explicit, conscious, very limited thought. What makes the expert better than the novice is not that the expert is thinking better but that everything is, as we say, "right at his fingertips." The consequences are immediately apparent. Everybody knows this, at some level. Certainly, research in chess has documented extremely well the fact that experts see the chessboard very differently from novices. They need only a few glances to reconstruct the whole board, whereas the novice cannot just do it from memory but has to copy the board by setting pieces. This is really the essence of what expertise is. When we understand how experience can be encoded so as to let us really see the implications of things, we will be a lot closer to understanding thinking.

What do you think about the economic outcome of connectionist research? Connectionism is a relatively new field—are there already any applications?

Yes, there do exist applications. What we are seeing in a number of areas is the incorporation of the connectionist learning algorithms into the tool kit of the Artificial Intelligence practitioner. I really don't subscribe to the notion that all you need to do is to take a back propagation network and throw some data at it and you will end up with a remarkable system. But we are at the point where several state-of-the-art methods for doing things like phoneme or printed character recognition are successfully drawing on connectionist insights. We have people who build highly structured modular systems for phoneme recognition that then use connectionist learning procedures to adjust the weights between the specific modules. In this way they are able to overcome some limitations in connectionist learning—for example, speed—while at the same time exploiting its ability to capture the ineffable interdependencies of different aspects of the speech signal.

It is not that applying connectionism is the answer to everything, but as a part of what an expert brings to the task of building a mechanism that would exhibit some intelligence, it is beginning to play a real role.

So in practice, the more traditional field of Artificial Intelligence and the field of connectionism come together? There are not two different directions—Artificial Intelligence on the one side, completely committed to symbol manipulation, and connectionism on the other?

Most people interested in application are extremely pragmatic. If they see something they can use they will try it out. This is one of the reasons why Carnegie-Mellon University is a place where different approaches can thrive well. At other institutions that are perhaps more philosophically oriented, there seems to be a clash. People here are exploring ideas in terms of what they are worth. Their worth is measured according to how they account for the facts and also by their ability to be actually used in an applied environment. Many computer scientists certainly have that bent. Newell very clearly has this as part of his goals. People with a more traditional approach will finally find a way to incorporate what they take to be a crucial new insight, or some piece of it, into their work. So we get a lot of hybrid systems around here.

What is the reason for the rise of connectionism in the last few years? The main ideas go back to the 1960s. Some people say that it has to do simply with the fact that there are more powerful machines now.

I've said that, for instance. My view is the following: the ability to explore things in a hands-on way, using computer simulation models, has had a tremendous impact. When Rosenblatt tried to do his work, he spent most of his time thinking and writing, and he got very, very limited access to computational resources. His book *Principles of Neurodynamics*[7] goes on for six chapters or so before he gets to his first simulation. He said something like, "Well, I got six hours of computer time between midnight and 6 a.m., and I got a hint that my ideas might work!" and that his findings were very preliminary.

Ultimately, the results of these explorations led people to have insights—they encode something, become more expert. They have insights that they can then write down in an elegant equation. The equation might lead to a breakthrough in what kind of training can occur in connectionist networks. It is not that there have not been breakthroughs but that what there has really been is the opportunity to do exploration. The fact that people have workstations on their desks has created the opportunity to explore these ideas and has led to key mathematical insights. The computational resources contribute to the development of a critical mass.

I wonder how one can fit into this field. You have to have knowledge about psychology, mathematics, computer science, and philosophy as well. How is that possible? Is it made possible by cooperation in groups? I would like to know more about this intellectual atmosphere.

This is a really good question, and it is something I think about a lot. My feeling is that each individual researcher is a limited "device" for incorporating expertise. What we have to do is create environments that bring together just the right combinations of things to allow people to make progress. It seems to me that the best way to do this is through a diversity of institutional arrangements in different universities. At Carnegie-Mellon University, we have the opportunity to combine connectionism with a focus on higher-level cognition. At another institution, they may have a much better chance to make a connection with linguistics, because there are linguists who have the right kinds of ideas about language and connectionists who have the right kinds of tools.

Who makes this policy—to hire the right people, to provide the environment? Is it coincidence? Why are there some centers around the country where people come together?

I think it is basically intellectual synergy between individuals, and maybe also a matter of fertility of the environment. Here at Carnegie-Mellon University, there is this sort of openness toward new ideas in spite of the people having developed their own ideas, which makes for the opportunity to explore new kinds of things. Other institutions are more constrained. But really what makes these things happen is when researcher A and researcher B, who have different but synergistically combinable backgrounds, see that they have something to offer to each other.

That is what happened at San Diego. Rumelhart had incredible mathematical modeling skills, and I brought an awareness of the details of the data in a particular area. I learned how to model from working with him, how to get this whole thing to work, how to account for the data I knew something about. Once I got this expertise, a guy named Jeff Elman said, "Hey, I can see how this relates to language," and he and I started to work together on language.[8] I really think that this is the only way in which such enterprises can be started: by people who have specific expertises, hooking up with each other when they see that what the other does is really relevant. This is the way I try to operate. I collaborate with individuals whose work really indicates that they have a deep understanding of their area.

Recommended Works by James L. McClelland

"The Case for Interactionism in Language Processing." In *Attention and Performance*, vol. 12: *The Psychology of Reading*, ed. M. Coltheart, pp. 1–36. London: Lawrence Erlbaum, 1987.

"Parallel Distributed Processing: Implications for Cognition and Development." In *Parallel Distributed Processing: Implications for Psychology and Neurobiology*, ed. R. Morris. Oxford: Oxford University Press, 1989.

(With D. E. Rumelhart.) *Explorations in Parallel Distributed Processing: A Handbook of Models, Programs, and Exercises*. Cambridge, Mass.: MIT Press, 1988. (Contains a floppy disk with several network simulations.)

(With D. E. Rumelhart and the PDP Research Group.) *Parallel Distributed Processing: Explorations in the Microstructure of Cognition*. 2 vols. Cambridge, Mass.: MIT Press, 1986. (James McClelland contributed to a number of chapters in these two volumes.)

(With M. St. Johan and R. Taraban.) "Sentence Comprehension: A Parallel Distributed Processing Approach." *Language and Cognitive Processes* 4 (1989): 287–335.

Notes

1. E.g., S. Sternberg, "Memory Scanning: Mental Processes Revealed by Reaction-Time Experiments," *American Scientist* 57 (1969): 421–57.

2. The Psychonomic Society hosts an annual meeting of psychologists committed to the experimental "measurement of the mind." The society publishes, among other journals, *Perception and Psychophysics* and *Memory and Cognition*.

3. G. E. Hinton, "Learning Distributed Representations of Concepts," *Proceedings of the Eighth Annual Conference of the Cognitive Science Society* (Hillsdale, N.J.: Lawrence Erlbaum, 1986), pp. 1–12.

4. J. L. Elman, "Finding Structure in Time," *Cognitive Science* 14 (1990): 179–212.

5. J. L. McClelland and D. E. Rumelhart, *Explorations in Parallel Distributed Processing: A Handbook of Models, Programs, and Exercises* (Cambridge, Mass.: MIT Press, 1988).

6. J. A. Fodor, *The Modularity of Mind* (Cambridge, Mass.: MIT Press, 1983).

7. F. Rosenblatt, *Principles of Neurodynamics* (New York: Spartan, 1962).

8. J. L. McClelland and J. L. Elman, "Interactive Processes in Speech Perception: The TRACE Model," chap. 15 of *Parallel Distributed Processing: Explorations in the Microstructure of Cognition*, vol. 2: *Psychological and Biological Models*, by D. E. Rumelhart, J. L. McClelland and the PDP Research Group (Cambridge, Mass.: MIT Press, 1986).

ALLEN NEWELL

The Serial Imperative

The field of computer science did not exist in the fifties. Computer science as a discipline does not show up until the early or midsixties. Before that, computers were viewed as engineering devices, put together by engineers who made calculators, where the programming was done by mathematicians who wanted to put in mathematical algorithms. Consequently, there was not really an intellectual discipline of programming. On the other hand, there clearly was the ferment of computer science in cybernetics. The postwar world was clearly in intellectual ferment. I would almost use the word *chaos*, but not in the sense that it was in trouble; it was just bubbling over. As the number of scientists was fairly small, lots of things that are regarded today almost as separate fields were thrown together.

So there was no computer science; people came from mathematics . . .
    . . . and from electrical engineering. The cybernetic ideas were very well known at that period. They involved strong interdisciplinary work, because cybernetics was not a separate discipline. It was all built around the notion that the electrical engineers understood the most about feedback. You have to remember that the whole field of feedback mechanisms had come into being during the war. After the war, it opened up: ideas on feedback circuits were dumped into the open intellectual world.

How did this lead to Artificial Intelligence?
    Back in the late forties and fifties, there were already speculations about computers, cybernetics, and feedback, about purpose in machines and about how computers might be related to how humans think. But these speculations went on at the top level without there even being a view that it actually was a coherent discipline.
    In psychology, there was this tremendous hiatus during the last four or five years of the war, where everyone went off and prostituted for war. A feature of a world war is that everybody in the society attends to the war. All the scientists leave the universities and do all kinds of things throughout the military command. The military becomes a different

Allen Newell one of the founders of the fields of Artificial Intelligence and cognitive science, died on 19 July 1992 at the age of sixty-five.

Newell's career spanned the entire computer era, which began in the early 1950s. The fields of Artificial Intelligence and cognitive science grew in part from his idea that computers could process symbols as well as numbers and, if programmed properly, could solve problems in the same way humans do. In cognitive science, he focused on problem solving and the cognitive architecture that supports intelligent action in humans and machines. In computer science, he worked on areas as diverse as list processing, computer description languages, hypertext systems, and psychologically based models of human-computer interaction.

His more recent work has centered on the development of "Soar." The goal of "Soar" is to provide an underlying structure that would enable a computer system to perform a range of cognitive tasks.

A native of San Francisco, Newell received a bachelor's degree in physics from Stanford University in 1949. He did graduate work in mathematics at Princeton University, then worked for the Rand Corporation as a research scientist from 1950 to 1961. While at Rand, he met Herbert A. Simon, then Professor of Industrial Administration at Carnegie Institute of Technology. Their discussions on how human thinking could be modeled led Newell to Pittsburgh, where he collaborated with Simon and earned a Ph.D. in industrial administration from CIT business school in 1957.

Newell joined the CIT faculty in 1961. He played a pivotal role in creating Carnegie-Mellon's School of Computer Science and elevating the school to world-class status. Newell, U. A. and Helen Whitaker Professor of Computer Science at the time of his death, wrote and co-authored more than 250 publications, including ten books. He was awarded the National Medal of Science and elevating the school to world class status. Newell, Helen Whitaker Professor of Computer Science at the time of his death, wrote and co-authored more than 250 publications, and ten books. He was awarded the National Medal of Science a month before his death.

place, because the military is in fact the civilian population. Almost all the—strongly behavioristic—psychologists went off and worked in the military in terms of training, operations research, engineering psychology, for example, worrying about instrument panels. All this does not change the discipline as long as it is all in the service of the war.

After the war, however, the scientists come back, and of course they are very different people now, having done different things. So you get this immense turbulence, this mixing effect that changes science. People in operational mathematics, cybernetics, computing, and information theory all talk to each other at the same time.

The shift in psychology that is now called the "cognitive revolution" really happens in the midfifties. From a historical perspective, that is still the postwar period. First came the war, then came this turbulence, and then big things begin to happen.

Could this historical turbulence be the reason why interdisciplinarity seems much easier here than in Europe?

There are several factors to be taken into account. This is one of them, at least in our part of the world, where things were carved up so differently from how they had been. But there are also some deep institutional factors. In the European, and especially in the German tradition, the academic world is built around the fact that the professor is king in his own domain and his own university. The idea that you would have four professors of theoretical physics at one university is not a conceivable notion.

But that is exactly what is true of American universities. Look at Berkeley: you cannot even count the professors of theoretical physics. Here in this department (the Department of Computer Science, Carnegie-Mellon University), there are three or four professors—Mitchell, Carbonell,[1] and myself—whose main interest lies in learning systems. This produces a whole environment for research. And this institutional factor has nothing to do with postwar development.

The other factor is mobility. What first amazed me in Europe was the idea that once you went to a place, you would stay there. Typically, a professor's career in the U.S. would be something like four years at Berkeley, three years at Michigan, twelve at Harvard. This produces an immense mixing that exposes professors to different pressures. The last factor is that Carnegie-Mellon University itself, unlike Berkeley, Stanford, or others, is a tremendously interdisciplinary place.

Was it difficult for you to work together with people from other fields?

I have never asked the question that way. My own education is such that I did not get my Ph.D. until I was a full professional. After the war I went back to graduate school to do mathematics. I felt I was

not a mathematician and so I left school. I went to the Rand Corporation in Santa Monica. It was only five or six years later that Herb
(Simon) convinced me to go back to graduate school. But when I came
back here, I took my degree in industrial administration—so I have a
business degree, really. I have considered myself as a physicist for a
long time, but at Rand I worked on experiments in organizational theory. We did social simulations of a whole organization and studied
how it behaved under a simulated load. Fundamentally, I was doing
experimental organizational sociology. It never occurred to me, and I
have never recognized any disciplinary boundaries. The real question
is why the environment did not stomp on me and force me to say what
I was.

This is due partly to this turbulence, which opens everything up. It
makes everyone appreciative of things going on in other areas. A place
like Rand is completely interdisciplinary. Operations research did not
exist before the war. It was created during the war and became a brand-
new field. It started out as narrow analysis—what the British called operational research. The extension to entire systems—systems analysis—
was pioneered by Rand. At the time I was there, in the fifties, the idea
was born that you could look at entire social systems, build quantitative models of them, and make calculations upon which to base policy
decisions.

Rand is a nonprofit corporation that got its money from the Air Force.
This is one of the cultural curiosities that happen in the United States for
which the boundaries in Europe are much too tight.

*The question of military-sponsored research is a matter of much debate
even in the United States, isn't it?*

The Vietnam War is a watershed point in terms of attitudes among the
intelligentsia. If you go back into the previous years, there is no such
feeling of confrontation or separation.

The scientists produced dozens of major inventions during the war.
So the government and the military came out of the war and, having
realized the huge impact of science, wanted to preserve it. A general initiated the creation of an institute to keep some part of academia working
on problems of national security. The statement of work in the contract
is that the Rand Corporation will do whatever is in the interest of national security. It says nothing about the Air Force. Also, it's located in
Santa Monica—to be as far away from Washington as possible. In the
early days, the only place where military people were allowed in was the
director's office. They were not allowed to walk around.

Then the Rand Corporation was shifted into the Congress budget,
where it has been ever since. Rand was the first FCRC—a Federal Contract Research Center. An FCRC is a nonprofit organization wholly sup-

ported by the government. It exists independently from the government, and its purpose is to provide competitive salaries for scientists, because the government does not pay enough money.

When did you come to Carnegie-Mellon University?

I went to Rand in 1950, and I came to Carnegie in 1955 to be a graduate student. I never ceased to be a Rand employee; I simply transferred. Then I was here full-time, first as a student, then as an adjunct faculty member but still a member of Rand staff, until 1961. In 1961, I shifted my allegiance to the university.

How did you get involved in computer science?

In the course of our organizational experiments, we were trying to understand what was going on in technological organizations, for example, in air defense. People there were reading information off radar screens, trying to build a common picture out of them and making decisions on them.

In studying this, we built models of the processing of information. We had to characterize why these groups were as effective as they were by understanding how they were actually processing information. We were deeply immersed in using computers as a simulated environment. In the middle of this emerges the computer as a device that can engage in more complex processes than we had ever imagined.

I even remember the exact moment Artificial Intelligence started for me: one afternoon at Rand, there were five of us talking together. Cliff Shaw was at Rand then. Herb was an economic consultant, but he became very interested in organizations. At this meeting in November 1954, Herb was present. Cliff was not able to listen, but I went into his office and told him the whole talk. As soon as I hear us talk, I absolutely *know* that it is possible to make computers behave in an essentially intelligent way. I started working on Artificial Intelligence that night, and I have never stopped since: I went home and designed what amounted to discrimination net schemes to simulate what was going on in our air defense direction centers. My first paper, in 1955, was about adaptation in chess machines.[2]

The so-called cognitive revolution came later?

The cognitive revolution starts in the second half of the fifties. If you want a landmark: the first work was the use of signal detection theory. Sweats and Tanner used the models of radars—how a radar makes a decision between signal and noise—to build a model of what an ideal observer should do. They applied that to how a human would detect whether he heard a faint sound or not. It was a direct transfer from the analysis being done for radars into models for the humans.

Miller's paper "The Magical Number Seven, Plus or Minus Two" was published in 1956,[3] as well as Bruner, Goodnow, and Austin.[4] In the same year, we wrote about the Logic Theorist.[5] Chomsky's first real paper, "Three Models of Language," was in 1956, too.[6] In some sense, the break with behaviorism happens across the board. My historical characterization is that Herb and I were part of it, but only one little part.

So you would see the cognitive revolution as a break with behaviorism?

Right, except that some of us, such as Herb and I, were never behaviorists. So our main motivation was not to react against behaviorism. It happens that a scientific group has as its principle the reaction against the previous one—in a dialectic way. You can see a piece of that in connectionism. For all people deeply involved in psychology, the break with behaviorism is a large part of the action, but less for people like Herb, who comes from the social sciences, or for me as a physicist. Therefore I have only peripheral connections with psychology. I certainly was not a behaviorist who then had to throw off this load; it was just not part of my mind-set.

But historically, this revolution is essentially a revolution against behaviorism. It is related to how you think about the inside of the mind, but it is also about theory, about how you postulate entities and a theoretical structure—whatever scientists want to do but behaviorists kept themselves from doing. All this happened in the second half of the fifties. It is, fundamentally, a postwar development, in spite of the apparent delay of one decade.

Other fields came into being, too. The business schools, which had been totally focused on business, underwent a revolution in which they hired social scientists and psychologists and economists. This revolution was, in large part, due to Herb. The first work on Artificial Intelligence on this campus took place in the business school: Feigenbaum wrote his thesis on EPAM[7] here in industrial administration.

When did you start with your work on the general problem solver (GPS)?

In 1957. The first internal document on GPS is in December 1957; the first external document is in 1959, for the IFIP congress in Paris.[8]

It was about problem solving by humans—without a context. At first, I think, you thought that one could isolate the problem-solving mechanism from knowledge about the world.

Those ideas were totally absent. We had this notion that one could build systems that were complex and could do complex intelligent tasks.

We called the area at first simply "complex information processing." The focus was not so much on intelligence but on intelligent tasks, like chess, geometry theorem proving, and logic. The first system we built was called Logic Theorist (LT) and was for theorem proving in logic. The reason it was not in geometry was that the diagrams would have increased the momentary burden on the system.

Herb and I have always had this deep interest in human behavior. The issue of always asking how humans do things and trying to understand humans has never been separate. So we produced LT, which is largely driven by pure task analysis. Although we wrote an article in which we talked about it as a theory of human behavior, the connections drawn to the brain are rather high-level nevertheless.

Then came a graduate student who started to look at protocols. So we used the protocols of how humans do tasks and logic. The GPS program is, in some sense, exactly the Logic Theorist and the evidence from the protocols taken together.[9] We thought, Let's write a program that does the tasks in exactly the way humans do it according to the protocol. The paper in Feigenbaum's and Feldman's *Computers and Thought* (it was the classical collection of papers) is the first paper on GPS that has to do with human behavior.[10] We just followed the protocol, looked at the mechanism, and put the mechanism into the program.

The question of context just was not asked. We had this evidence of human behavior, we could see the mechanism, and, given what we then knew about the Logic Theorist, we could build a new version.

You were also excited by the success of the first model . . .

Sure. It was quite clear to everybody, including ourselves, that with LT, we had really made a great step forward. First of all, LT really dealt with symbolic material. Later on, one would get concerned that the symbols were formalized things and not commonsense reasoning or language. But at that point, this distinction was not so important. It was important that the representation was not numerical but symbolic. It was a truly symbolic representation of a task.

The criticism concerning commonsense came later?

Right, in the seventies. One of the things that are true for every science is that discriminations that are very important later on simply have no place in the beginning. They cannot have a place, because only when you know something about the problem, you can ask these questions.

On the one hand, you write that problem solving is one of the major intellectual tasks of humans, and we think about intelligence as the ability to solve problems. On the other hand, we succeeded in building

*machines that could do pretty well in problem solving, but we did not
succeed so well in pattern recognition, walking, vision, and so on. What
is the reason for this? Formerly, recognizing a face was not seen as an
intelligent task.*

This particular discrimination goes way back. In the Feigenbaum/
Feldman collection of 1963, you find an article by Selfridge and Neisser[11] that is exactly about this question: why is it easy to do logic but
difficult to do pattern recognition? First of all, this is not nearly as true
as it was: we know lots about how to do vision, and there are successes
in robotics. In retrospect, part of the issue is that the volume of data that
has to be processed from a two-dimensional retinal array in order to
deal with real vision is in the millions of bits. Until the computational
capacity is up to dealing with millions of bits per millisecond, you cannot do any of these processes. If you can do visual processing only at a
rate of one bit per minute, you will never even find out if you are on the
right track.

So it was a question of advancement in technology, too?

That is partly a retrospective comment. It has to be clearly understood
that as soon as you deal with vision or speech, there is this tremendous
amount of data processing. It comes coupled with the fact that there is a
lot of noise, so that the processing deals with the extraction of signals
from noise. It is easy to be confused about what the problem is—if it is
lack of computational power or if we don't understand pattern recognition. If you cannot experiment, there is no way of knowing which problem it is. We are now at a point where the computational processing is
getting under reasonable control. Think of cognitive reasoning in terms
of computational demand, and then of speech, which is one-dimensional
with noise, and of vision, which is two-dimensional. The amount of data
processing for speech is hundreds of thousands of bits per millisecond,
and for vision it is millions of bits.

As long as we are speaking of computers in the fifties and sixties, they
do not have enough capacity to do either of these—you are always struggling with the amount of processing. Speech comes under control before
vision, and now, in fact, the signal processing for speech is no longer a
bottleneck. There are still serious problems with respect to actually extracting the signals, but we do have speech recognition systems. We also
have visual recognition systems, but there the computation is still just
marginal.

*Today, people try to get around this bottleneck by parallel processing. I
wonder how much your physical symbol system hypothesis is bound up
with serial computation. You once wrote that one of the problems with
humans is that they have to do one step after the other, with the excep-*

*tion of the built-in parallelism of the eye. What do you think about con-
nectionism, and how is it related to the physical symbol system hypothesis? Would you say that connectionism is also part of cognitive science?*

Connectionism is absolutely part of cognitive science. The issue is that
it is also part of neuroscience. And yet there is a certain way in which
cognitive science and neuroscience are separate: they come out of different bases of inspiration. There is a need for seriality in human behavior
in order to gain control. If you worked totally in parallel, what you think
in one part of your system could not stop what you are thinking in another part. There are functional reasons why a system has to be serial
that are not related to limitations of computation.

The physical symbol systems hypothesis, the nature of symbolic systems, is not related to parallelism. You can easily have all kinds of
parallel symbol systems. You can think of connectionist systems as not
containing any notion of symbols. So you can, in one sense, say that
connectionist systems are systems that are nonsymbolic and see how far
you can go when you push them to do the same tasks that have been
done with symbol manipulation without their becoming symbolic systems. There is certainly an opposition between the physical symbol system hypothesis and connectionism.

*Could one combine the two hypotheses, for example, by saying that on
a lower level, processing is more connectionist, but on a higher level, you
need symbol manipulation?*

That is the standard way of doing it.

*Do you think it is correct? That is, to say that vision or speech
recognition is better captured by connectionist models?*

No, that is the cheap way out. What the connectionists are doing is
not just trying to deal with speech and hearing. They are trying to deal
with what central cognition does—with the learning of categories, or
with the organization of behavior. If you try to put connectionism down
only at the senses, only at perception, that is not really the domain they
are working in.

The usual way of putting them together is to say, Clearly, a symbolic
system must be realized in the brain. So it has to be realized by one of
these parallel networks. (We can argue about how close the connectionist systems model the brain. They may not model it very well at all. There
is a lot of internal discussion in connectionism about that.) Therefore,
symbolic systems must be implemented in systems that are like parallel
networks. That is the implementational point of view.

A lot of connectionists would not buy that, in part because they believe that connectionist systems *do* higher-level activities in ways that
would not look at all like what we have seen in symbolic systems. If you

say that it is nothing but an implementation technology, you take them away all these possibilities. Many people would reject this.

There are numerous funny contradictions in this effort. Look, for example, at Soar,[12] our latest system: Soar is built around a production system memory. A production system memory is a massively parallel memory. But if you look at the literature, you will find that production systems are viewed as the essence of symbolic systems. So all of a sudden you get one of these conceptual intransitivities: when connectionists characterize physical symbol systems, they take production systems as their prototype. Yet production systems are massively parallel systems, not serial systems. This distinction, which is also meant to be a distinction between serial and parallel, vanishes.

So in your view, the physical symbol systems hypothesis is not challenged by connectionism?

My personal assessment is: no, there is no challenge. If you ask me as a commentator on the scene: yes, there is a challenge. It is quite possible that as the underlying technology of using distributed representation, weight adjustment, and so forth gets better, and if we then try to build systems of that kind that become symbol systems, they could be so different from the kind of symbol systems that we've got now that, in fact, the pragmatic content of the criticism is really there. In that respect it is a challenge.

It is important to explore these technological alternatives. Humans use stories that they tell themselves in order to get themselves to work on this or that. These stories often deal with confrontation between areas and ideas. From some point of view, it is almost always the case that these high-level stories are relevant only as motivation and not really relevant to what eventually happens in terms of technical understanding.

From a certain point of view, connectionist systems are not a point in technology space with a different set of arguments. They are not outside this world, they are part of computer science. We analyze them in the same way by means of complexity analysis; we need to understand their properties, and as we do so they enrich our set of possibilities for modeling how the human mind works.

Recommended Works by Allen Newell

Unified Theories of Cognition. Cambridge, Mass.: Harvard University Press, 1990. (The
    William James Lectures for the Department of Psychology at Harvard, which A. Newell
    delivered in 1987, were the basis for this book.)
"Unified Theories of Cognition and the Role of Soar." In *Perspectives on Soar*, ed.
    J. Michon and A. Anureyk. Norwell, Mass.: Kluwer, 1991.

(With W. Harvey, D. Kalp, M. Tambe, and D. McKeown.) "The Effectiveness of Task-Level Parallelism for Production Systems." *Journal of Parallel and Distributed Computing* 13 (1991): 395–411.

(With J. Laird and P. Rosenbloom.) *Universal Subgoaling and Chunking.* Norwell, Mass.: Kluwer, 1986.

(With J. F. Lehman, T. A. Polk, and R. Lewis.) "The Role of Language in Cognition: A Computational Inquiry." In *Conceptions of the Human Mind,* ed. G. Harman. Hillsdale, N.J.: Lawrence Erlbaum, 1992.

(With H. A. Simon.) *Human Problem Solving.* Prentice-Hall, 1972.

Notes

1. See R. Michalski, J. Carbonell, and T. Mitchell, eds., *Machine Learning: An Artificial Intelligence Approach* (Palo Alto, Calif.: Tioga, 1983).

2. A. Newell, "The Chess Machine: An Example of Dealing with a Complex Task by Adaptation," *Proceedings of the 1955 Western Joint Computer Conference* (1955): 101–8.

3. G. A. Miller, "The Magical Number Seven, Plus or Minus Two: Some Limits on Our Capacity for Processing Information," *Psychological Review* 63 (1956): 81–97.

4. J. S. Bruner, J. Goodnow, and G. Austin, *A Study of Thinking* (New York: Wiley, 1956).

5. A. Newell, and H. A. Simon, "The Logic Theory Machine," *IRE Transactions on Information Theory* IT-2, no. 3 (September 1956); A. Newell A., J. C. Shaw, and H. A. Simon, "Empirical Explorations with the Logic Theory Machine: A Case Study in Heuristics," in *Computers and Thought,* ed. E. A. Feigenbaum and J. Feldman (New York: McGraw-Hill, 1963), pp. 109–33.

6. N. Chomsky, "Three Models for the Description of Language," *IRE Transactions on Information Theory* IT-2, no. 3 (1956).

7. EPAM – Elementary Perceiver and Memorizer; see E. A. Feigenbaum's "An Information-Processing Theory of Verbal Learning" (Ph.D. diss., Carnegie Institute of Technology, 1959).

8. A. Newell, J. C. Shaw, and H. A. Simon, "Report on a General Problem-Solving Program," *Proceedings of the International Conference on Information Processing* (Paris, 1960).

9. See K. A. Ericsson and H. A. Simon, *Protocol Analysis: Verbal Reports as Data* (Cambridge, Mass.: MIT Press, 1984).

10. A. Newell and H. A. Simon, "GPS: A Program that Simulates Human Thought," reprinted in *Computers and Thought,* ed. E. Feigenbaum and J. Feldman (New York: McGraw-Hill, 1963).

11. O. Selfridge and U. Neisser, "Pattern Recognition by Machine," originally published in *Scientific American* 203 (1960): 60–68.

12. J. E. Laird, A. Newell, and P. S. Rosenbloom, *Soar: An Architecture for General Intelligence* (research report, Carnegie-Mellon University, 1986).

Stephen E. Palmer was born in 1948 in Plainfield, New Jersey. After attending the public schools in his hometown of Westfield, New Jersey, he entered Princeton University, where he received a B.A. degree in psychology with highest honors in 1970. He did graduate work at the University of California, San Diego, where he worked with Donald A. Norman and David E. Rumelhart, and received a Ph.D. in 1975. Upon graduation, he took a position in the Psychology Department at the University of California at Berkeley, where he has remained ever since. His field of specialization is visual perception, where he takes an information process-ing approach to understanding how people perceive the structure of objects and scenes. He is currently Director of the Institute of Cognitive Studies at Berkeley and head of the undergraduate degree program in cognitive science.

STEPHEN E. PALMER

Gestalt Psychology Redux

My training in cognitive science began as a graduate student in psychology at UC San Diego. With Donald Norman and David Rumelhart we began doing work in cognitive science before there was actually any field by that name, in that we, as psychologists, were constructing a large-scale computer model of how people understand language and retrieve information from their memories. That was in the early seventies. During that time we came to know a number of important people in computer science and Artificial Intelligence, because Don organized a series of conferences. He brought people down to talk to us. Those people included Allen Collins and Ross Quillian,[1] and Gordon Bower and John Anderson, who were doing the HAM model at that time.[2] We went up to Stanford and visited Roger Schank's[3] laboratory. Danny Bobrow[4] came down from Xerox PARC (Palo Alto Research Center), and a number of other people came, too.

We were doing two pieces of cognitive science, in the sense that we were worried not only about human psychology but also about how to do computer simulation models. That was before computer simulation models were well known in psychology. It was one of the first big psychological computer simulation models. If it were done today, it would be recognized immediately as cognitive science, but at that time it was just some strange psychologists doing what computer scientists normally do.

So psychologists started cognitive science at UC San Diego? Is it right that Norman and Rumelhart were the founders of the Institute?
First there was the Center for Human Information Processing, founded by George Mandler in the late sixties. It consisted primarily of psychologists. Later on it broadened. Roy D'Andrade[5] and Aaron Cicourel, both anthropologists, and eventually some linguists became associated with the center. I don't think that there were any computer scientists. Don and Dave had a couple of graduate students in computer

science, because there was nobody at UC San Diego doing Artificial Intelligence at that time. The Center for Human Information Processing was the beginning of cognitive science at UC San Diego.

UC San Diego had the first institute of cognitive science. Did it have a special research program then? Today, it looks like the center of connectionism.

As I said, I think that cognitive science probably got started at UC San Diego, almost sooner than anyplace else, except maybe MIT and Carnegie-Mellon. The connectionist days came later, actually after I was gone. Back then the major project that would be called cognitive science was the memory model project of Lindsay, Norman, and Rumelhart.[6]

By the time I came to Berkeley I had already started to become interested in perception more than in language. I had started out in the memory project, and I came to perception through a kind of back door, via language. I was part of a group within the LNR research group that was studying the meaning of verbs. We developed a theory in which verbs were broken down into structures of primitive components. We had a nice theory of the structure of the meanings of many of these verbs, but at that time, nouns had no interesting cognitive structure in the theory. They were just represented as unitary nodes in a network.

I decided we needed a better representation for nouns, and that was how I got interested in perception. Because once you start to worry about nouns and concrete objects, you start to think about the representation of their physical structure, what they look like, and how you identify them. I did my dissertation on perceptual structures, actually, and it became clear that my interest in perception pointed to the direction that Gestalt psychologists had been interested in. I did not have any training in Gestalt psychology because most of the psychologists at UC San Diego were relatively young at the time and did not know the Gestalt literature.

How can one combine Gestalt psychology and cognitive science? Isn't this a contradiction? Dreyfus, for example, criticizes cognitive science from a Gestalt approach.

In a way, I am sympathetic to that. I see a tension in cognitive science between the very formalistic approach that, for example, the MIT school takes toward cognition, and the opposite pole, which is kind of Gestalt: the idea that you cannot analyze things into atomic parts or logical symbols and do symbol manipulation processing on them in very logical and formally derived ways. Real human cognition is more fluid, more flexible, more context-sensitive than classical cognitive science has been able to capture in the kinds of models they use and the kind of paradigm they developed.

That is the reason, in fact, why I am so interested now in the connectionist movement. I see it as a very central and interesting part of what is going on in cognitive science. Part of what makes it so interesting is that it seems to have these properties of being more fluid, more the way people think: less formal, less rigid, and maybe ultimately more humanlike.

In your opinion, then, the MIT approach is more formal, with the main idea of the physical symbol systems hypothesis and symbol manipulation? And the other side could be Gestalt psychology?
Yes, Gestalt psychology is the other pole.

Is the difference between the East and West Coasts a coincidence?
Some people even claim that it is correlated with the different lifestyles on the two coasts. The East is where things are very busy and uptight and crammed together. People in the West are more laid-back and open to new ideas. I don't know whether there is anything to that or not. In fact, prior to the connectionist work, most of the work at UC San Diego, including the staff I was part of—the LNR memory model—was very much classical Artificial Intelligence. We were using standard propositional network representations and processes that operated on networks, which is more in the Quillian mode than, say, Newell's and Simon's production systems. The family resemblance between what we were doing and the symbolic systems approach that characterizes the more formal East Coast approach to cognitive science was very strong. It was only later, when McClelland, Rumelhart, and Hinton started the connectionist work, that it got a particularly Gestalt flavor.

How was it possible to develop approaches differing from the then dominant MIT approach?
UC San Diego is far away from MIT, so I was not very much under that influence. Maybe that is part of the geographical reason for the difference between the East and West Coasts: the farther you are from MIT, the more it becomes possible to do something different. For me it was easy to get interested in Gestalt theory, because I was phenomenon-driven. I think it is a general characteristic of the people—at least, at Berkeley—that they tend to be more sensitive to the phenomena of human cognition and less concerned about the formal status of theories.

How would you outline your critique of the more formal approach and the advantage of a Gestalt approach?
I got interested in Gestalt theory because I was interested in phenomena that Gestalt psychologists had studied, especially contextual phenomena. The idea of holism is essentially that there is global interaction

among all the parts and that the whole has characteristics that are quite different from the characteristics of the parts that make it up. In perception, those phenomena are everywhere—you have to try very hard to ignore them. You can think of Gestalt psychology as having discovered these phenomena and thus having identified a very important set of problems that need to be solved theoretically.

The Gestalt psychologists managed to overthrow the prior prevailing view of perceptual theory, which was the structuralist theory promoted primarily by Wundt[7] and his followers. Their idea was borrowed from British empiricist philosophers, namely that there are sensory atoms and that complex perceptions were constructed by associating these sensory atoms together. The Gestaltists' demonstrations of principles of perceptual organization—grouping, figure-ground relationships, reference frames, and so on—were effective in showing people that the atomistic approach to theory was not going to work very well.

Eventually, they proposed their own way of theorizing about those things, but these theories were not well accepted. When Köhler[8] proposed that electrical fields in the brain were responsible for the holistic interactions, it allowed people to do physiological experiments to try to determine whether these fields in fact existed and whether they had causal properties in perception. And it turned out they did not. Because their physiological hypothesis was wrong, all Gestalt theoretical ideas were discounted.

In my view, the Gestalt psychologists were proposing a more abstract notion, namely what they called "physical Gestalten": physical systems that operate according to holistic principles. They thought that the brain was a physical Gestalt, that it somehow was a dynamical system that would relax into a state of minimum energy. The electrical fields they proposed were one way in which this theory could be implemented in the brain, and it turned out to be wrong.

Now we have an alternative. Connectionist networks that have feedback loops, like Hopfield networks, are in fact another example of physical Gestalten. They have all the same properties: they are dynamical systems that evolve over time, they relax or settle into a minimum energy configuration. Hopfield defined the energy function that was minimized by the dynamic properties of these networks.[9]

I view recurrent networks as a new kind of Gestalt theory. That does not say that it is going to be the right theory, but it has given Gestalt theory a new lease on life. It was more or less banished from psychology on the basis of those experiments. There are no electromagnetic fields in the brain that cause perceptions, but there might be neural networks that are working in the same way, which are responsible for the very same kinds of phenomena arising in visual perception.

In one of your articles,[10] you define five underlying assumptions of information processing, one of which is the recursive decomposition assumption. How does that fit with Gestalt theory? How does Gestalt theory fit into cognitive science under these assumptions?

The recursive decomposition assumption is the assumption that you can characterize the human mind in terms of a functional black box, in terms of a mapping between inputs and outputs, and that then you can recursively decompose that black box into a set of smaller and simpler black boxes that are connected to one another by a flow diagram of some sort. Each of those black boxes can itself be decomposed recursively into smaller and smaller black boxes. The standard assumption in cognitive science is that this recursive decomposition can take place in small steps, that the decomposition process is going to be "smooth." In its extreme form, the Gestalt psychologist would say that you start out with one black box, which is the mind, and then immediately you have to decompose it into a zillion neurons that are connected to one another in probably very funny ways.

My own view is that something in between those views is probably right. The mind is probably a nearly decomposable system in the sense that Simon defined, down to some level. You can get modules inside modules down to some level, but at some point you cannot get any farther and, all of a sudden, you have to go down to the level of individual neurons. The decomposition is smooth only for a few levels, and then you get these extremely strongly interacting sets of neuronlike elements that define that module, and it does not make sense to define intermediate levels or structures within that module. This is the point where things get very Gestaltlike. It might be that you have a module for recognizing faces that works as a Gestalt, that is, a single black box in there that just takes its input, does something mysterious with it, and in this massively interactive and parallel way produces an output. But those modules are probably put together in ways that are understandable, so that you can in fact do a certain amount of decomposition. At some point you stop being able to decompose, and you have to go all the way to the functional level, which corresponds to neuronlike elements—which are, in fact, the basic units of a connectionist model.

In your opinion, then, the decomposition assumption is right, but only to a certain level.

That is my current guess. The strong form of the classical information processing approach in psychology supposes that you can recursively decompose all the way down to the bottom, however you define your functional primitives. That may well be wrong.

What do you think about the physical symbol systems hypothesis?

I think it is approximately true. The view that I like best is Smolensky's notion that you can build an information processing system out of connectionist modules that would look approximately like symbol manipulation if you look at it on a grand scale.[11] When you try to model the transitions between different states in higher-level cognition, where the time scale is large and the operations take fairly long, it does a really good job. All the work Newell and Simon have done indicates that it is possible to model the microgenesis of human problem solving nicely using those kinds of models.

The fact is that we have never been able to get very good models of human perception, pattern recognition, or human learning. We have never been able to get contextual effects to arise naturally out of physical symbol system models. These are exactly the domains where connectionism is strong.

Do you think an algorithm like back propagation is a satisfying model?

Unlike most people, I am less enthralled with the learning aspects of the connectionist paradigm than I am with the dynamic system aspects. Back propagation is interesting as a kind of learning, but it strikes me that it remains to be demonstrated convincingly that it is the kind of learning people do. It does have nice properties: it is gradual in a way that human learning is, and it may be able to model things like small quantitative changes producing large qualitative changes over a long term, which is consistent with what we know about human learning. But it does not convince me as a general model for human learning.

So in your opinion, neither symbol manipulation nor connectionism is the whole story—they need to be combined?

I don't know how to combine them. But it does seem to me that their strengths are so complementary that there has to be a breakthrough either in how you can get formal symbol systems to do things the way connectionist processes do—multiple constraint satisfaction, and so on—or in how you can get connectionist systems to do serial symbollike kinds of processing for one or the other of them to dominate the field. My guess is that when you do get such a system, it will look like a hybrid—maybe like a hybrid of the kind Smolensky envisions: when you look at it from far away, it looks like symbol manipulation; but when you look at it up close, what is actually going on is some sort of connectionist system.

In one of your articles[12] you deal with cognitive representation. Is the representation of huge amounts of world knowledge—the so-called commonsense problem—for you an in-principle problem, or is it just a problem of time or scale?

I really don't know. One of the attractive things in the connectionist paradigm for me is the idea that information can be represented in a system in a very opaque way. The standard view of representation is always that you have something in the system that is a symbol, and it corresponds in a fairly clean way to something, a relation, a property, or an object in the world. The relationship between the element inside the system and the external elements is conceptually transparent. In the connectionist schemes you use back propagation or one of the other learning algorithms to develop their own internal representations. This has had a very liberating effect on the way we think about what representations might be like, because there is not necessarily any transparent relationship between the elements in the systems and what is out there in the world. You get distributed representations where you can make sense out of what the nodes are doing only by looking at the pattern of activation over all of them.

This suggests that some of the problems involved in commonsense knowledge, like ideas we have about causality or about the physical world in terms of "If you push against something, how hard is it going to push back?" may be encoded in these networks in ways that are conceptually opaque. We will not be able to say which nodes correspond to my concept of gravity and how they are connected to the nodes corresponding to my notion of force. That may be fine for talking about how a physics student learns something about a domain of formal academic physics, but it probably does not capture anything like the kind of knowledge we have about the world we live in.

Maybe the connectionist paradigm will provide an answer for how commonsense knowledge might be represented, but we don't know that yet. As far as I know, no one is really working on it.

What do you think about the notion of background and the idea that it is not representational at all? Do you think cognitive science should be concerned with it?

I think connectionist networks are in principle compatible with the idea of the background. Connectionist networks have given the ideas about the background respectability in the sense that before them, you did not have a viable alternative theory. All you could say was: the symbolic view is insufficient to talk about the kind of knowledge we call the background. But what connectionism is in the process of doing is providing an alternative where you can begin to think about what the background might be in terms of a different kind of representation, namely, some sort of distributed representation that is conceptually opaque.

You can also talk about the background in terms of the issue of recursive decomposition. The idea is that there may be some level down to which you can decompose, and what is below that is the background, which can only be captured by some very messy neural network or connectionist kind of scheme. But I still think that psychologists and cognitive scientists have to be concerned with it, because it ends up being the foundation on which so much higher-level cognition is built. There may be certain things you can do without it, but at some point you have to understand what it is.

You write[13] that the broadest basis of cognitive science is the information processing paradigm. What is psychologically interesting in information processing? Searle says that even if you could describe a river in terms of an information processor, it does not lead to new psychological insights. What is the psychologically relevant idea of information processing?

That is about the hardest question I know. I do not have a good answer for it. The interesting question is: what is the difference between an information processing system that we would say has cognition and one that we would say does not? We would not say that the river has cognition just because we can describe what it does as information processing.

One of the examples I always use in my courses is the thermostat example. We can describe what a thermostat does as information processing, but we would not call it cognition.

But there are some people who do.[14]

There is a principled distinction to be made here. The central issue is the nature of consciousness. What people are really talking about when they say that people and animals have cognitive systems but rivers do not is the issue of consciousness. A good example would be the human immune system. The human immune system does a lot of information processing: it has to recognize foreign tissue, and when it does, it mobilizes its resources, and so forth. But the immune system does not have consciousness.

Anyway, what all this stuff is about, and what Searle's Chinese Room Argument is about, is consciousness. I do not think that the information processing approach has anything terribly profound—or anything at all—to say about consciousness yet. I wish it were otherwise, but there are some deep reasons why it does not. The deep reasons have to do with the extent to which you want to make science a third-person as opposed to a first-person endeavor.

What the information processing approach *can* do is divide a cognitive system into a part that can be accounted for by information processing and a part that cannot. The part that can be accounted for is all the stuff an automaton could do if it were doing the same information processing. What it cannot account for are the conscious experiences that we have and that we think other people and at least some animals have.

We may be able to provide an information processing account of what things are conscious and what things are not, and maybe even of what the function of consciousness is by talking about what things can be done only if they have the property of being conscious—whatever that means.

But we cannot get into what experiences are actually like. The relation between the first- and third-person view of science is that to say what these experiences are like is a first-person statement, and as long as the science we are doing is restricted to third-person statements, we cannot get into that. It is methodologically closed to the kind of science we do currently. We might come to some extended notion of science that would allow those things to be part of it. And it does not mean that the science that we do is not informed by our own first-person experiences: there have been many important advances in psychology that have been argued principally on the ground of first-person experiences.

When I read your articles, I thought that you believed the information processing paradigm to be the most promising approach in cognitive science. But now you say that there is no notion of consciousness in it . . .

I *do* think that information processing is the most promising approach in cognitive science, because there is not *any* approach I know of that would give us an account of consciousness. The phenomenological perspective does not actually give the kind of account of consciousness I am talking about.

For me, this is a contradiction. On the one hand, the goal of cognitive science is to explain how the mind works, and on the other hand, it leaves out consciousness, the most essential feature of the mind.

Probably not everybody would agree with me that it does. I just do not see how information processing is going to give us a complete account of consciousness. It can give you an account of a lot of aspects of consciousness—for example, how two experiences are related to each other— but I do not think it can answer the nitty-gritty question of why there are experiences and what they are like individually. I do not see how information processing can ever tell the difference between a person who has conscious experience and a complete computer simulation that may or may not. It is just neutral about it. It has no features for saying whether a person or a creature or a computer is conscious or not.

Let us return to the enterprise of cognitive science. What is the future of cognitive science? Should it become a discipline by itself, and, if so, is an interdisciplinary discipline not a contradiction in terms?

The problem with cognitive science as a discipline is that the methods of the component disciplines are so different. To a certain degree, the component disciplines depend on the topic they are studying. If you are working in vision as I am, then the principle component disciplines of a cognitive science approach to vision would be psychology, computer science, physiology, and maybe some philosophy. Linguistics does not enter into it, except maybe indirectly if you want to study the interface between language and perception. But if you are interested in language, then of course linguistics is a very central part, and you would have to know a fair amount about anthropology and sociology as well.

Some people think linguistics has to be a major part of cognitive science.

Historically, linguistics has been a very important part of cognitive science, largely because of the relationships between modern linguistics and computers through computer languages and automaton theories. This was one of the foundations of cognitive science. But it is not necessarily an important part of cognitive science for every cognitive scientist. For example, I do not think it is important at all for me to know government and binding theory[15] in order to do interesting work in perception.

Actually, I think that vision is probably the best example in which cognitive science is really working, because we know more about the biology of the system than we do in the case of language. We can actually look at the relationship between biological information processing and the computational models that have been constructed to do vision. Within vision, there is this one very restricted domain, which I think is the very best example of cognitive science anywhere in the field, and that is color perception.

You can tell a very interesting and extended story about color vision that goes all the way from the biological aspects, like the three different cone types that we have in the retina, and how that representation is transformed into an opponent red/green, black/white, and blue/yellow representation and so forth, through the psychological structure of color space (hue, saturation, and brightness) and how they relate to one another, all the way up into anthropology, where you can tell the Berlin and Kay story about basic color terms in different cultures[16] and how those basic color categories are grounded biologically in terms of the underlying processes and structures of the visual system. It is a beautiful story that connects all the way from biology through psychology and up to anthropology. It is the best example we have of how the different disciplines of cognitive science can all contribute to the understanding of a single topic in a way in which each one illuminates the others. There are really very nice examples of synergy here.

How should cognitive science establish itself? On the one hand, we get and need training in our respective disciplines. On the other hand, we see now that we have to work together with other fields. Isn't this a contradiction?

It is a real problem. It is the reason why we at Berkeley have gone very slowly in the direction of a cognitive science program. The problem is, as I said, that the methods of the different disciplines are quite different. As a psychologist, I learn how to do experiments and how to analyze empirical data. A computer scientist learns how to write programs in LISP. A linguist learns how to construct formal grammars of a language. A biologist learns how to record from individual neurons in the brain. A sociologist or anthropologist learns how to study people in different cultures. That is why, when we first started cognitive science at Berkeley, we had just an extended set of tutorials from the different fields for the first two years.

The initial phase has to be educational to learn the vocabulary of the other fields and to obtain what I would call an "outsider's view" of what those fields are like. But to do cognitive science, you have to be grounded in the methods of at least one discipline. My own view is that, at Berkeley, we are currently in the zeroth order of approximation to cognitive science, that is, where you start out with a bunch of people who are each in their own discipline and who learn about what the other people in other fields are doing. They started to move into the first-order approximation to cognitive science at places like UC San Diego, where they are designing a graduate program in cognitive science from the ground up.

In San Diego they are doing it right: they are starting their department from scratch; they are figuring out what the curriculum should be, how the courses should be structured, what people should know about the different disciplines. And from that will emerge the first generation of *real* cognitive scientists, who we hope will have a selection of tools that is truly cross-disciplinary. By this I mean that the students coming out of this program will probably know how to write LISP code, how to do computer simulation, *and* how to do experiments and analyze data so that they can work as psychologists as well as computer scientists. And they probably will have some knowledge of other fields as well.

The danger, of course, of being jack of all trades and master of none is that it does not necessarily help you to be a little bit good at all the techniques but not really good at any one of them. It is currently an open question how much integration is going to occur through programs like the one in San Diego.

You mentioned the question of different methods as one of the main problems. What are these different methods?

In psychology, the principal methods are empirical. As a graduate student, you learn how to do experiments, how to analyze data, and how to make inferences from the results. When you see people in other fields doing experiments, they very often do bad ones, experiments from which you cannot draw any conclusions because they do not include all the proper control conditions or they do not set the experiment up so that the results can be analyzed properly.

But in psychology, there are several theories about, say, vision, and they all have their own experiments.

I am talking about experiments at a more abstract level. People who believe in different theories would have the same beliefs about what constitutes a proper psychological experiment in terms of its design and analysis. They might think that the actual conditions that are studied in a certain experiment are fairly uninteresting in trying to understand vision, but still, because they have been trained in the same methods, they will understand the difference between a good experiment and a bad experiment.

The techniques a psychologist learns are very different from those a computer scientist learns. You can talk about Artificial Intelligence research as being empirical in the sense that they get an idea, they build a system to see whether it can do what they think it can do, and if it cannot, they revise it and go back to do something else. It is empirical in that sense, but the tools that they use to perform experiments are completely different from the tools we use to do psychological experiments. They do their experiments by programming computers. You can do psychological experiments without using a computer at all or even knowing what a computer is. After all, people did psychological experiments for decades before there were any computers. So the methods and techniques of these two disciplines are very different.

Philosophy and linguistics are much more analytically oriented. The knowledge you get in graduate training in philosophy is about logic and logical argument. This is training we do not get in psychology, for example.

Is this an advantage or a difficulty?

I do think it *is* an advantage. It is why cognitive science is still afloat, but it is getting larger, so it is working. It is a difficulty in trying to train a student to be a complete cognitive scientist, because he or she has to know all the techniques of all the disciplines. There is a certain overlap, but if you want to be master of all these trades, it would take you a long time to learn them. One way is to be master of one methodology, and to have a working knowledge of other methodologies so that you can understand what is going on and be informed by what is going on in those disciplines, at least enough to influence the work you do in your own methodology.

When I do experiments in vision, knowing what is going on in computer vision might in fact help to define the experiment I want to do. I might try in my experiment to test some theory that was constructed by a computer scientist in vision. That certainly constitutes cognitive science research, in my view.

For lay people it is very difficult to decide which theories are right, for example, in psychology: each theory has its own data and its own interpretation of them. Is there any confusion?

People with different theories tend to carve out different parts of the domain. They are pretty good at accounting for the kind of experiments they are running and other groups are not. It is very hard to do a critical experiment in psychology in the same way as a critical experiment in, say, physics. In psychology there are no such experiments where the results send a whole field into a quandary over how to solve some very deep problem. It is because the underlying mechanisms of cognition and the human mind are much more complicated. There is always more than one way to test any given idea, and when you test it in different ways, there is always the possibility that you get different answers, because there are some factors that you have not taken into account.

A lot of the confusion is well justified on the grounds that there often are different experiments that seem to be testing the same thing and come out differently. We do not understand enough about the mechanisms that produce it to know what all the relevant factors are. Maybe twenty years down the road, we can look back at the experiments and explain the difference in the results that, for example, in one experiment people could see the surroundings of the display, and in the other the surroundings were dark; and maybe we know that this allows some mechanism to come into play that we did not understand at the time.

Let's look at it the other way around: what should be a criterion for success in cognitive science? Would you think that the Turing Test is a useful test idea?

Conceptually, the Turing Test is the appropriate one: that you can program a machine that should be indistinguishable, in its behavior, from a person. Methodologically, this is a behaviorist viewpoint, appropriately aligned with the information processing approach, in the sense that if you get a machine to behave like a person, doing the same kind of things in the same amount of time and with the same kinds of errors, then you can claim that you capture the critical information processing capabilities of the organism that you study.

You can test the behavior only. But the central claim of cognitive science is that it wants to explain the "inside" of the black box.

How can we cognitive scientists or cognitive psychologists determine what is inside the black box? We do experiments that try to distinguish between having this kind of internal structure or that one. We can do the very same experiment on a computer simulation that we do on a person. If the answer comes out the same way, we make the inference that the computer may have the same information processing structure as the person. If in all possible experiments the simulation behaves in the same way that people do, then we have to make the same inferences about what is inside, about the structure of the process inside the person. The way we make the decision that they have the same internal structures is by doing methodologically behavioral experiments, but it is entirely compatible with cognitive science in that sense.

But outside behavior is not a guarantee for the internal structure. I could simulate an analog watch or a sand clock by constructing a digital watch—I get the same function, but the structure is completely different. Is it possible to conclude from equal functioning to equal structure?

Only after the point that you have tested it in your experiments. The analog watch and the digital one will probably break in different ways, for example, or they will react differently when the battery runs down. The errors that computers make have to be the same as people make: a breakdown must occur under the same conditions, and the amounts of time that processes take have to be the same, and so forth. The idea is that all these constraints together plus something like Occam's razor (i.e., that there is a simplest account relative to the constraints) will be sufficient to say: up to the constraints inherent in the information processing approach, these two things work in the same way, and when we understand the structure of the simulation, we will understand the structure of the person's mind.

So it's crucial to find the right experiments.

Take Weizenbaum's ELIZA:[17] as long as you kept within certain parameters, it really did sound like a Rogerian therapist. But that was because you just were not doing the right experiment. As soon as you stepped out of the normal mode of therapy, it did something bizarre. This is an example of doing the right experiment to find out that it does not behave like a person. It is a very simple example, however, and as simulations get more and more like real people, the experiments get more and more complex.

Another example from color vision demonstrates the level up to which an information processing account can go and beyond which it cannot. We know what the structure of color vision is. We can characterize normal people's color experience in terms of a three-dimensional color space with the dimensions of hue, saturation, and brightness. It is a sort of double cone structure. We can characterize it on behavioral grounds, by asking: "Which is more similar to red—orange or green?" And people say "orange." I could ask you all kinds of questions about similarities of colors and could show you patches and have you make decisions. On the basis of all these data we can construct a spatial representation of what your color experiences are like. If you have normal color vision, yours will be almost identical to mine. Within an information processing account, that is the best we can do. We cannot say that my experience of red is like your experience of red, because your color space could be inverted, and my experience of red could be like your experience of green.

An information processing account does not even give us a reason to say that you *have* conscious experiences, it only gives us the right to say that you *say* you do. Someday, we may figure out a way of going beyond that, maybe when we—as Searle would claim—get to the real biology. Maybe there we will be able to say what conscious experiences are about. I actually doubt that myself. I think what experiences are qualitatively like is going to be forever beyond us. We may be able to say whether there is some biological property that corresponds to something being conscious or not. But I do not see how we can get beyond information processing constraints, without changing something about the way we do objective science.

I, for one, would be loath to give up this kind of science. We had problems with this in psychology long ago when introspectionist views were predominant. There was no way of deciding who was right. One group said that thought was a flow of sensory images, and the others said that thought was imageless. How can you decide who is right on the basis of introspective evidence?

Let's get back to the Turing Test. On the one hand, you say that we cannot say anything about experience or consciousness. On the other hand, you consider the Turing Test to be a useful test idea. If one assumes that, then in order to be intelligent it is necessary to have consciousness, intentionality . . .

There is no evidence that for something to be intelligent it is necessary for it to be conscious. There are programs now that apparently can play chess at the level of masters or even grand masters. Are we going to claim that this is not intelligent behavior? It is clear to me that, as far as behavior goes, this is intelligent. One of the most interesting things that we have found out so far in cognitive science with computer simulations is that the high-level processes seem to be a lot easier to do than the low level ones. We are much closer to understanding how people play chess than we are to understanding how people recognize objects from looking at them.

Suppose we have a machine that not only plays chess but can make breakfast. Let's further assume, with John Searle, that it is not conscious, although I do not think we have any reason to say either way. Are we going to deny that its capabilities, which are the same as ours, are intelligent just because it is not conscious? My view is that intelligent behavior should be defined without reference to consciousness.

Some people may claim that it is an empirical fact that only conscious beings achieve those criteria, but I do not think so. We already have convincing examples of intelligence (though not general-purpose intelligence) in nonliving things, like the chess computers. The reason why we do not have general purpose intelligence is that so much of what is required for this is background knowledge, the low-level stuff that seems so difficult to capture.

In your opinion, what is the future prospect of cognitive science and, in cognitive science, of Artificial Intelligence? What could be the outcome in the next few years?

My own principal interest is, What is going to happen with the connectionist movement? How is it going to relate to, or integrate, or fail to integrate with the symbolic view that has dominated the field until now? I do not really have a clue how it is going to come out—whether it will turn out to be one of these great ideas that helps solve unsolved problems in cognitive science, or whether it will turn out to be a waste of time.

What are these unsolved problems in cognitive science, at the moment?

All the low-level processes, like perception, action, and learning, in which we have not made very much progress and where there are interesting prospects coming up from the connectionist perspective.

Do you think that the connectionist approach will continue to fit into cognitive science?

The people who say that connectionism is not really cognitive science because "they are only building biological analogues" are the hard-line symbolic systems people. They seem to take a very narrow view of what cognition is, and if it turns out to be wrong, they will have to end up saying that there is not any cognition. I think this is silly. There clearly is cognition; it is just a question of what it is like. They might end up being right in that it can be characterized in terms of physical symbol systems; but they might be wrong, which does not mean that cognition does not exist. Cognitive science is the science that studies cognition, and we know that cognition exists, so it is not a field that is going to disappear. It is not like alchemy, which disappeared when it turned out that it is impossible to change one element into another by chemical means. There are phenomena of cognition, and there is, therefore, a field of cognitive science defined as the interdisciplinary study of these phenomena.

It was for this very reason that I wrote the article about the information processing approach to cognition. The symbol systems people were talking about cognitive science and information processing and computationalism—they used all these names to mean what they did and what they believed, and they tried to pass it off as being all these things. I am an information processing psychologist, but I do not believe all the things the hard-line symbolic people believe. When Kimchi and I analyzed that, we came to the conclusion that information processing is a distinct stand from this symbolic computational viewpoint. It is a weaker stand. We could be right on issues where they could be wrong.

The symbolic systems hypothesis is a stronger view than the information processing approach. It is therefore both more interesting and more likely to be wrong. It is not yet a well-understood topic, but it may turn out that symbolic and connectionist views are distinct and mutually exclusive cognitive theories. But it might turn out that they are not—we don't really know yet. The physical symbol systems hypothesis is fairly well characterized now, but we are just beginning to understand what connectionism is about and what its relationship is to symbolic systems. It will take a lot of people a lot of time to figure it out.

Will there be a great breakthrough, or will progress be slow?

The biggest breakthrough was the breakthrough that allowed cognitive science to happen. It was the breakthrough that included the Chomskian revolution in linguistics, and Newell's and Simon's work in computer science. Since cognitive science was born, there has been only one revolutionary change, and that is the advent of connectionism. This has changed cognitive science more than anything else since it started.

Why did it happen so late?

It is somewhat similar to the story about Gestalt psychology. The early perceptron research died a premature death. Its death was caused by a very important book written by Minsky and Papert—*Perceptrons*.[18]

They made some very well founded arguments against a certain class of perceptron-type theories, showing that these theories could not do certain things. They did not prove that those same things could not be done by a more complex class of perceptrons, which was the principal basis of the more recent advances. And they did not always address the question of how important these disabilities were. Some important new results were obtained by the PDP (parallel distributed processing) group in San Diego, but what we don't know yet is how far it is going to carry us.

I doubt whether there will be another breakthrough soon, because you usually get breakthroughs under conditions where there is dissatisfaction with the way things are. Part of the underlying conditions for the connectionist movement to happen was that there was a bunch of people who were not terribly happy with certain shortcomings of the symbolic systems approach. They found a way around some of these problems.

Now you have a division largely into two camps: the symbolic systems people and the connectionists. The symbolic systems people are happy with the standard view and will continue to work within it. The chances that they will produce a revolution are fairly small. Most of the people who had problems with that view are either neutral or going to the connectionist camp. These people are busy just working now, because there is so much to be done within the connectionist framework. It will keep them busy for a long time, and it will be a long time before they get to the point where they become dissatisfied in a fundamental way with the connectionist paradigm.

And only under the conditions of deep-seated dissatisfaction is another revolutionary breakthrough likely. As far as the current situation goes, the next decade is going to be spent working on what the connectionist paradigm is and how it relates to the symbolic systems approach.

Recommended Works by Stephen E. Palmer

"Fundamental Aspects of Cognitive Representation." In *Cognition and Categorization*, ed. E. Rosch and B. L. Lloyd, pp. 259–302. Hillsdale, N.J.: Lawrence Erlbaum, 1978.

"Gestalt Principles of Perceptual Organization." In *The Dictionary of Cognitive Psychology*, ed. M. W. Eysenck. Oxford: Blackwell, 1991.

"Modern Theories of Gestalt Perception." *Language and Mind* 5 (1990): 289–323.

"The Psychology of Perceptual Organization: A Transformational Approach." In *Human and Machine Vision*, ed. J. Beck, B. Hope, and A. Rosenfield, pp. 269–339. New York: Academic Press, 1983.

"Visual Perception and World Knowledge: Notes on a Model of Sensory-Cognitive Interaction." In *Explorations in Cognition*, ed. D. A. Norman and D. E. Rumelhart, pp. 279–307. San Francisco: Freeman, 1975.

(With R. Kimchi.) "The Information Processing Approach to Cognition." In *Approaches to Cognition: Contrasts and Controversies*, ed. T. Knapp and L. C. Robertson, pp. 37–77. Hillsdale, N.J.: Lawrence Erlbaum, 1985.

Notes

1. See e.g., A. M. Collins and M. R. Quillian, "Experiments on Semantic Memory and Language Comprehension," in *Cognition in Learning and Memory*, ed. L. W. Gregg (New York: Wiley, 1972), pp. 117–37; and "How to Make a Language User," in *Organization of Memory*, ed. E. Tulving and W. Donaldson (New York: Academic Press, 1972), pp. 309–51.

2. HAM = human associative memory. See J. R. Anderson and G. H. Bower, *Human Associative Memory* (New York: Holt, 1973).

3. R. Schank, "Conceptual Dependency: A Theory of Natural Language Understanding," *Cognitive Psychology* 3 (1972): 552–631.

4. See, e.g., D. G. Bobrow and D. A. Norman, "Some Principles of Memory Schemata," in *Representation and Understanding*, ed. D. G. Bobrow and A. M. Collins (New York: Academic Press, 1975), pp. 131–50.

5. see R. D'Andrade, "The Cultural Part of Cognition," *Cognitive Science* 5 (1981): 179–95.

6. D. E. Rumelhart, P. H. Lindsay, and D. A. Norman, "A Process Model for Long Term Memory," in *Organization of Memory*, pp. 197–246.

7. Wilhelm Wundt (1832–1920) is considered to be the founder of experimental psychology; see E. G. Boring, *A History of Experimental Psychology* (New York: Appleton-Century-Crofts, 1950).

8. Wolfgang Köhler (1887–1967), co-founder and leading member of the Gestalt school of psychology; see *Gestalt Psychology* (New York: Liveright, 1929); *The Mentality of Apes* (New York: Humanities Press, 1925); and *The Task of Gestalt Psychology* (Princeton: Princeton University Press, 1969).

9. J. J. Hopfield, "Neural Networks and Physical Systems with Emergent Collective Computational Abilities," *Proceedings of the National Academy of Sciences* (1982): 2554–58; reprinted in *Neurocomputing: Foundations of Research*, ed. J. A. Anderson and E. Rosenfeld (Cambridge, Mass.: MIT Press, 1988).

10. S. E. Palmer and R. Kimchi, "The Information Processing Approach to Cognition," in *Approaches to Cognition: Contrasts and Controversies*, ed. T. Knapp and L. Robertson (Hillsdale, N.J.: Lawrence Erlbaum, 1985), pp. 37–77.

11. See P. Smolensky, "Information Processing in Dynamic Systems: Foundations of Harmony Theory," chap. 6 in D. E. Rumelhart, J. L. McClelland, and the PDP Research Group, *Parallel Distributed Processing: Explorations in the Microstructure of Cognition*, vol. 1 (Cambridge, Mass.: MIT Press, 1986); and "On the Proper Treatment of Connectionism," *Behavioral and Brain Sciences* 11 (1988): 1–74.

12. S. Palmer, "Fundamental Aspects of Cognitive Representation," in *Cognition and Categorization*, ed. E. Rosch and B. L. Lloyd (Hillsdale, N.J.: Lawrence Erlbaum, 1978), pp. 259–302.

13. S. E. Palmer and R. Kimchi, "The Information Processing Approach to Cognition," in *Concepts and Controversies*, pp. 37–77.

14. J. McCarthy, "Ascribing Mental Qualities to Machines," in *Philosophical Perspectives in AI*, ed. M. Ringle (Brighton: Harvester Press, 1979), pp. 161–95.

15. N. Chomsky, *Lectures on Government and Binding* (Dordrecht: Foris, 1981).

16. B. Berlin and P. Kay, *Basic Color Terms* (Berkeley and Los Angeles: University of California Press, 1969).

17. J. Weizenbaum, "ELIZA: A Computer Program for the Study of Natural Language Communication between Man and Machine," *Communications of the ACM* 1 (1966): 36–45.

18. M. L. Minksy and S. A. Papert, *Perceptrons* (Cambridge, Mass.: MIT Press, 1988).

HILARY PUTNAM

Against the New Associationism

I got involved in cognitive science in two ways, because I worked for many years in mathematics as well as in philosophy. I am a recursion theorist as well as a philosopher. So the theory of Turing machines is one that I have known, as it seems to me, for all my adult life. In the late fifties I suggested a philosophical position that I named "functionalism." It became very popular. I no longer believe in it, to the regret of many of my former students. But the idea is that the mind or the mind-brain is, basically, a computing machine, and instead of focusing attention on the question of whether the substance of our mind is or is not material, we should be focusing attention, rather, on the different question of whether the organization of our mind is or is not the same as that of a Turing machine.

People are already speaking of "Putnam I" and "Putnam II." What was the reason that you abandoned this position, that you completely reversed your view?

I would not say that it is a complete reversal. The question of whether the brain can be modeled as a computer is still wide open. But what I don't think anymore is that our mental states are simply isomorphic to computational states. If you take a simple mental state, for example, believing that there are roses in Vienna, this mental state might correspond to one mode of functioning in my brain and to a quite different mode of functioning in someone else's brain. The idea that the program or some feature of the program must be exactly the same for us to have the same thought now seems to me untenable.

What was the reason for changing your view?

It came about because I was working in two different fields at the same time: in philosophy of language as well as in mathematical logic and computer science. At a certain point, it seemed to me—and indeed to

Hilary Putnam was born in 1926 in Chicago. He studied at the University of Pennsylvania and at the University of California at Los Angeles, where he earned a Ph.D. in 1951. He taught at Northwestern University, Princeton University, and MIT until he became Walter Beverly Pearson Professor of Modern Mathematics and Mathematical Logic at Harvard University in 1976, a position he still holds.

some of my students it still seems that way—that the ideas of computational philosophy of mind fit well together with contemporary philosophy of science.

But as I worked more deeply into this subject, I realized that the direction in which I was going in the philosophy of language does not really point to a computational model of the mind. The mind may be a computer in the sense that a computer may be the "body" of the mind, but the states that we call mental states are not individuated in the way computational states are individuated. The way we tell whether someone is or is not in the same mental state is not the way we tell whether a computer is or is not in the same computational state.

One point that made functionalism so attractive was that we have to study only the structures and not the underlying matter, and therefore the implementation does not matter. What is, in your opinion, the main failure in functionalism? Functionalism is, after all, still very strong.

It is nice to produce something that goes on. . . . I still agree with the point of functionalism you just mentioned, that is, that at least in principle the mental states don't have any necessary connection to one particular form of chemical, physical, or metaphysical organization. I like to say that our mental states are compositionally plastic, by which I mean that a creature, for example, a Martian or a disembodied spirit, could have a very different matter from what I have and be in the same mental state. *That* point of functionalism I have not given up.

What I now say is that our mental states are also *computationally* plastic. Just as two creatures can have different chemistry and physics and be in the same mental state, it must be considered that computers can have different programs and be in the same mental state. This is a view that some computer scientists have also come to—like Newell, who distinguishes not only a computational level but also a more abstract, symbolic level. When you describe a program by saying what it does, for example, that this program integrates elliptical functions, that statement is not in a language of computer science. It does not say how the program works—in BASIC or in LISP or in any other language. It gives, as it were, an intentional description of the program. Newell himself has emphasized in recent years that this level of description, which I call the intentional level, is a higher-level description than the computational description. Rather than two levels of description—computational and material—we need a minimum of three: the intentional or psychological description, perhaps a computational description, and then a physical description.

What could be the reason why people did not follow you in this abandonment of functionalism?

A number of people see the problems, although they are not quite apparent. Let me say a little more about what the conflict with philosophy of language is: in modern philosophy of language, there have been two important directions. One, especially in this country, has been the emphasis—which began with my paper "The Meaning of 'Meaning' "[1]—that meanings are not in the head. To see what somebody means, we cannot look inside the brain or inside the mind. We also have to see what sort of community he lives in, what the social interactions are, and even at some of the physical substances and things that they interact with. Our philosophy of language has moved in an antisolipsistic direction. In 1928, Carnap called his position "methodological solipsism." And by and large, solipsism has become increasingly unpopular. Jerry Fodor still defends it, but even he is now looking mainly for a theory of what he calls "broad content," which is to say, a theory of the nonsolipsistic aspect of meaning and reference. His next book is going to be an attempt to solve the great Kantian problem of how language relates to the world.[2] That is not a solipsistic problem.

That is part of what pushed me away from functionalism: here I had a functionalistic picture of meaning, that meaning was entirely in the head, and there I had a philosophy of language that said that meaning was not in the head. Something had to give way . . .

The other great tendency in contemporary philosophy of language, which appears in German as well as American philosophy, is a tendency toward what has been called holism. This is something Jerry Fodor does not like at all, whereas Daniel Dennett does (if people in the computer science community tend to disagree with me, they also disagree very much with one another). Holism says that words don't really have meanings in isolation. The meaning of a word is a function of the role it plays in a whole symbolic system. The French carry this to what I regard as absurd lengths—to the extreme of denying that there is a world or such a thing as truth at all.[3] But even those of us who are not so extreme have been forced to recognize this very holistic character of meaning.

So you have these two phenomena: that meanings are not in the head, and the holistic character of meaning, which does not fit well with the idea of a determinate computer program or feature as the meaning of *this* word and another determinate feature as the meaning of *that* word. In many ways, from the viewpoint of a philosopher of science like Quine or a philosopher of language like Wittgenstein, that looks retrograde, like a move away from their insights.

What do you think about attempts like Searle's to find a way for intrinsic intentionality?[4]

That is an attempt to go still farther back. Searle seems to be saying, Since functionalism did not work, let's go back to the old kind of brain-mind-identity theory that we had in the fifties, before functionalism, with authors like J.J.C. Smart and the neuroscientist Place[5] who said that mental states are just brain states. I think that Searle is just going back to identity theory without having answered any of the objections to it. By intrinsic intentionality, as far as I can understand, he means that certain brain structures have the intrinsic power to refer to external things. He has not explained either philosophically how this makes sense or scientifically how he would meet the well-known objections to identity theory.

He says that the problem of intentionality is just like the liquidity of water. But the property of being liquid can be defined in terms of fundamental physics. If he thinks the property of referring is something we can define in terms of fundamental physics plus chemistry plus DNA theory, let him do it . . . and if he is *not* talking about that kind of reduction, his language is extremely misleading, as it constantly suggests that he thinks a reduction is possible. If no reduction is possible, the analogy to the property of liquidity is broken. He can't have it both ways—that referring is a property of the brain tissue in just the way that liquidity is a property of water, but that it is not definable in terms of physics and chemistry. What makes liquidity a reducible property is that we can actually give a reduction, we can actually give necessary and sufficient conditions of an explanatory kind in terms of fundamental physics for something to be liquid. We cannot give necessary and sufficient conditions for something to refer to something in the language of fundamental physics.

It seems to me that his Chinese Room Argument is a very strong argument against strong Artificial Intelligence. Do you think that argument is correct?

I have certain problems with the details of the argument, but I share Searle's intuition. I disagree with the systems reply to Searle, which says that the room is a speaker of Chinese and that the intuition is wrong. But again, Searle draws too strong a conclusion. The conclusion is that having a certain functional organization is not *sufficient* for consciousness. He concludes that it is not *necessary* for consciousness. And that does not follow. If a certain property is not sufficient for consciousness, it does not follow that it is not necessary.

What do you think is the main problem in Artificial Intelligence and cognitive science at the moment? Is functionalism a problem?

I think the main problems have very little to do with whether functionalism is right or wrong. The main problem with Artificial Intelligence at the moment is the enormous gap between what has actually been accomplished and the sort of talk about it. The situation is as if someone at the time of Newton had wanted to discuss the question, Can we have atomic energy? It turned out that we can, but no one at the time of Newton could have said anything sensible about it. It is nice as a speculation—I like science fiction, and I like philosophy. We like to speculate about computers that talk to us, and we may do that even if functionalism is false. It may be that one day we shall produce computers we can talk to, and it is very likely that we won't understand their mental states better than we understand our own. The assumption that if we design a system we will understand it, is not necessarily right. There are already systems that have been produced by groups of people over periods of time that no one really understands.

So you don't think that Artificial Intelligence is a useful approach to understanding the human mind?

I would say that at present it is not. At present, it is a useful form of engineering. It seems to me that Artificial Intelligence is so far able to replicate certain features of the mind—to remember and to retrieve things, for example. Expert systems seem to me the greatest achievement of Artificial Intelligence. But I don't believe that intelligence is just a matter of having a huge library of already-preset repertoires and some very fast methods of searching it to come up with the right repertoire.

The problem that stopped philosophers when they were trying to reduce intelligence to a set of rules was the so-called problem of inductive logic. It already began with Bacon and his slogan of the New Organon.[6] That was supposed to culminate in a second type of logic, because people recognized that deductive logic, by itself, is not all of reason. John Stuart Mill, in 1843, published his *A System of Logic, Ratiocinative and Inductive*.[7] No one has really made any progress on the second form of logic since, let alone the third form that the American philosopher Peirce talked about, which he called abduction, meaning how to think of good hypotheses.[8] The problems of induction and abduction are just intractable. We have not yet had a genius who came up with at least a first idea of how to tackle them. It is as if we tried to talk about atomic power at the time of Newton—if somebody comes up with an idea, no one will be more delighted than I.

I am disturbed by the following, which is an undeniable sociological fact of Artificial Intelligence: no branch or sub-branch of science in the twentieth century has engaged in the kind of salesmanship that Artificial Intelligence has engaged in. A scientist in any other field who tried to sell his accomplishments in the way Artificial Intelligence people have consistently done since the sixties would be thrown out. This is something very disturbing. There are valuable engineering achievements, but the claim to have made something like a breakthrough or even a really new approach in thinking about the mind and psychology, seems to me fraudulent.

What about cognitive science in general? Do you think it is different from Artificial Intelligence as an approach?

I don't think one should construe cognitive science as the name of an *approach*. It seems to me that cognitive science really began with the conferences that were funded by the Sloan Foundation.[9] In a sense, the Sloan Foundation created cognitive science. I saw that as a—very successful—attempt to get linguists, philosophers, recursion theorists, and cognitive psychologists, of which there are many varieties—Eleanor Rosch,[10] for one, and many others—to talk to one another. It seems to me like semiotics: if it is to be the name of science, you would have to ask: what science? If you think of semiotics as an attempt to thematize certain concerns and to say that the present-day academy with its present departments is not well adapted for discussing these concerns, then you have something. Cognitive science is similar. To me, it simply is a way of recognizing that the department structure that we have inherited from the middle ages is not sufficient.

But is that not an exciting development, to cross the boundaries of disciplines?

That is exciting, but as soon as someone tries to say what all cognitive scientists believe, or that we all have to agree with Chomsky or Minsky, then it is going to become a millstone around our necks rather than something good. What is good about cognitive science is that debates like this one can take place beneath the general umbrella of cognitive science.

One reason cognitive science has become strong, I think, was that it is an attack against behaviorism. But the other reason could be that the computer is a very powerful metaphor to study the human mind.

Cognitive science is like philosophy. Philosophy includes its opponents. If somebody says, as Wittgenstein did, that philosophy is no longer possible[11] or raises the question, as Habermas did, of whether philosophy has a future[12]—that is doing philosophy. Similarly, cognitive science includes its own opponents. I am an opponent. I say that there is no evidence at all that the approach is useful.

If we take the approach in the narrow sense, I see no serious evidence that it is useful. As soon as I ask what the evidence is, either I am told about expert systems or the Chomskians turn out their little argument that everything must be innate.[13]

Marvin Minsky holds a very different point of view. In his opinion, what we are doing now is creating minds, and in a couple of years, we will be able to construct machines that act like human beings.

Yes, he has said that for twenty years, and we are not any closer than we were. I have known Marvin for a long time, and he said the same thing in 1960. I don't see that we are closer to this goal than we were in 1960. Nor do I see any fundamental idea. His speculation about the mind as a society[14] does not make essential use of the computer metaphor. A nineteenth-century philosopher could equally have said, Look, the mind is like a society.

I want people, when they look at this kind of literature, to consider the following possibility: forget the question of whether, in principle, Artificial Intelligence could work (or even my arguments about functionalism, although I hope that people buy my book *Representation and Reality*[15]). I want people to decide for themselves whether they are encountering any new ideas or whether they are constantly encountering the sort of informed, optimistic scientist who hopes for what he may be able to do some day. All *I* see is the informed, optimistic scientist who says: "Well, if all goes well, we will have machines that think, machines that talk, machines that write poems, machines that fall in love . . ."

Where are the scientific results now? I remember being on a symposium with Seymour Papert many years ago, in the sixties, at Berkeley. And he said: "Now we have programs that read children's stories." I got up and said: "What Papert did not tell you is that the program can answer questions about *one specific* children's story, and you would have to write a new program if you changed the story." And he dropped the example. That is an example of what I call the horrendous over-sell—both the frightening Madison Avenue aspect, and the aspect of a religion of the machine—that people hope that machines will succeed us. This is an antihuman side that Weizenbaum can tell you much more about than I can.

I want to focus on only the scientific side. Scientifically, the idea that maybe there could be an algorithm for being intelligent is an idea that has been given no concrete flesh and bones.

What do you think about connectionism?

There are two things to say about the so-called new approach of connectionism. First, as to the brain side: as Gerald Edelman has pointed out, there are very good physical reasons why the particular relaxation algorithms that current connectionism is based on cannot be the algorithms that the brain uses. I think the brain does use algorithms. I am more inclined to think that the kinds of algorithms Edelman describes in *Neural Darwinism*[16] are much more relevant to brain functioning than the actual connectionist algorithm is.

As far as the connectionist algorithm goes, it is important to realize that about 98 percent of what that algorithm does has been replicated using paper and pencil and ordinary desk calculators, using only regression analysis, that is, standard statistical tests for regression. In other words, what that algorithm does is find correlations between sets of variables that it is given. Things that cannot be expressed in terms of correlations between given variables cannot be found by the algorithm. That is how Steven Pinker at MIT was able to show that you cannot claim that neural nets learn the past tense in English.[17] In principle, they cannot, because there are features in the past tense—for example the division of verbs into classes that form their past tense differently—that cannot be expressed simply as correlations between variables given to the machine as input.

Now connectionists are saying that maybe they would need higher-level connection engines. But then they will probably turn out to be universal Turing machines. Let's imagine that we discovered a fast algorithm for doing elementary statistical analysis, which is more of an analog device than a digital device—if people said of it that it learns to speak a language by itself, that would be false.

But connectionist models have some nice features—so-called graceful degradation for example, or that with parallel processing you can now tackle the question of real-time problem solving. A Turing machine could, in principle, compute all these states, but connectionists claim that in their models you don't have to represent them symbolically.

We have known for a long time, though, that correlations can often be found faster by analog devices than by digital devices. The Guttman scale,[18] which is just a matter of shifting things around on a board, is the way most sociologists look for correlations: you play around with the patterns, and you quickly find which variables are correlated. It is certain that the brain uses devices of this kind, perhaps more of the Edelman kind than of the connectionist kind. I have no doubt that the brain uses pattern recognition algorithms. I have never shared the view that

Artificial Intelligence is never going to say anything about pattern recognition. Hubert Dreyfus had some kind of mysticism about pattern recognition that I never understood.

He is now more optimistic about connectionism.

Because he always focuses on pattern recognition. But the problem for me is not pattern recognition; it's thought. Thought is probably where you do need recursions (which is, by the way, where connectionism stops). I don't think we have, at present, even a suggestion as to what it would mean to produce something that can go into an open situation and learn. We have no real theory of learning.

If the outcome of Artificial Intelligence and cognitive science is not so relevant, what can be the reason for its being so attractive and powerful?

My guess is that we have been looking for some kind of reductionist account of thought for centuries. In its time, the association of ideas was more popular than any idea in the social sciences has been before or since. The great scholar in England, Walter Jackson Bate,[19] wrote a very early book in which he showed that the effect of this new psychological theory—the association of ideas—of the time of Hume was to change totally the way of thinking. Romanticism would have been impossible without the overthrow of the old picture. That overthrow and the new picture of taste as much more random, much more arbitrary, much more unreasonable came in with the association of ideas. Bate even did a study about how many books were written on the new theory of the association of ideas—it was just staggering: in relation to the European population at that time, the number was far greater than the number of books written on Artificial Intelligence or any other theory. For two or three centuries, psychologists continued to talk about nothing but the association of ideas—even though Kant and others had already pointed out that it had, in a way, no substance.

My view is that we are in the same position. We think that because Newton somehow reduced the physical world to order, something similar must be possible in psychology. Just to conclude, as we say in the United States (although I am not from Missouri): "I'm from Missouri—show me!"

Recommended Works by Hilary Putnam

Philosophical Papers. 3 vols. Cambridge: Cambridge University Press, 1975—83. (A collection of earlier articles.)
Realism with a Human Face. Cambridge, Mass.: Harvard University Press, 1991.

Renewing Philosophy: The Gifford Lectures, 1990. Cambridge, Mass.: Harvard University Press, 1992. (See esp. chaps. 1—4.)

Representation and Reality. Cambridge, Mass.: MIT Press, 1988.

Notes

1. H. Putnam, "The Meaning of 'Meaning,'" reprinted in *Philosophical Papers*, vol. 2: *Mind, Language, and Reality* (Cambridge: Cambridge University Press, 1975), pp. 215–71.

2. See J. A. Fodor, "Methodological Solipsism Considered as a Research Strategy in Cognitive Psychology," *Behavioral and Brain Sciences* 3 (1980): 63–109. The new book mentioned here is *"A Theory of Content" and Other Essays* (Cambridge, Mass.: MIT Press, 1990).

3. E.g., the "deconstructionist" school of Derrida; see J. Derrida, *La Voix et le Phénomène* (Paris: PUF, 1972), published in English as *Speech and Phenomena, and Other Essays on Husserl's Theory of Signs* (Evanston, Ill.: Northwestern University Press, 1973).

4. J. R. Searle, *Intentionality: An Essay in the Philosophy of Mind* (Cambridge: Cambridge University Press, 1983).

5. J.J.C. Smart, *Philosophy and Scientific Realism* (New York: Humanities Press, 1963); "Sensations and Brain Processes," *Philosophical Review* 68 (1959): 141–56; and U. T. Place, "Is Consciousness a Brain Process?" *British Journal of Psychology* 47 (1956): 44–50. See also C. V. Borst, ed., *The Mind-Brain Identity Theory* (London, 1970).

6. F. Bacon, "New Organon," in *Selected Writings of Francis Bacon*, ed. H. G. Dick (New York: Modern Library, 1955).

7. J. S. Mill, *A System of Logic Ratiocinative and Inductive, Being a Connected View of the Principles of Evidence, and the Methods of Scientific Investigation*, 2 vols. (London: J. W. Parker, 1843); also in *Collected Works of John Stuart Mill*, ed. J. M. Robson, vols. 7 and 8 (London: Routledge and Kegan Paul, 1973–74).

8. C. S. Peirce, *Collected Papers* (Cambridge, Mass.: Harvard University Press, 1931–35).

9. The Alfred P. Sloan Foundation played a decisive role in launching cognitive science by organizing the first meetings and establishing a research program. See *Proposed Particular Program in Cognitive Science* (New York: Sloan Foundation, 1976); and H. Gardner, *The Mind's New Science: A History of the Cognitive Revolution* (New York: Basic Books, 1985).

10. E.g., E. Rosch, "Human Categorization," in *Advances in Cross-Cultural Psychology*, ed. N. Warren (New York: Academic Press, 1977); and "Principles of Categorization," in E. Rosch and B. B. Lloyd, *Cognition and Categorization* (Hillsdale, N.J.: Lawrence Erlbaum, 1978), pp. 27–48.

11. In *Tractatus logico-philosophicus* (London: Routledge and Kegan Paul, 1922).

12. In *Der philosophische Diskurs der Moderne* (Frankfurt: Suhrkamp, 1988).

13. See also the discussion on innate ideas between Putnam and Chomsky in *Readings in Philosophy of Psychology*, ed. N. Block (Cambridge, Mass.: Harvard University Press, 1981), vol. 2.

14. M. Minsky, *The Society of Mind* (New York: Simon and Schuster, 1985).

15. H. Putnam, *Representation and Reality* (Cambridge, Mass.: MIT Press, 1988).

16. G. M. Edelman, *Neural Darwinism* (New York: Basic Books, 1987).

17. See S. Pinker and A. Prince, "On Language and Connectionism: Analysis of a Parallel Distributed Processing Model of Language Acquisition," *Cognition* 28 (1988): 73–193.

18. A standard scaling method in empirical social science; both the position of the items and the attitude of the interviewees are represented ordinally on a joint scale. See *Lexikon zur Soziologie*, ed. W. Fuchs, R. Klima, et al. (Reinbek: Rowohlt, 1975), vol. 2, pp. 615–16.

19. W. J. Bate, *From Classic to Romantic* (London: Harper Torchbooks, 1946).

From Searching to Seeing

I did my graduate work in the study of memory and discovered that there were too few constraints on what we do. Around 1970 I started trying to understand long-term memory: how is knowledge stored in memory, and what is stored? One thing we knew from psychology was that mainly meaning is important. People can remember things that are meaningful and will forget things that are not. As a graduate student I worked in the area of mathematical psychology. I thought that the tools of mathematical psychology were by and large too weak. So I became a model builder: I began to think that computer simulation was an important way to go. When I tried to figure out what kind of information might be stored in memory, and that it would have to do with meaning, I began to look around for people who knew anything about meaning, who knew what meaning is like. And there I saw several things.

One thing I saw was the work of people in Artificial Intelligence, people like Ross Quillian, who invented the semantic network.[1] I saw the work in linguistics, for example, that of Charles Fillmore.[2] At that time he had his case grammar approach, which looked like a beginning of an approach to do meaning. I also saw work in philosophy, such as Frege's.[3] I attempted to put all those things together. This led to a kind of computer simulation approach that had many elements from Artificial Intelligence work, elements taken from Fillmore's ideas and also from some of the philosophical work. It was what my colleague Don Norman and I called an "active semantic network."[4] The three of us, Peter Lindsay, Don Norman and I, began work on a general project trying to characterize what is stored in long-term memory and what knowledge is like.[5] This was in 1970.

It grew naturally, in my view, into the ideas that have since come to be called cognitive science. In the book that Norman and I wrote in 1975[6] we commented on the emerging field of cognitive science. That book was a summary of what we and our students had done during the previous five years. The first time I noticed the term *cognitive science*

David E. Rumelhart received a Ph.D. in mathematical psychology from Stanford University in 1967. He was Assistant Professor, Associate Professor, and then Full Professor of Psychology from 1967 to 1987 at the University of California at San Diego, until he became Professor of Psychology at Stanford University. While at San Diego he was a co-founder of the Institute of Cognitive Science. During the last ten years Rumelhart has concentrated his work on the development of "neurally inspired" computational architectures.

was in another book, by Bobrow and Collins, published in 1975 as well, where *cognitive science* was used in the subtitle.[7] I think I began working in this area when I tried to characterize the nature of memory and knowledge.

As you look back now, what do you think was responsible for the shift from serial processing to PDP?

In my own mind, I had been intrigued by parallel processing ideas for a long time. I was very much impressed by the work of Reddy and his colleagues at Carnegie-Mellon when they developed the so-called HEARSAY speech recognition system,[8] where the idea of cooperative computation was central. I wrote a paper in 1975 on an interactive model of reading,[9] as I called it, which was an application of the cooperative computation idea to characterize the nature of the reading process. These ideas haunted me for really a long time. I attempted to model using conventional parallel processing ideas.

For the next several years, I tried to build computer models that were appropriate for the kind of semantic information processing I was interested in. Story understanding was another area I was working in: how understanding the meaning and the structure of stories works. I found that it was very difficult, using the formalisms that were available to me, to try to create a cooperative computational account that I thought was right. During the period from 1975 to 1980, a close colleague of mine, James McClelland, developed what he called the "cascade model," which was a particular model of reaction time.[10] It occurred to me that we could put together the particular kind of model he was developing with this interactive model that I had been trying to make in order to make a more "neurally plausible" model.

One of the complaints people raised about my initial work was that brains could not do it. It requires too much processing and could not be done by a real brain. We decided to build a model that *could* be done by a real brain. Whether it is or is not is another question, but it could be. McClelland and I developed the interactive activation model of reading or word perception. We started it around 1978, and it came out in 1980 or 1981.[11] It was my first foray into building neurally inspired models of some process. This was an attempt to understand cooperative computation of a kind that we thought brains could really carry out.

At the same time I had another colleague, a postdoc at UC San Diego, where I was at that time. That was Geoffrey Hinton, who came from England.[12] He had a background in the area of associative memory. He had worked with Higgins and Hillshaw in Britain and had an understanding of associative memory systems, which were also neurallike memory systems.

As I began to see the relationship between associative memory and our activation models, we tried to connect them. Hinton, McClelland, and I set out to do that. In a series of meetings we established what we called the "PDP Research Group," in late 1981, I think. We decided that we would try to sort out the different brainlike models that we knew about. We would spend several months sorting them out and writing a book, putting together what we had learned. It took us quite a while to do this, longer than we had thought. In the end it turned out to be a two-volume book.[13] We thought we would spend just six months working together and use the next six months to write it down, but overall it took five years to get the book out. By then, virtually all the work we published was work we had done in the meantime; it was not exactly what we'd originally had in mind.

It was really an attempt to understand two things: one was the nature of cooperative computation. The idea that human beings really are parallel processing processors originally stems from the work of Reddy and his colleagues on HEARSAY. We were not asking more of our models than what we knew could go on in the head of someone. When we tried to put together those things, they drew me to the body of work that had accumulated since 1980 in this area. So many interesting and open questions intrigued me that I moved most of my research into trying to answer them, into understanding what these models were good for and how they worked.

At first, wasn't it difficult to get funding and attention for the PDP approach? The idea is an old one, but what did it take to have it accepted after the book of Minsky and Papert?[14]

We were fairly well situated. In 1976 or 1977, Norman and I founded the Institute of Cognitive Science at UC San Diego. In the first instance, this institute received money from the Sloan Foundation. They paid for the postdoctoral program for Geoffrey Hinton and others. McClelland and I had already done a fair amount of work on the study of reading, so we could be funded for our work without much consideration of what it might have to do with perceptrons and Rosenblatt,[15] and Artificial Intelligence for that matter. We also then received money from a private foundation, the System Development Foundation, which supported the work for essentially the five years starting from 1981 through 1985 or 1986. Not being a government agency, they were able to risk the money. The people at SDF were interested in supporting computational ideas of various kinds, including parallel processing. I think our work fit nicely in that general area. We actually did not suffer a lack of money in this period. We were pretty much given our head and could go ahead.

In your opinion, what is now the relationship between the serial approach and the parallel approach?

There are different intuitions that underlie the two approaches. In his recent work, for example, Allen Newell provides us with a kind of argument that will help us understand the domains. Normal computation models conscious reasoning or "recipe following," which is something human beings can do. And it is clearly very serial. When we consciously think something through, we go from step to step. We will see that, at a very slow time scale, there is seriality in human thought. But at a faster time scale, I don't believe that serial processing is the dominant form. The brain is simply not fast enough to be like a digital computer in this respect. A digital computer can overcome this by doing things very fast and serially and pretending it is doing them all at once. We have to imagine that the serial part of human processing is quite slow. And this slow serial part, the part that dominates our consciousness, is an important part of human cognition.

From a biological and psychological point of view, I believe that this serial processing is dependent, ultimately, on an underlying parallel processing system, namely the brain. But it is also approximately describable in its own right. When we look at things that happen at a very slow time range, we see things that look familiar from a symbol processing type of perspective. When we look at things that happen at a faster time range, we see things that look more parallel, less symbol-oriented. In one of the chapters of our book we try to sketch how it might be that this longer-term serial processing results from fundamental parallel processing systems.[16] But if you don't want to bother talking about where it comes from, you can probably talk about it in a serial way.

That is what is right about the serial processing accounts. What was wrong about the computer metaphor idea, in my view, is the attempt to push it all the way down to phenomena that occur very quickly. At that level it is not correct. But it gives an approximately correct solution or a vehicle for modeling these slower processes.

What becomes of the main idea of functionalism, that is, that the hardware does not matter?

The hardware clearly does matter. It matters less and less the more abstract the process is. Given time enough, the human being can act like almost any machine. But the faster we get, the more we press the system, the more important it becomes what kind of machine we have. This is true in general. If you don't care about the efficiency of an universal machine but only about its output, then the architecture doesn't make any difference. But when you make efficiency considerations, you have to ask what the machine is made out of and what its fundamental processing characteristics are. That's when the architecture begins to matter.

It's the same thing with brains. Of course, the brain is a parallel processing system; no one really doubts that. But when we work slowly enough, it may act as if it were a serial system. It could simulate one, so to speak—I could act as a Turing machine. To the degree I did that well, I would be simulating it. I could also simulate the computer on my desk, but it would take me forever to do even the simplest computation, because the serial steps that a conventional computer takes are so small and we are so slow that we could not keep up with it. That is why it is nice to have computers. When we do serial processing, we have to pay the price of going slowly, very slowly, by the standards of conventional computers.

In the long run, then, PDP will replace the serial approach, even if we can simulate some processes with serial processing?

At a theoretical level, it will. At a practical level, it is just as in physics: we might have a quantum theory that talks about quanta. Then we might have a macrotheory that talks about objects and aggregates of things. For this macrolevel it is useful if we can ignore their quantal properties. When building a bridge, it might be sufficient to ignore the microstructure, because we cannot possibly keep track of all the quanta. We try to get macroaccounts of what is going on. This might be true in psychology or cognitive science as well. Sometimes, a macroaccount will be the more parsimonious one, even if we have to give up some detail when we go for such an account. A microaccount might be too complicated for us to express at this higher level—although I believe that the microaccount captures more of what is going on.

In your opinion, which sciences constitute cognitive science or should play an important role in it?

There are many areas. A short answer, I suppose, would be: people who try to understand cognition. What does that mean? Obviously, this includes fields like Artificial Intelligence, cognitive psychology, linguistics, computer science, philosophy, cognitive anthropology, cognitive sociology, neurobiology. These people are all interested in cognition.

These are a lot of sciences, and from our educational background we are used to working in one discipline. Is it difficult to overcome the barriers between disciplines?

I should think that not all people in each of those areas would view themselves as doing cognitive science. I think that the real cognitive science is a particular mode of understanding, and this involves the use of a computational theory: the idea that ultimately the appropriate theory will be a computational one.

The consequences of this are not quite clear to me now. But if you look empirically at people who call themselves cognitive scientists, they almost always champion a kind of computational account—in some cases a connectionist account, in other cases a symbol-processing account—but there is always a computational orientation. I believe this is a central aspect of cognitive science. So some linguists could be considered or consider themselves cognitive scientists, whereas others might not. Similarly in philosophy or, clearly, in computer science. Those are areas where some are cognitive scientists and some are not.

At least in philosophy, it is not clear at all what "computational account" means.

That is correct. These things remain to be sorted out.

Nobody knows what it means when a program on one machine is equivalent to another program on another machine.

Those are technical questions to be answered. The working scientists are never bothered by them. They have a program of research that is reasonably clear. It involves the development of computer simulations as a mode of expressing or exploring theories, or sometimes more mathematical forms of expression. The questions of what is computation and what is not, and so forth, are semantical questions that eventually need to be resolved, but they are not the key. If it turns out that computation is not what we intuitively think it is, then we have just used the wrong word. These are issues that don't worry the working cognitive scientist very much.

Let's come back to the difficulties of different disciplines. In forming the PDP group, were there difficulties to overcome at first—of language, of concepts? How did this group work?

You have to understand that the Institute of Cognitive Science had been around for a while, and we had been doing interdisciplinary work of some sort already. Norman and I had forged rather close ties to Artificial Intelligence. Some people thought we weren't cognitive psychologists at all but simply Artificial Intelligence people. But of course we were trained in psychology.

At the same time, we had made efforts to interact with linguists, with philosophers, with anthropologists, with cognitive sociologists. We already had a good deal of experience in interdisciplinary work.

One of the things that came out of the Institute was a series of interdisciplinary workshops. And out of one of the workshops came the book entitled *Parallel Models of Associated Memory*, by Hinton and J. A. Anderson. The workshop was in 1979, and the book came out in 1981.[17]

Could you explain in more detail how an interdisciplinary workshop works?

It is not easy. First of all, you have to select the people carefully. Usually, the way we ran our workshops was that for each speaker, we allocated an hour and a half. During that time, the speakers would have their say. There would be a reasonably long discussion, and there would be a break in which the participants could talk informally to one another. We would have maybe two speakers, one in the morning and one in the afternoon. We also had the philosophy that the speaker should try to be understood by everyone, which meant that we would try to probe and understand the point of the speech instead of pressing on to the next talk. The main point was communication.

The hardest part in interdisciplinary work, in my view, are the differing goals that people have. Each discipline has its goals, its questions, and also its methods. As far as the goals are concerned, they usually go unspoken. We don't know what one another's goals are, yet we use the same language. So we think we are talking about the same things, but it turns out that our idea of what a right answer would be like is different. One important thing in any interdisciplinary effort is to sort out what you are trying to find out, what the goals are.

The second important issue is that of methods. Cognitive science, I think, is largely methodologically eclectic. By that I mean that we believe there is no absolute path to truth, but every method has its strengths and weaknesses. What we usually mean in part by "discipline" is exactly that there is a methodological discipline, a method that is the accepted method for this discipline to proceed.

Cognitive science does not yet have a "discipline" in that sense. What it has is a collection of methods that have been developed, some uniquely in cognitive science, but some in related disciplines. The difficult part here is learning to have respect for the disciplines of other fields. The linguist, for example, feels that the psychologist's mode of taking some undergraduate student and asking him or her about language is a ridiculous way to find out anything about how language works. Whereas the psychologist thinks that the linguist's method of using their intuitions is a ridiculous way to proceed in terms of learning how people really work.

It is clear that we have to learn to appreciate one another's approaches and understand where our own are weak. Sometimes we use observational approaches, sometimes experimental or introspective approaches.

Sometimes we use computer simulation, sometimes theorems, sometimes philosophical argumentation. Psychologists are notoriously suspicious of philosophical argumentation: it is not their way to get at truth; in their view scientists should go and observe. Whereas philosophers feel that by argument you can actually resolve things. These are things you have to learn. We work very hard to get these ideas across to one another.

When we started our interdisciplinary postdoctoral program, we also had a philosopher. The five postdocs and Norman and I met as much as ten hours a week. And during this time, we struggled with one another's methods and goals, trying very hard to sort out: what are you after? How do you find out? Once students are raised in an interdisciplinary environment, it will be second nature for them. They will recognize the validity of alternative methods.

Do you think that cognitive science should be a discipline of its own? Isn't an "interdisciplinary discipline" a contradiction in terms?

In the best of all worlds, it should form its own discipline. However, the universities are very conservative, at least in the United States, and it is not so easy to start a whole new department. Departments are very powerful in the universities, and once one is started, you cannot get rid of it very easily. Change will be very slow, but we already see various interdisciplinary institutes or centers of one kind or another, which are much easier to begin. They are created everywhere, and I think they are quite important. They are the way cognitive science will proceed for the time being. We will need people in the different departments like me, who encourage students to be broadly educated and to take interdisciplinary perspectives. It is not easy because there is a lot of inertia in the system.

That is a conservative strategy, and universities are very conservative. But I think cognitive science is reasonably important now, and it should increase. Still, it is not easy to hire someone in a traditional department who does not have the traditional credentials. I encourage my students to get a degree in a traditional discipline also, for practical purposes. Interestingly, business is less conservative. They don't care so much what field you were in.

What role does Artificial Intelligence play in cognitive science?

My view is that, roughly speaking, Artificial Intelligence is to cognitive science what mathematics is to physics. Artificial Intelligence, as I see it, consists of three different branches: theoretical Artificial Intelligence, experimental Artificial Intelligence, and applied Artificial Intelligence. Theoretical Artificial Intelligence I take to be the development of algorithms and procedures that are inspired by considerations of human or biological intelligence.

Experimental Artificial Intelligence is when you test these algorithms out by writing programs and systems to evaluate them. And applied Artificial Intelligence is when you try to take them and use them for some purpose.

The techniques of experimental Artificial Intelligence are the ones that are important for cognitive science. I see theoretical Artificial Intelligence as really a branch of applied mathematics. They develop abstract algorithms and procedures without consideration for whether they really account for how people work. But a cognitive scientist might use the methods that are being developed by Artificial Intelligence and propose that they are theories of cognition as embodied in some biological organism. When they do this, the algorithms become an empirical area. Artificial Intelligence is not empirical. It is much like mathematics in this respect, which is not an empirical science, but a formal science. But cognitive science itself has both the formal part, which is Artificial Intelligence, and the empirical part, in which assertions can be correct or incorrect descriptions of the world. Those are the cognitive science theories of how things might work in actual cognitive systems.

In your approach to intelligence, do you think that the Turing Test is a useful idea?

Sure. It does not mean that anybody solved anything, but it is a stringent enough criterion; it is a first cut within an area.

Some people claim that the Turing Test is useless—that if a machine passes the test it does not mean that its behavior is intelligent. Others claim that it is an artificial communicative situation that has nothing to do with everyday life.

These arguments are both true; I do not deny either of them. On the other hand, I think it is a challenge to understand the nature of cognition and intelligence sufficiently well to be able to build a machine that can mimic them at a level at which it is indistinguishable from a person. It is not the bottom line, but there are not many machines, if any, that can pass it within any breadth, and I would say it is a perfectly reasonable thing to shoot for.

Even though it can only test behavior?

That is correct. As I say, it is not the ultimate answer, but it is a goal we could achieve and that would be a significant step forward, even though it would not mean that we then understood everything.

What are now the biggest problems to solve in cognitive science and Artificial Intelligence?

I think the most challenging problem, the problem that motivates me the most, is that of language: how does it work? What is the nature of meaning? How does language interact with thought? Those are the problems on my stack in terms of difficult and central questions I would like to have answers for. But they are the kind of questions I suspect will hang around for a long time. We won't get satisfactory answers for these for some time.

Do you think that the so-called commonsense problem is a big problem for cognitive science and Artificial Intelligence? Some claim that the conventional approach will never solve the commonsense problem.

I guess what you mean is what I call "everyday thinking" or "everyday reasoning." I think that if we can understand understanding, we will move toward it, because reasoning is mostly understanding. My view is that to understand a problem is to solve it—a sort of Gestaltist view. A lot of what we do when we approach a problem is to make an attempt to understand it. Once we understand it, the solution is implicit in our understanding.

The process of understanding becomes, with experience, the process of seeing. So we come to perceive the solutions of problems. We look and we see. There is this movement from an abstract problem that we can solve maybe only by trial and error, or by some formulaic method like logic or algebra, to the point where we can understand it. At that point we have moved from the problem solving activity as search to problem solving as comprehension. Then we move on from problem solving as comprehension to problem solving as seeing or perceiving. We are always moving along this line.

With respect to everyday reasoning, I think that very little of it is in the form of search. Most of what we know about symbolic Artificial Intelligence techniques is about search. Mostly this is not relevant to understanding, which is what we usually do, or perceiving, which is even farther from what these search methods are about.

What will be the outcome of cognitive science research in the next few years? Machines, expert systems, a better understanding of cognition— what will be the practical and theoretical results?

This is very difficult. I don't like to predict too much. I think that the PDP approach is maturing, and it will allow us to understand better the relationship between mind and brain.

We already see applications of the connectionist approach in engineering and other areas. Similarly, we see that conventional Artificial Intelligence approaches are being applied in many different areas of practical use. It is really difficult to say what will come out of them eventually—certainly a clearer picture of mind and cognition. . . . That is what I hope for, and I hope for a better understanding of the relationship between mind and brain.

We will see steps toward this, but certainly no sudden insight. Slowly, the results will begin to accumulate. But we won't see a flash of light.

Recommended Works by David E. Rumelhart

Introduction to Human Information Processing. New York: Wiley, 1977.

(With M. A. Gluck.) *Neuroscience and Connectionist Theory.* Hillsdale, N.J.: Lawrence Erlbaum, 1990.

(With J. L. McClelland.) *Explorations in Parallel Distributed Processing: A Handbook of Models, Programs, and Exercises.* Cambridge, Mass.: MIT Press, 1988.

(With J. L. McClelland and the PDP Research Group.) *Parallel Distributed Processing: Explorations in the Microstructure of Cognition.* 2 vols. Cambridge, Mass.: MIT Press, 1986.

(With D. A. Norman and the LNR Research Group.) *Explorations in Cognition.* San Francisco: Freeman, 1975.

(Ed., with W. Ramsey and S. Stich.) *Philosophy and Connectionist Theory.* Hillsdale, N.J.: Lawrence Erlbaum, 1991.

Notes

1. M. R. Quillian, "Semantic Memory," in *Semantic Information Processing*, ed. M. Minsky (Cambridge, Mass.: MIT Press, 1968), pp. 216–70.

2. C. Fillmore, "The Case for Case," in *Universals in Linguistic Theory*, ed. E. Bach and R. Harms (Chicago: Holt, Rinehart, and Winston, 1968), pp. 1–90.

3. G. Frege, "On Sense and Meaning" (1892), reprinted in *The Philosophy of Language*, ed. A. P. Martinich (Oxford: Oxford University Press, 1985), pp. 200–212.

4. D. E. Rumelhart and D. A. Norman, "Active Semantic Networks as a Model of Human Memory," *Proceedings of the Third International Joint Conference on Artificial Intelligence* (1973): 450–57.

5. D. E. Rumelhart, P. H. Lindsay, and D. A. Norman, "A Process Model for Long-Term Memory," in *Organization of Memory*, ed. E. Tulving and W. Donaldson (New York: Academic Press, 1972), pp. 197–246.

6. D. A. Norman, D. E. Rumelhart, and the LNR Research Group, *Explorations in Cognition* (San Francisco: Freeman, 1975).

7. D. G. Bobrow and A. Collins, eds., *Representation and Understanding: Studies in Cognitive Science* (New York: Academic Press, 1975).

8. D. R. Reddy et al., "The HEARSAY Speech Understanding System," *Proceedings of the Third International Joint Conference on Artificial Intelligence* (Stanford, 1973).

9. D. E. Rumelhart, "Notes on a Schema for Stories," in *Representation and Understanding: Studies in Cognitive Science*, ed. D. G. Bobrow and A. Collins (New York: Academic Press, 1975), pp. 211–36.

10. J. L. McClelland, "On the Time Relations of Mental Processes: An Examination of Systems of Processes in Cascade," *Psychological Review* 86 (1979): 287–330.

11. D. E. Rumelhart and J. L. McClelland, "Interactive Processing through Spreading Activation," in *Interactive Processes in Reading*, ed. C. Perfetti and A. Lesgold (Hillsdale, N.J.: Lawrence Erlbaum, 1981).

12. G. E. Hinton, "Implementing Semantic Networks in Parallel Hardware," in *Parallel Models of Associative Memory*, ed. G. E. Hinton and J. A. Anderson (Hillsdale, N.J.: Lawrence Erlbaum, 1981), pp. 161–88.

13. D. E. Rumelhart, J. L. McClelland, and the PDP Research Group, *Parallel Distributed Processing: Explorations in the Microstructure of Cognition*, 2 vols. (Cambridge, Mass.: MIT Press, 1986).

14. M. L. Minksy and S. A. Papert, *Perceptrons* (Cambridge, Mass.: MIT Press, 1988).

15. F. Rosenblatt, *Principles of Neurodynamics* (New York: Spartan, 1962).

16. In vol. 2, chap. 14 of *Parallel Distributed Processing*, pp. 7–57.

17. G. E. Hinton and J. A. Anderson, eds., *Parallel Models of Associative Memory*.

John R. Searle was born in Denver, Colorado, in 1932. He attended the University of Wisconsin from 1949 to 1952 and studied at Oxford University, where he received his B.A., M.A., and Ph.D. He taught as Lecturer in Philosophy at Christ Church in Oxford from 1957 to 1959 and since then has been Professor of Philosophy at the University of California at Berkeley. He has also been a visiting professor at many universities both in the United States and abroad, including Venice, Frankfurt, Toronto, Oslo, and Oxford.

JOHN R. SEARLE

Ontology Is the Question

How did I get involved in cognitive science? Well, it happened over a dozen years ago. The Sloan Foundation decided to fund major research projects in the new area of cognitive science, and they asked various people in related disciplines if they would be willing to come to meetings and give lectures. I was invited, and I accepted. That way I met a lot of other people, and I was asked to participate in the formation of a cognitive science group in Berkeley. It is one of those things where outside funding actually makes a difference to your life. The fact that a major foundation was willing to fund a group and provide us with travel expenses, clerical support, and computers made a big difference to my participation. So, the short answer is that I got involved by way of the Sloan Foundation.

But what was the reason? As I was told, the Sloan Foundation planned at first to give the money to MIT and Stanford University, and Berkeley came in later.

They gave the biggest grants to MIT and Berkeley. I asked the vice president, who was in charge of donating the money, why they would give it to those two institutions and in particular to Berkeley, where there was not a typical cognitive science group. His answer was: "Because you submitted a good application." I think he was right: we have some good people at Berkeley, and they do good work. The Sloan Foundation is not so ideologically strict as to exclude a nonconformist group like ours. They let all these different flowers bloom.

How would you outline the recent developments and connections between sciences in the area of cognitive science? Is cognitive science a cooperation of different sciences, or is it a new science by itself?

I am not sure whether there is such a thing as cognitive science. But it is often useful to get together with people who have related interests and to work together. I do not take these departmental categories too seriously: are you really a philosopher or a linguist or a historian or a psychologist or a cognitive scientist? Such questions are not really very interesting, and I think they are for university deans to worry about—mostly useful for budgets. When you work on a real intellectual problem, you just work on the problem. I remember when I was a young philosopher, people used to tell me, "What you are doing is not 'real philosophy'; it is only linguistics." But nowadays it is "real philosophy." People change their opinions, so I would not worry too much about how to categorize cognitive science.

However, much of what is interesting about cognitive science is that it is partly a reaction against behaviorism in American intellectual life. It expresses an urge to try to get inside the mind to study cognitive processes instead of just studying behavioral responses to stimuli. It grew out of cognitive psychology, but it needed more resources than were available in psychology departments. It needed resources from computer science, linguistics, philosophy, and anthropology, to mention just the most obvious cases. That is the official ideology: cognitive science is that branch of the study of the mind that succeeds or tries to succeed where behaviorism failed, in that it tries to study inner contents of mental processes.

But having said this, I have to make two qualifications immediately. The first is that, in fact, many cognitive scientists repeat the same mistake of behaviorism. They continue to think that cognition is not a matter of subjective mental processes but a matter of objective, third-person computational operations. And that leads to the second comment. I think that cognitive science suffers from its obsession with the computer metaphor. Instead of looking at the brain on its own terms, many people tend to look at the brain as essentially a computer. And they think that we are studying the computational powers of the brain.

Why is that so? What is the role of computer science or Artificial Intelligence in cognitive science?

In the history of cognitive science, the computer has been crucial. In fact, I would say that without the digital computer, there would be no cognitive science. And the reason for that is that when people reacted against behaviorism and started doing cognitive psychology, they needed a model for cognitive processes. And the temptation was to think, "Well, we already have a model of cognitive processes. Cognition is a matter of information processing in the implementation of a computer program." So the idea was, as you know, to treat the brain as a digital computer and to treat the mind as a computer program. Now, what was wrong with that? Well, first, of course, strong Artificial Intelligence, which is the view that all there is to having a mind is implementing a computer program, is demonstrably false. Because minds have

semantic contents and because programs are purely syntactical, programs could never by themselves be sufficient for minds.

I demonstrated that in my Chinese Room Argument. But even weaker
versions of Artificial Intelligence, such as the view that says that minds
are partly constituted by unconscious computer programs, are also confused. The confusion is this: such notions as implementing a program,
being a computer, and manipulating symbols do not name intrinsic
processes of nature in the way that gravitation, photosynthesis, or consciousness are intrinsic processes of nature. Being a computer is like
being a bathtub or a chair in that it is only relative to some observer or
user that something can be said to be a chair, a bathtub, or a computer.
The consequence of this is that there is no way you could discover unconscious computational processes going on in the brain. The idea is
incoherent, because computation is not discovered in nature; rather,
computational interpretations are assigned to physical processes.

So, to summarize my answer to this question very briefly: without the
computer, there would have been no discipline of cognitive science as we
know it. However, one of the most important results of the development
of cognitive science has been the refutation of the computational model.

In your opinion, has cognitive science changed since, let's say, 1980?
There are many changes. I think, in fact, that today very few people
defend strong Artificial Intelligence. Of course, they do not say that they
have changed their mind, but they have. I do not hear as many extreme
versions of strong Artificial Intelligence as I used to. But in the hard core
of cognitive science, the biggest change has been the advent of connectionism, parallel distributed processing, neural net models of human
cognition. I am more sympathetic to these than I am to traditional Artificial Intelligence, because they are trying to answer the question of how
a system that functions like the brain might produce intentional and intelligent behavior. If we think of them as weak Artificial Intelligence—as
attempts to model or simulate but not actually duplicate—some features
of human cognition, they are quite useful. They provide us with models
of how systems that are constructed not exactly like, but something like,
the brain might actually function. Now, let's think about these a minute.
One exciting thing about PDP models, I think, is not that they produce
so much by way of positive results. The results are relatively modest. But
much of their popularity derives from the fact that traditional Artificial
Intelligence is a failure. It is the failure of traditional Artificial Intelligence rather than the positive successes of connectionism that has led to
much of the popularity of the connectionist models. However, as I said,
I am very sympathetic to the connectionist project, and I will be interested to see how far they can get.

The idea of PDP is an old idea that comes from the sixties, from the work of Rosenblatt.[1] Do you think that its popularity nowadays is only due to the failure of traditional Artificial Intelligence?

There has been some progress since Rosenblatt: the early perceptrons were attacked by Minsky and Papert quite effectively,[2] but the more recent versions of connectionist nets have hidden nodes; they have properties that the original versions did not have. So, they largely evade the earlier objections. The main point that I was trying to make is this: connectionist models have been, in many ways, successful, but I think that the popularity exceeds the success. And the reason for that popularity is the failure of traditional, old-fashioned Artificial Intelligence models.

In your opinion, what are the failures of traditional Artificial Intelligence?

Well, they did not get very far. So let's discuss some of their problems. I am assuming that strong Artificial Intelligence is a failure, for the reasons I have stated: the syntax of the program is not by itself sufficient for the semantics of the mind. I am now talking just about Artificial Intelligence efforts at simulating human cognition. Within traditional Artificial Intelligence, you have to distinguish between those programs that are just for commercial applications and those that are supposed to give us information about psychology. For example, most of the expert systems that I know anything about, such as those that are used for medical diagnosis or for oil prospecting, are in fact essentially commercial devices. They are useful as tools, but they do not even attempt to give you real insights into human psychology, and the methods of information processing that they use are, in general, rather simple retrieval methods. Maybe there are some programs I do not know about, but as far as I know, the so called expert systems are not very interesting from a psychological point of view. So when we get to the real issue in Artificial Intelligence, we are talking about those programs that are supposed to give us psychological insights. Now, the basic problem is very simple: we did not get much by way of psychological insights from traditional Artificial Intelligence. Progress was very, very slow. In fact, there has been, to my knowledge, virtually no recent progress of any theoretically significant kind. Occasionally, the problems become apparent within the Artificial Intelligence community. For example, discussions of the "frame problem" exhibit the limitations of traditional Artificial Intelligence. I think that the frame problem is just an instance of what I call the "background problem," namely that all intentionality functions against a background of commonsense knowledge, practices, behavior, ways of doing things, know-how, and such preintentional capacities, both cultural and biological.[3] There is a theoretical obstacle to programming the background capacities: they are not themselves representational. So there is no way that their essential features can be captured in representations, but by definition, traditional Artificial Intelligence uses representations. You might get the same results using representational methods for something that a human brain does without representational methods, but it does not follow that you have gained any insight into how the human brain does it.

I will give you an example: I am sitting here with you, drinking coffee. I reach out and pick up the cup to take a sip of coffee. If I tried to describe that activity in information processing terms, it would involve a prodigious number of representations: hand-eye coordination, solving of vector calculations as to how exactly to move my arm and how to grip the cup and how much energy to use in lifting it. It would look as if I were solving a very large number of difficult differential and quadratic equations. If you tried to build a robot to do it using existing technology, you would have to have a prodigiously powerful mathematical system for the robot to go into a room it has never been in before, to pick up a cup it has not been programmed to deal with before, and to take a sip of coffee from the cup. Now, according to me, I can drink this coffee without solving any differential or quadratic equations. I do it without solving any equations. I do it just as an exercise of my capacities, my abilities. But if that is true, then you cannot explain my abilities in terms of information processing, equation-solving capacities. In my view, the greatest single failure—or class of failures, I should say—in traditional Artificial Intelligence is the failure to come to terms with the nonrepresentational character of the background.

Let's go back to the strong Artificial Intelligence position. You published the Chinese Room Argument back in 1980.[4] Why is there still a discussion about it? In general, the arguments are the same as the ones published with the article back in 1980, aren't they?

I can only speculate about what keeps it going. What keeps traditional Marxists going, given the failure of communism everywhere in the world? The faithful have careers built on it; they have commitments. There are organizations, there are budgets and jobs, professorships and directorships. Similarly, people have built their professional lives on the assumption that strong Artificial Intelligence is true. And then it becomes like a religion. Then you do not refute it; you do not convince its adherents just by presenting an argument. With all religions, facts do not matter, and rational arguments do not matter. In some quarters, the faith that the mind is just a computer program is like a religious faith.

That does not explain why new people now coming into the field are stuck with the same arguments. Perhaps you cannot reduce the question to a number of individual people with individual interests. Perhaps there is some sort of mainstream thinking in our society about how science should function.

I think you are right about this. There are deeper impulses. On the surface, I think there is this sheer professional commitment and the fact that young people get socialized into a certain professional ideology. But the deeper question is the one you are pointing at now: what is it in our society that makes this particular religion attractive? After all, a religion has to survive in a given society. I think the answer is that it provides us with what a lot of people think is a scientific vision of the mind. Our overall scientific vision, going back to the seventeenth century, is that all of scientific reality is objective. You do not have to worry about subjective, conscious mental states, because reality is physical, and physical reality is objective. The beauty of strong Artificial Intelligence is that it enables you to recognize the special character of the mind while still remaining a materialist. You can say that the mind is a nonphysical formal, abstract mechanism. But at the same time, you do not have to become a Cartesian to say this. You do not have to say that there is some special mental substance. And, on this view, the key to solving the mind-body problem is the computer program. Because a computer program can be defined purely formally or abstractly, it satisfies the urge to think of the mind as nonphysical, but at the same time, the solution is completely materialistic. The program, when implemented, is always implemented in a physical system.

Another part of the mistake—and it is in a way one of the leading mistakes of our civilization—is to think that the crucial question for any system is "What does it do?" Thus if you get a program that does what the mind does it is tempting to think they must at some level be the same thing. I want to urge that the crucial question is not "What does it do?" but rather "What is it?"

Speaking of computer programs and their philosophical significance, what do you think about the Turing Test?

If the Turing Test is supposed to be philosophically significant and conclusive to show that behavior constitutes cognition and intelligence, then it is obviously false. The Chinese Room refutes it. You can have a system that passes the Turing Test for understanding Chinese, but it does not understand Chinese. I am not really sure that that is how Turing intended the test, though. If it is a test for shortcutting philosophical arguments and saying, "Well, look, if the machine can do this, then let's treat it as if it were intelligent," then I have no objection. So, the Turing Test depends on how it is used: whether it is used just as a practical device for deciding when we should treat a machine as having succeeded in simulation or whether it is supposed to be a philosophically significant device showing the presence of conscious intelligence. If it is used as a practical device, with no theoretical significance, I have no objection. If it is supposed to be a theoretical philosophical claim, it is obviously false. And its falsity is symptomatic of the deep mistake I was just calling attention to: we think the question "What does it do?" is the crucial question. But for science and philosophy that is an incidental question. The crucial question is always "What is it? What is it intrinsically?"

Let us reexamine the Turing Test and the Chinese Room. One could argue that the Chinese Room Argument is the Turing Test from the inside. The Turing Test is an objective view, just as science is seen in our society from a third-person point of view, and the main idea of your Chinese Room is the first-person view. And therefore your argument is conclusive only if the thought experiment starts from the first-person view: I have to imagine that I am in the room.

I think it is easiest to *see* the Chinese Room Argument from the first-person point of view, and there is a deep reason for that—namely, that the ontology of the mental is essentially a subjective, first-person ontology. But the actual form of the argument says that from the fact that you have the syntax of Chinese it does not follow that you have the semantics. And that is a valid argument as it stands, given that syntax is not sufficient for semantics. The philosophical depth of the argument, however, derives from the fact that it reminds us not just that syntax is not sufficient for semantics but that actual semantics in a human mind, its actual mental contents, are ontologically subjective. So the facts are accessible only from the first-person point of view. Or, rather, I should say that the facts are accessible from the first-person point of view in a way that they are not accessible from the third person.

Let me add something to this discussion of the Chinese Room parable. From my point of view, as I see it now, I must say that I was conceding too much in my earlier statements of this argument. I admitted that the computational theory of the mind was at least false. But now it seems to me that it cannot even be false, because it does not make sense. I mentioned this point briefly earlier; now let me explain it a bit more in detail. The natural sciences describe features of reality that are intrinsic to a world that is independent of any observer. That is why gravitation, photosynthesis, and so on can be subjects of natural science—because they are intrinsic features of reality. But such features as being a chair or a bathtub or being a nice day for a picnic are not intrinsic features of reality. They also have features that are intrinsic to reality because they are physical entities with physical features. But the feature of being a nice day for a picnic exists only relative to observers and users. Now, computation is defined in terms of symbol manipulation, but "symbol" is not a notion of natural science. Something is a symbol only if it is used or regarded as a symbol. The Chinese Room Argument showed that semantics is not intrinsic to syntax; this argument shows that syntax is not intrinsic to physics. There are no physical properties that symbols have that determine that they are symbols. Only an observer, user, or agent can assign a symbolic interpretation to something, which becomes a symbol only by the act of interpreting it as a symbol. So computation exists only relative to some agent or observer who imposes a computational interpretation on some phenomenon. Of course you can ask: "Can you assign a computational interpretation to those brain processes that are characteristic of consciousness?" But this question is trivial, because you can assign a computational interpretation to anything, even to a river flowing down a hill. But if the question is whether consciousness is intrinsically computation, then the answer is that nothing is intrinsically computational; it is observer-relative. This now seems to me an obvious point. I should have seen it ten years ago, but I did not. It is devastating to any computational theory of the mind.

So, in your view, the Turing Test and the Chinese Room Argument
are basically concerned with consciousness? But consciousness is not
a notion that is often used in cognitive science.

No. I originally presented the Chinese Room Argument as an argument about intentionality, and I made no reference to consciousness. For the purposes of the argument about the Chinese Room and the Turing Test, the issues can be kept quite separate from the problem of consciousness. However, it is true that until quite recently many workers in cognitive science and neurobiology regarded the study of consciousness as being somehow not a concern of their disciplines. They thought that it was beyond the reach of science to explain why warm things feel warm to us or why red things look red to us. But I think that it is precisely a question that neurobiology should find an answer for. Maybe a kind of residual dualism prevented people from treating consciousness as a biological phenomenon like any other.

By consciousness, I simply mean those subjective states of awareness or sentience that begin when you awake in the morning and continue throughout the day until you go to sleep or become "unconscious," as we would say. Consciousness is a biological phenomenon, and we should think of it in the same terms as we think of digestion, or growth, or aging. What is particular about consciousness that distinguishes it from these other phenomena is its subjectivity. There is a sense in which each person's consciousness is private to that person. You are related to your pains, tickles, feelings, or thoughts in a way that other persons are not. You could also say that conscious states have a certain qualitative character, and for that reason they are sometimes described as "qualia." For consciousness to be a biological phenomenon, it must be related to the brain. How should we think of that relationship? This is the famous mind-body problem. It has a long history in philosophy, but I think that the solution—or maybe, better, its dissolution—is rather simple. Here is how it goes: conscious states are caused by lower-level neurobiological processes in the brain, and they are higher-level features of the brain. As far as we know anything now about how the brain works, variable rates of neuron firings in the brain cause all the enormous variety of our conscious life. But to say that conscious states are caused by neuronal processes does not commit us to a dualism of physical and mental things. The consciousness that is caused by brain processes is not an extra substance or entity. It is just a higher-level feature of the whole system. These sorts of relations are not unusual in nature. Think, for example, of the liquidity of water or the transparency of glass and of their relations to the microstructures of the substances in question. Of course, these are only analogies. But the point I want to make is this: Just as one cannot take a molecule of water and say, "This one is wet," so we cannot pick out a single neuron and say, "This one is thinking about tomorrow's lecture." Just as liquidity occurs at a much higher level than that of molecules, we know that thoughts about lectures occur at a much higher level than that of single neurons or synapses. But there is nothing metaphysical about water being wet, just as there is nothing metaphysical about consciousness.

Now let's go back to the Turing Test. The Turing Test inclines us to make two very common mistakes in the study of consciousness. These mistakes are to think either that consciousness is a certain set of dispositions to behavior or that it is a computer program. I call these the behavioristic mistake and the computational mistake. Now, behaviorism has already been refuted, because a system could behave as if it were conscious without actually being conscious. There is no necessary connection between inner mental states and observable behavior. The same mistake is repeated by computational accounts of consciousness. Computational models of consciousness are not sufficient for consciousness, just as behavior by itself is not sufficient for it. Nobody supposes that the computational model of a thunderstorm will leave us wet, so why should the computational model of consciousness be conscious? It would be the same mistake in both cases.[5]

In the light of your criticism, what do you think will be the future of cognitive science and in particular of Artificial Intelligence?

That is a very hard question to answer, because, of course, one is only guessing. I think that cognitive science will continue for some years—and for the simple reason that there is money. And there is what I call

"Searle's Second Law": that is, that "Brains Follow Money." If there is enough money to support this, people will do it. Additionally, it has become fashionable, and universities have set up courses and so on. There is no doubt that for the next decade or so cognitive science will continue. What happens after that, who knows? Then it depends on results. But right now we know that it is going to continue because there is money available. What I think will happen to Artificial Intelligence is that it will be immensely useful commercially. Computers are generally useful commercially, so various efforts to simulate the processes and results of human intelligence will be useful commercially. But my guess is that traditional Artificial Intelligence, as a scientific research attempt to understand human cognition as opposed to being just a straight commercial attempt, has run its course. I do not think we will get any more new and exciting results from it. The ones that you see now are really rather feeble, when compared with the exaggerated predictions made by its founders, such as Minsky or Simon and Newell.

Connectionism is more interesting. I could not attempt to predict its future, because it is still too early. We are now just exploring the properties of the various connectionist systems. I think that commercially, again, there will be a lot of possibilities for parallel machines. For me the most interesting results are not the commercial results but the scientific results.

So you think that the metaphor of the mind as a computer program will disappear after a few years, as did the metaphor of the mind as a steam engine?

I think it is going to be with us for a while, because we do not know how the brain works. The brain is the most important organ. It is where we live, so to speak. And we would really like to know how it works. We do not know how it works, so we fasten onto metaphors. In Western history we have thought, successively, that the brain must be a catapult or a steam engine or an electric circuit or a switchboard or a computer. And God knows what people will come up with next. Until we get a science of the brain, we will always have these desperate and pathetic metaphors. And I just do not know how close we are to having a science of the brain. Maybe we have still a long way to go.

What, in your view, would constitute a breakthrough in this science of the brain?

The most important scientific discovery of our time will come when someone discovers the answer to the question "How exactly do neurobiological processes in the brain cause consciousness?" In broad outline we now think that it must be a matter of variable rates of neuron firing relative to specific neuronal architectures. But, of course, this is only a tentative hypothesis, as any causal hypothesis must be. It might turn out that we overestimate the importance of individual neurons and of synapses. Perhaps it will turn out that the crucial unit is a column or an array of neurons. But what is important is that we are looking for causal relationships. Many people object to my solution of the mind-body problem with the argument that we do not have an idea of how neurobiological processes could cause conscious phenomena. But that is not so different from the situation we were in with other "mysterious" phenomena, like electromagnetism. In terms of Newtonian mechanics, electromagnetism had no place in the conception of the universe and seemed mysterious. But with the development of a theory of electromagnetism, the metaphysical worry dissolved. I believe it will be the same with consciousness. The problem is to figure out how the system, the brain, works to produce consciousness; and that is an empirical-theoretical issue for neurobiology.

Recommended Works by John R. Searle

Intentionality: An Essay in the Philosophy of Mind. Cambridge: Cambridge University Press, 1983.
Minds, Brains, and Science: The Reith Lectures, 1984. London: Penguin Books, 1984. (A small and readable book, also containing John Searle's critique of Artificial Intelligence.)
The Rediscovery of the Mind. Cambridge, Mass.: MIT Press, 1992.

Notes

1. F. Rosenblatt, "Two Theorems of Statistical Separability in the Perceptron," *Proceedings of a Symposium on the Mechanization of Thought Processes* (London: Her Majesty's Stationery Office, 1959): 421–56; and *Principles of Neurodynamics* (New York, Spartan Books, 1962).
2. M. L. Minksy and S. A. Papert, *Perceptrons* (Cambridge, Mass.: MIT Press, 1988).
3. J. R. Searle, *Intentionality: An Essay in the Philosophy of Mind* (Cambridge: Cambridge University Press, 1983).
4. J. R. Searle, "Minds, Brains, and Programs," *Behavioral and Brain Sciences* 3 (1980): 417–57.
5. The theses advanced here are more fully and more recently laid out in J. R. Searle, *The Rediscovery of the Mind* (Cambridge, Mass.: MIT Press, 1992).

Terrence J. Sejnowski studied physics at Case Western Reserve University (B.S.) and Princeton University (M.A., Ph.D. in 1978). As a research fellow at Princeton University and Harvard Medical School he specialized in biology and neurobiology. His professional career began at John Hopkins University, where he served as Assistant Professor and later Associate Professor from 1982 to 1987. In 1988, he was appointed Professor of Biophysics at Johns Hopkins, before joining the University of California at San Diego in the same year, as Professor of Biology and Physics. Since 1988, he has also been a senior member of the Salk Institute, San Diego.

TERRENCE J. SEJNOWSKI

The Hardware Really Matters

Cognitive science came late in my career. It represents a synthesis of many of my other interests. The essential question for me has always been how the brain works. In the process of trying to understand how the brain works, I have come to appreciate the complexity not only of the structure of the brain but also of its cognitive abilities. My feeling now is that the study of the brain is going to require collaboration between at least two major areas of science—the cognitive sciences on the one hand and the neurosciences on the other hand—to make a synthesis possible.

Which sciences would you include?

The neurosciences, at least in the United States, are an amalgamation, a combination of many disciplines that were formerly isolated in different departments—neural anatomy, neurophysiology, neuropharmacology, psychiatry, and physiological psychology. These were all different faculties. About fifteen years ago in the United States, many research workers in these areas who were all working on different aspects of the brain started a new society—the Society of Neuroscience. It is one of the most successful new societies in the whole world, its membership now reaching twelve thousand and continuing to grow linearly, with about eight hundred new members annually. Each year it holds a large meeting at which over ten thousand people present results in literally hundreds of sessions.

What has been important is the realization that the structure and function of the brain at all the different levels, from the molecular to the behavioral, are intertwined and cannot be separated. The levels have to be put into contact. If one is studying behavior, it can often be very important to understand some of the pharmacological substrates that can alter behavior. And equally, if one is interested in pharmacology, in the chemical composition of the brain, one needs some guidance as to which chemicals are important, which ones should be the focus. Neurosciences as a whole have provided the scientific foundations on which we can build toward the cognitive sciences.

That is one hand. On the other hand, I see in the cognitive sciences a similar coming together—perhaps not on this scale, but it is already clear if you look at the charta of the cognitive science society in the United States: it includes psychology, linguistics, and Artificial Intelligence, and it also includes neuroscience. Historically, neuroscience has always been included, but it has not played a major role in the development of the cognitive sciences, partly for historic reasons but mainly for deeper scientific reasons.

It has to do with the approach. The easiest way to explain this is to go back to the early days of Artificial Intelligence and try to understand some of the fundamental assumptions that were made in the late fifties. One of the most influential statements was made by Allen Newell, when he outlined a research program that would carry Artificial Intelligence into the next decade. I only heard this from Newell and do not know where it is printed, but he said that the present knowledge in neuroscience was not detailed enough to take advantage of it. Therefore we had to look elsewhere for inspiration if we are interested in understanding how human beings think. The key is the realization that knowledge about the brain was potentially an important resource and constraint on human cognition. However, it was rejected on the grounds—justifiable at the time—that not enough was known about the brain. The difficulty of working on single neurons—which are incredibly microminiaturized—were beyond the techniques and capabilities of that time.

Things have changed dramatically over the last forty years. It is now no longer the case that neuroscience is behind. In fact, in many respects the techniques are so developed that it may, over the next decade, become one of the leading resources of constraints and information. I am not saying here that neuroscience by itself is going to provide all the constraints that are needed, only some constraints that would be unavailable with any other approach. That has to do primarily with the timing characteristics—constraints that come from thinking fast: being able not only to solve a problem but to solve it on a time scale of seconds, which of course is essential for survival.

But to get back to the Artificial Intelligence story: somewhere along the line over the last forty years the emphasis shifted from "neuroscience is not ready" to "neuroscience is not necessary." You still often hear from Artificial Intelligence researchers who are not aware of the revolution that has occurred in neuroscience statements like "Knowledge of the brain is not relevant," and "The details of how biology solves the problem are not important for me, because I want to solve the problem on a more functional, general level. Whether the brain is made out of neurons or of chips or of tin cans really is not relevant to the level I want to address." That the mind should be studied independently from the brain has become almost a dogma with certain people, but I don't think it was a dogma with the founders of Artificial Intelligence, certainly not with Allen Newell. It has become a dogma in part because of a very misleading metaphor—the metaphor of the digital computer.

One dramatic change is that we use the term *computer* now in a very different way from how it was used, let's say, before 1950. When I was in school learning physics, we computed with a slide rule. It was very efficient for a limited range of computations. Nonetheless, it was computation, and people did computation. They used log tables, they used mechanical devices, they used analog computers. These were all computers, but now the prototype of the computer has become the digital computer. In fact, our abstract definitions almost preclude what formerly was called a computer. If you assume that computation necessarily means sequential operations and instructions, you exclude analog computation. I want to go back to a much broader definition and try to imagine what sorts of computation might be going on in the brain, if we all assume that computation is a term that extends to brains as well.

Digital computers also have the property that software is separated from hardware. This has almost become part of the definition of computation. If we think of software as a program and of hardware as an arbitrary implementation—then it becomes more plausible that the brain is not important, once you identify the brain with the hardware. That is where the fundamental mistake occurs, because the brain is not just the hardware; the brain is the hardware *and* the software. In fact, it is almost impossible to separate the two. The very concept of software has to be modified, I think, when you come to the question of how the brain is programmed. The brain is clearly not programmed by lists of instructions. The fact is that most of the things we do are carried out by routines that we have learned by experience: by going to school, or by watching and mimicking. There are many sources of learning and experience, but a large fraction comes to us through ways we are not even aware of. What happens in this process is that the actual circuitry in the brain is being modified; it is as if the hardware itself is changing. There is a point where hardware and software come so close that it does not make sense any more to separate them.

The digital computer is no longer a good metaphor or model for the brain, then?

Certainly not for the human brain, although some continue to argue that it is still a good metaphor for human cognition. I think we have to question that, too. We have to go back to basic psychology and ask whether there may be a more fundamental way of looking at human cognition, of which the traditional view is only an approximation. It need not be wrong, only approximating something else, a process that is very different.

If one assumes that human cognition consists in formal symbol manipulation, then the conclusion is that hardware does not matter; whereas your opinion is that the hardware matters, too. Is that right?

That is right. It could matter even in a very crucial way. Hardware has been largely ignored both in cognitive science and in Artificial Intelligence. This is ironic because of a consideration that is obvious and that any engineer recognizes as essential—the time it takes to do the computation.

We evolved and survived because we were able to solve problems quickly. When you are threatened by a predator and cannot react within a fraction of a second and do the right thing—recognize what and where the danger is coming from and take evasive action—you will not survive; your genes will not be propagated, and this would be the end. Most digital computers would not survive long without care and feeding by humans.

The essential part is that it is not enough simply to show in a formal sense that a program can solve a problem. You also need to show that you can solve the problem in the same time, namely problems having to do with planning, making decisions about when and where to move, even very high-level cognitive acts like aesthetics: people decide on the spot whether they like something or not. They like a painting, a performance: they don't need to go off-line and do a lot of symbol crunching to make those decisions. It is a feeling but certainly a very high-level feeling, one that involves the highest cognitive abilities humans have: artistic sense.

This draws near the question of commonsense. It is clear that commonsense is something we have on-line, that we have access to, that we can apply and call up without thinking, without waiting or doing calculations. We immediately know what the appropriate response is when someone asks us, for example, what to do with a check, even if we have never had this particular experience, never gone to a bank, before. But we have generalized from previous similar experiences, and it has been incorporated in the knowledge structure that we have in the brain, such that we can apply it within a few clock cycles. That is the essential missing constraint that we need to include if we want to come up with a scientifically plausible explanation of human thought.

Most Artificial Intelligence programs require millions of steps before coming to even the simplest conclusions. We know that the brain does not have the benefit of all those steps. We know, for example, that the time scale for signals to be processed by neurons is measured in milliseconds. To be precise: it takes about five milliseconds for a signal to go from input at a synapse to output. If I respond to a question in half a second, which is not difficult at all—if I ask you, "In what continent is Germany?" the answer comes to you immediately; and I could ask you questions from any knowledge area—that allows for only around one hundred iterations or steps. These are very elementary steps having to do with one path through a network in order to retrieve and to make the appropriate response and even plan it in the case of a motor response. All that has to be going on and has to be over in one hundred clock cycles or less. The only Artificial Intelligence program I know that uses this as a constraint is Allen Newell's Soar.[1] If a program is meant to be a model for cognition, this is something that really has to be taken into account.

The problem of time is one that people in Artificial Intelligence have to confront even on their own grounds: a LISP machine has fundamental limitations on how many inferences per second it is able to do: a clock cycle of nanoseconds still requires so many operations to solve a problem that even if you grant a billion operations, it is still not fast enough to solve it. Computers today are a million times faster than the first generation, and new machines will be faster, but they are still orders of magnitude away from real-time solutions. A LISP machine will never "see" in real time, even if such a program could be written.

You claim that even with bigger and faster machines they won't succeed?

Not if they are doing it strictly sequentially. If they are following the traditional view of computation as being one operation at a time, then it will be very difficult. But it is dangerous to predict that something is not possible.

However, here is something I am more confident about and that has to do with hardware constraints: any good programmer on a conventional digital machine knows that if you want a program to be fast, you have to take advantage of the architecture. You can vectorize if you know that the computer has a vectorizing pipeline, like a supercomputer. The more effort you put into taking advantage of the hardware, the faster the program is going to run.

This illustrates a very general point: when speed is a factor, the link between hardware and software becomes closer and closer. In the case of biology, the difficulty is to solve problems fast. That means that nature had to close the loop and take advantage of the physics of the neurons itself. There is a much closer relationship between the solution of the problem—the algorithm—and the actual hardware itself.

Here is another way of looking at things: if I have a computer that runs certain things very, very fast and others incredibly slowly, it is almost certain that I will search for algorithms that take advantage of that machinery. If you have to solve a practical problem, you are almost forced to do that. In retrospect, a great part of the reason why cognitive science and Artificial Intelligence have focused on symbol processing was that there were machines available that were very efficient at symbol processing. That is what digital computers are good at, but they are not good at sensory processing, motor control, or a number of other things that humans do very easily. This is well known.

In a sense, the programs in Artificial Intelligence have been optimized for digital computers. I would argue that, far from being machine-independent, the research in the field has been entirely dependent on that architecture. Having this architecture has strongly influenced the realm of computation where people have searched for algorithms. If a radically different architecture were available, however—for example, computers with a billion processing units—they would not be suitable for running a LISP program, a traditional symbol processing program, but they would be very good at signal-processing primitives. Then there would be much more motivation for studying and discovering algorithms on that architecture.

In other words: hardware, far from being irrelevant, is the most important guiding influence on the way people think about algorithms. Parallel algorithms are now being discovered that were totally overlooked but are now practical, because the connection machine with sixty-four thousand processing units can efficiently implement them. The hardware really influences your thought process. The effort of finding new algorithms is now focused on a different part of computational space. This is ironic, because one of the original motivations for Artificial Intelligence and cognitive science was the belief that this way of approaching computation was machine-independent. Indeed, in a formal sense this is true: a Turing machine can compute any computable function, but not necessarily in a second or less—and therein lies the problem: any practical machine has a finite cycle time and therefore will not necessarily be able to solve the problem in a second.

Another factor that may be even more important than these practical considerations is inspiration—how you get ideas. In other words: I, and I think most human beings, use concrete images or exemplars as ways of getting new ideas, even abstract ideas. And the exemplar of the Turing machine has been very fruitful; it has led to an enormous amount of very good research.

Just over the last few years, we are beginning to develop a new model, a new exemplar for computation, based on principles that are very different from the Turing machine—or the von Neumann architecture, to be more precise. That is the neurally inspired architecture developed by the PDP group within psychology, by the connectionist group within Artificial Intelligence, and now within neuroscience, in the area that I represent: computational neuroscience, which explores the use of computer models to try to understand neurons, brain circuits, and processing systems within the brain.

Could you explain for a nonexpert public what the main differences are between a von Neumann architecture and PDP? Isn't a PDP architecture a combination of a number of von Neumann–type processors, still with a common cycle that controls input and output of every unit?

This is certainly one type of parallel architecture. The connectionist architecture is somewhat different. The processing units are much simpler than von Neumann architectures. They are typically nodes that are capable of only elementary arithmetic: adding, subtracting, multiplying, and so on, but with limited memory. The connectivity between the processing units accounts for much of the computational power.

Ideally, these processing units are connected to hundreds or thousands of other units, whereas typical circuit elements on chips have two or three fan-outs, and even contacts between boards are ten or less. There is a good reason for that: wire is expensive on a chip, because you have to charge it. If you want to connect a circuit board to a hundred others, you have to have lots of currents passing through all the wires to charge all the supports. This directly affects the speed, how well the whole computer can synchronize, and so forth. We are dealing here with fundamental hardware constraints. The brain does have a technology that allows it to support a very high degree of interconnectivity. It is by no means completely interconnected, but there are thousands to tens of thousands of connections per neuron. As I said earlier, this constraint will change the nature of the algorithms you are going to explore.

Suppose I wanted to speed up a commonsense calculation: I want to know, for example, whether this glass could be used for bailing out a flooded area in my kitchen. In a traditional Artificial Intelligence program you would have to represent "glass" by a symbol and all its properties by other symbols, and these representations would be stored at different locations. Then you have to be able to do a lot of matching between the symbolic expressions having to do with glass, with containers, with water. . . . Even assuming that all the structures were there, you would have to go through many steps in order to decide that the glass is too small, that the water on the floor is a thin layer, that the glass is not suitable. You also have to be able to decide that, if the water in my kitchen were in the sink, the glass would be a perfectly suitable instrument.

How would I do a similar calculation but with a billion processing units, each of which is so tiny it could certainly not hold a whole symbol? How do I take the concept of a glass, a container, and distribute that concept over a million units in a way that I could use this representation of the glass to compute whether or not it is appropriate to contain water that is on the floor? The advantage of the distribution is that a large fraction of the units have a piece of the representation that they can combine locally with other pieces of other representations, because the other representations are being stored in the same population in a distributed way. Much of this can be done locally, so that the information need not travel through the whole system. In only a few steps or iterations it may be possible to come up with a solution to the problem, because the computation has been distributed over many processing units, and the information is being combined everywhere in parallel. The results are being accumulated, a decision is made, and you can act—all in less than a second.

What would a representation distributed over thousands of parts look like?

That is the current focus of the research in PDP and connectionist models. What can be represented, and how can it be represented? We are coming up with many counterintuitive answers that would never have been guessed based on the model of memory found in a digital computer, that is, a very specific and precise location for each fact. If you have to distribute this fact over a million units, you have to do it in a way that is very different. In a sense, the representation of "glass" contains partially the information you need to solve the problem, because the representation already contains information about its properties and uses. It is a very rich representation.

This is one of the major differences that scientists using this new approach are now finding. This illustrates the point I made earlier: the assumptions about the hardware have repercussions (that you could never have imagined) on how you approach a problem, how you think about issues like representations, how you come up with practical solutions.

I should mention here that the field is spreading and has a very strong component from the engineering side that is outside Artificial Intelligence. There is, for example, a journal that has just started, published by MIT Press and entitled *Neural Computation*. It includes papers from the VLSI community that are concerned with actually building chips. Why are the chip people interested in neural computation? Because computer scientists have come to realize that problems like vision and speech recognition are so computation-intensive that in order to solve them in real time you need to design special chips that can solve those problems in parallel. What they have found is that a great deal of inspiration about how the circuits should look has actually come from neuroscientists who are concerned with the circuitry in the brain.

In the first issue, for instance, there is a paper by Lazzaro and Mead[2] about sound localization. They have designed a chip that converts sound waves into frequency-filtered analog signals, which are then combined in a way that is similar to the way they are combined in the brain of a barn owl. That allows the chip to compute the localization of the sound. Barn owls can do that much better than we can, and they use very specialized neural circuits. I would not say that the hardware duplicates their brain, because the degree of microminiaturization in the brain exceeds anything that can be done today with VLSI. But some of the computational principles having to do with how the information is represented, how it flows through the circuits, the algorithms that are used to extract the information—those appear to have very close similarities to the way the brain computes.

Similarly, there are groups concerned with speech recognition. It would have an enormous impact on interfaces if a cheap, practical, efficient system could be designed that allowed us to feed information into the computer through the voice. So far, there has only been progress with limited vocabulary and single speaker systems, which are still very expensive and not nearly as good in their performance as humans. We know that the problem can be solved, because humans solve it; but we have to deliver enormous computing power to do it. One way to deliver that enormous computing power is to go through a massively parallel system and by using the connectionist approaches.

One approach that has been successful and looks promising is the idea of learning algorithms applied to speech data. The idea here is, rather than to program the speech problem, to teach a computer to recognize individual speakers one by one. Perhaps it is possible to train the computer by giving it many examples of utterances and trying to arrange the circuitry so that the algorithm—in this case, the learning algorithm—can generalize from its experience, in the same way that I am able to understand you although you have an accent and I have never heard you before. But I don't have trouble understanding you, because I can generalize from other people I have heard who have similar accents. As I listen to you, I can tune up my auditory system in order to understand you better.

That adaptability is a task that our brain is very good at on all levels: from the early level of signal processing—there are circuits that adapt to the sound level, to the frequency bands, to the background noise. This happens in real time—as I am listening to you. There is adaptation also on the highest level of computation, having to do with the semantic components of understanding. As I understand that you use words in a slightly different way from how I do, I can adapt and compensate— again, all in real time.

I want to emphasize the factor of real time, because it really concentrates the mind. You have to solve the problem, because the signal is there now, and it is not going to be there anymore in a second. Your short-term memory is very limited; it is a very small window, and you have to be able to extract the information now or lose it forever. You can extract and save only a small fraction of all the information that is flowing in, and when conditions change, you have to adapt to them. And in particular, as information flows in, you have to extract the parts that are important.

You mentioned three notions: neural computation, PDP, and connectionism. Are these three different names for the same thing, or is there a difference of approach?

Different disciplines have used different words to describe similar approaches. Within psychology, the application of massively parallel computation of networks and of learning procedures has come to be known as PDP, or Parallel Distributed Processing, primarily because of an influential book that was published in 1986 by Rumelhart and McClelland.[3] Within the Artificial Intelligence community, this approach—how to program a massively parallel computer to solve difficult programs in vision and analogy engineering—has become known as connectionism, or the connectionist approach. There is an enormous amount of overlap, obviously, because there are similar algorithms and similar architectures, but perhaps a different emphasis. The emphasis in computer science is on solving practical problems, whereas in psychology the problem is to understand humans. Within the neuroscience community, there is yet another term that is used for massively parallel computation: computational neuroscience. Again, very similar, often identical algorithms are being applied now to the problem of understanding specific brain circuits.

What is common to all these approaches and what unifies the research workers who are exploring these problems from a mathematical, psychological, engineering, or biological angle, is the focus on this new architecture. This new architecture has, in a sense, offered an alternative to the von Neumann architecture as a basic model. It changes some of the fundamental assumptions about the way you approach the problem. As I mentioned earlier, the hardware, far from being irrelevant, could be the driving force. The explorations we are engaged in, looking at massively parallel architectures from all these different disciplines, are going to change the way we think not only about computation but even about the organization of other distributed complex systems at the level of societies or economies.

Why are the sciences coming together at this time to form neuroscience? Why now and not, say, fifty years ago?

Part of the reason is social and historical. Departments in the United States are set up along the European model, where you separate anatomy and physiology and so forth. That has been the case, and it continues to be the case even today. However, there are some problems that cannot be separated, for which you have to know about both the anatomy and the physiology in order to understand the function. Previously, experts in one or the other area had to collaborate, because no one person knew enough about both areas.

But today, because of the existence of the neuroscience society, neuroscience programs, and neuroscience departments, it is possible for students to become experts and to receive instruction on anatomy and physiology, and psychology, and pharmacology. All these areas are important if you want to study the brain. The major shift has been in the training and education of students, which is leading to a new generation of researchers who have at their disposal techniques from all these disciplines and who are, as a consequence, able to approach problems that were totally inaccessible before, such as: what is memory; what types of memory are there; and how is it accomplished in the brain? Which brain areas are storing which types of information, and how are they storing it? What are the mechanisms and how do they affect the way information is retrieved?

All these questions are now in the forefront of the research, and some of them could, in the next decade, actually be answered. We may know, within the next ten or twenty years, the fundamental neural mechanisms that form the basis of some aspects of human memory. Having that knowledge will have major implications not just for medical treatment of memory disorders but also for designing massively parallel computers that are built along similar principles.

There is a whole list of outstanding problems that no one can solve with conventional computers. I don't say that it cannot be done, but it is possible that you need so much computational power that it is just not feasible with conventional technology—unless you go massively parallel. And then the question becomes, Are there algorithms within the limits of the massively parallel architecture that can be used to solve the problem? Apparently, biology has discovered some, because we can solve the problems.

In the beginning I mentioned the turning point in Artificial Intelligence, which was the decision not to pay attention to neuroscience. This dogma is now being reexamined, first on technological grounds: we can build massively parallel computers; we can build chips with millions of transistors on them. Would it not be nice to use those transistors in a parallel way rather than a sequential way? How to do it is a research problem. We can build computers with sixty-four thousand processors today, maybe with a million processors in a few years. How do we program them? There is a technological force that is pushing us toward massively parallel architectures.

But there is another force, which may be even more influential in the long run: the neuroscience revolution—the fact that many of the biological mechanisms that before were totally out of reach and therefore could not be of any help for someone trying to understand cognition can now be studied. Some of these questions are going to be answered in a mechanistic way, in a way we can take advantage of in forming our models, our theories. It is going to be a shift of immense consequence.

Connectionism actually started in the sixties, but only now is it beginning to play an important role. Why this time lag?

I could give you two reasons. One of them is technological: it was not possible to build massively parallel architectures in the sixties. The second reason is that neuroscience was still in its infancy. One example: back in 1949 Donald Hebb[4] postulated that neurons could change the strength of coupling, the so-called synaptic strength. When the presynaptic neuron and the postsynaptic neuron were active together—when there was a simultaneous coincidence—then, he postulated, the strength of the connection should increase. That is now called a Hebbian synapse. That hypothesis has been around for forty years. But only in the last few years have neuroscientists been able to test it. What we have discovered in the last few years is that there are specific synapses in the hippocampus—a part of the brain that is intimately involved in memory functions—that display Hebbian synaptic characteristics. Here is a fact about the brain that we did not know in 1949, when it was only hypothesized. In fact, over the last twenty years, almost all neural network models that incorporate memory functions have used one or the other variation of Hebb's synapse as the fundamental local learning rule, without any experimental evidence whatsoever.

Now we have the experimental evidence. Furthermore, it is turning out that the Hebb's synapses in hippocampus have some very remarkable special properties, having to do with the temporal time scale. Those properties can be put back into the models and can change the performance of the models. The focus of my own lab is to try to combine the facts and insights from neurobiology with the models that are being investigated by the PDP and connectionist community, and to couple neuroscience into the research program in a way that was not possible in 1960.

Was this a gradual progression, or were there steps and stages such that now it is possible to make these experiments? Are there new technologies?

The intellectual development has been fairly gradual, but it is accelerating now. What has not been gradual is the policy. For a long time, the researchers who were actually building the field were doing it outside the major research centers and in isolated groups. But you need a critical mass and resources in order to be able to make major advances. And that has not been available—it is just becoming available now.

The answer to your question is that there has been a gradual increase in knowledge but that you need to reach a threshold before you can really succeed to solve practical problems. We are just reaching it now. In the last few years, we have computers that are fast enough to do the simulations, and we have hardware like analog VLSI chips that can deliver the architecture to the field in an efficient package. That may account for what appears, from the outside, as a major change. From the inside, I don't think there has been a major change; it is just the awareness and the number of people involved in the field that have increased. The actual ideas and the basic fundamentals have been developed much more gradually.

What is, in your opinion, currently the biggest problem that has to be overcome by cognitive science, Artificial Intelligence, and neuroscience?

The field as a whole—I am speaking now for all these areas—has to confront problems of scaling: most of the models that have been explored so far are small ones, typically having a few hundred to a few thousand processing units. To go from there to a million units will require some new research strategies. Scaling means also temporal scaling, that is, being able to design systems or multiple networks that are able to integrate information over time and, in the limit, approach the processing capacity that sequential von Neumann architectures are so good at.

In other words: what I envision happening over the next decade is a development in which computer science and neuroscience will become more unified rather than being separate, as it appears now. I look at them more as a continuum with two ends. There is going to be a gradual fusion toward the center as the problems of the approaches start to merge, so that the insights and strengths of both approaches can be taken advantage of and combined.

There will be a practical outcome, too?

There will be both practical and intellectual outcomes. First, it will be possible to build new generations of computers based on these principles. This is inevitable but will take time, because it is a new technology, and it is still being explored. For example, it will be possible to build chips that use principles of neural computation, that are lightweight and can be used in robots for seeing, hearing, and motor control. The higher levels of cognition will take longer to emulate with this architecture.

Let's look at this from the biological perspective of evolution. Stand back and ask yourself: where has the brain come from? In the case of mammals, the brain has evolved over hundreds of millions of years. But it has only been very recently, within the last few hundreds of thousands of years, that we have developed the ability to use language symbolically and to create artifacts that enhance our cognitive powers. Think of pencil and paper: it is an enormous leap over what you can remember! That has been only in the last few thousands of years—a pinprick in the history of time.

The focus of the research in Artificial Intelligence has been on that pinprick, on a tiny part of the biological evolution. In some respect, maybe the most difficult problems are not the ones which have been the focus of Artificial Intelligence—like chess, theorem proving, or logical deduction. The most difficult part of computation is perhaps the part that took hundreds of millions years to evolve: the sensory-motor representation; how we come up with a solution so quickly, and so on. Those might be the most computation-intensive problems where we have to wait the longest to come up with the right answers.

The second factor is the bias toward language. Given that in Artificial Intelligence there has been a focus on symbol processing, language, and logic, people have looked for intelligence in those terms. Take the Turing Test: it is stated in the form of a language problem, of querying another entity using language and using the answers as your assay for intelligence. Can you use that assay for a prelingual child? Everyone knows children can be incredibly intelligent without being able to speak, in terms of solving problems and dealing with the world in a very sophisticated way. Are you going to exclude prelingual children from intelligence because they don't pass the Turing Test? That is a mistake. But that opens up the more general question of intelligence in the animal kingdom, and especially the issue of consciousness, which by no means is a settled question. Some progress, at least, is made by recognizing that consciousness is not a binary predicate but a graded continuum. There are degrees of consciousness—even amongst humans, by the way, and it is true among animals as well.

One of the things you come away with after thinking about the problem of intelligence from a biological perspective is that we may be myopic in our insistence in identifying intelligence with language. Not everybody has that opinion, but nonetheless it has biased the way we think about the problem. Perhaps with a new type of architecture we will gradually come up with a new perspective on intelligence that has more to do with the interaction of internal representations with the real world in real time. Real time is a part of the equation that is essential on the biological side. This constraint on intelligence has not yet been fully considered.

Will this new research and technology lead to new social impacts?

Any new technology carries with it the problem of readjustment. Society has to readjust right now because of the widespread use of computers. If in fact people were programmable Turing machines, this would not be a problem—you put in a new tape. That illustrates that human beings are not Turing machines but creatures created by genes and experience. The reason why you cannot change a typist even from one keyboard to another overnight is that we cannot do arbitrary mappings. We can do only some mappings, and these have to be learned.

One of the strengths in the new research area is that the circuits we are dealing with here are fundamentally adaptable. In principle, they should be able to adapt to human beings, and not the other way around. In the past, you had to train human beings to adapt to the machine. The machine was the constraint. Here is a new concept that could shift the balance: if the machine itself can adjust to the human, you can imagine, in the future, machines that are personalized, that adapt to your particular circumstances, your individual handwriting, your personal way of saying things. Similarly, there are many problems in manufacturing and control, in running large businesses, that would look very different if the process, the organization itself, were adaptable. That is a fundamental way in which the brain works, and I don't see why it should not be incorporated in machines as well.

Recommended Works by Terrence J. Sejnowski

(With D. Ballard and G. E. Hinton.) "Parallel Visual Computation." *Nature* 306 (1983): 21–26.
(With P. S. Churchland.) *The Computational Brain*. Cambridge, Mass.: MIT Press, 1992.
(With P. S. Churchland.) "Neural Representation and Neural Computation." In *Mind and Cognition*, ed. W. G. Lycan, pp. 224–51. Oxford: Blackwell, 1990.
"Perspectives on Cognitive Neuroscience." *Science* 242 (1988): 741–45.
(With C. Koch.) "Computational Neuroscience." *Science* 241 (1988): 1299–1306.

Notes

1. See J. E. Laird, A. Newell, and P. S. Rosenbloom, Soar: An Architecture for General Intelligence (research report, Carnegie-Mellon University, 1986); and J. Michon and A. Anureyk, *Perspectives on Soar* (Norwell, Mass.: Kluwer, 1991).

2. J. Lazzaro and C. Mead, "A Silicon Model of Auditory Localization," *Neural Computation* 1 (1989): 47–57.

3. D. E. Rumelhart, J. L. McClelland, and the PDP Research Group, *Parallel Distributed Processing: Explorations in the Microstructure of Cognition*, 2 vols. (Cambridge, Mass.: MIT Press, 1986).

4. D. O. Hebb, *Organization of Behavior* (New York: Wiley, 1949).

HERBERT A. SIMON

Technology Is Not the Problem

My original area is organization and management. I approached organization and management from the standpoint of decision making. I had training in economics and understood the economist's model of decision making. But that seemed to me very far from what was going on in organizations. So I tried to develop alternative theories, theories that I generally label by "bounded rationality." This is the view that people are not as rational as economists think. That led me into trying to understand human problem solving, because it is part of what is involved in decision making. This research started back in 1939 and went on through the forties and fifties.

Meanwhile, I had contact—almost by accident—with early computers. I used punched card equipment for various research activities. I got fascinated, and so I kept my ears open to what was happening. I heard about those wartime computers, and by 1952 I had learned how to program a computer. But that was just for the interest in it. In 1954, I began to be interested in the possibility of chess playing by computers. There had already been attempts: Turing had published an article; Shannon had also published an article; von Neumann talked about how hard it was. About that same time, I met Allen Newell.

Gradually, we began to think that it would be a practicable research project to write a program that would do intelligent things. I wrote a little paper about chess programs, about how to do combinations. Newell wrote a more elaborate paper published in 1955 on chess playing programs,[1] but we had never programmed one.

Then he came back here (to Carnegie-Mellon University) to get his doctoral degree. We started working together in 1955. At first we thought of building a chess machine, but for a variety of reasons it turned out that the first thing to do was build a program that would discover proofs in symbolic logic. We worked with a third person— J. C. Shaw. We saw that there was no way of programming this without

Herbert A. Simon received his Ph.D. from the University of Chicago in 1943. He has also received numerous honorary doctorates from universities all over the world. He was Assistant Professor and then Associate Professor of Political Science at the Illinois Institute of Technology, before serving as Professor of Political Science at that institution from 1947 to 1949. From 1949 to date, he has been at the Carnegie Institute of Technology, as the Richard King Mellon University Professor of Computer Science and Psychology at Carnegie-Mellon University. The importance of his work was acknowledged by the receipt of the Nobel Prize in Economics 1978.

the appropriate languages. So the ideas of list processing gradually emerged.

By Christmas 1955, we had a program called LT, the Logic Theorist, which is capable of discovering proofs for theorems by heuristic search. Within a month, we had the basic ideas for list processing. By August of the next year, 1956, we had a running program.[2]

When we actually began in 1957 to compare the way this program did its business with protocols we had been gathering from human beings, we saw that human beings use means-end analysis. That gave us the idea for the general problem solver (GPS), which was built between 1957 and 1959.[3]

Do you think that the idea of the GPS and your concern about human problem solving have anything to do with what is called the "cognitive revolution" at the same time, which is usually described as a reaction to behaviorism?

That had not started yet. You might say that we started it. About the same time, people who, out of wartime experience, were doing human factors research were going back to higher mental functions. George Miller[4] is an example in the United States, Broadbent in England.[5]

So behaviorism was still the leading school?

Behaviorism was dominant in the United States until the seventies. In the book by Miller, Galanter, and Pribram—*Plans and the Structure of Behavior*[6]—they almost apologize to behaviorism for talking about the mind. The word *mind* was almost a dirty word. They came out of psychology, so they had to be worried about the response. In our first psychological paper in *Psychological Review*, in 1958,[7] we spent a lot of time arguing that what we were doing was just a natural continuation of behaviorism and Gestalt psychology. We still had to make these arguments until the seventies.

In your opinion, when did cognitive science take off?

Artificial Intelligence came first. Artificial Intelligence is often dated back to the Dartmouth Conference in 1956 or to the IRE meeting at MIT in September 1956.[8] Allen and I went to Dartmouth and presented our LT. It was the first running program; nobody else had one.

The term *cognitive science* really grew up in the seventies, when people began to take more seriously the application of Artificial Intelligence to psychology. The word *cognitive* got back into psychology through a book called *Cognitive Psychology* written by Neisser, in the middle of the sixties.[9]

Most of us now make a fairly consistent distinction between Artificial Intelligence and cognitive science, in terms of the purpose of the research and the way research is tested. Either the purpose is to make computers do intelligent things—that is Artificial Intelligence. Or else one might be interested in how human beings think, and one would use the computer as a model of human thinking, which is to be tested against real instances of human thinking. That is what I am concerned with, and that is what is called cognitive science. I have done both, but I don't really care whether we have more efficient computers in the world. On the other hand, I do care for understanding what the human mind is like.

In your opinion, then, building intelligent machines has nothing to do with cognitive science?

It is not exact to say that they don't have anything to do with each other, because, in fact, over the last thirty-five years, there has been a very useful flow of information back and forth between the two directions. When we started out, the first thing we had to do was build appropriate programming languages. We built the first list processing languages. In doing that, we borrowed ideas from associative memory in psychology.

Similarly, today we use certain techniques in our computing, and then we ask whether they have psychological counterparts. Production systems, for example, first arose in a computing context, and now we think that they are a good way to describe languages for modeling human thinking.

The word cognitive science *exists in German, but there has been, until now, no established interdisciplinary research program behind it.*

There does not have to be a real thing for every noun—take *philosophy*. The reason why the term *cognitive science* exists over here and why we have a society is that people found it convenient to have conversations across disciplinary boundaries: psychology, philosophy, linguistics, and Artificial Intelligence. A few anthropologists are wandering in now. Cognitive science is the place where they meet. It does not matter whether it is a discipline. It is not really a discipline—yet. Whether cognitive science departments survive or whether people want to get their degrees in psychology and computer science remains to be seen. Here (at Carnegie-Mellon University) we decided that we will still give degrees in computer science and cognitive psychology, not in cognitive science. These are just labels for the fact that there is a lot of conversation across disciplines.

In Europe, labels are treated like boundaries.

We have this danger here, too. Nouns are tyrants. People think that if there is a noun, there must be *an* idea behind it. The world is much more fluid than that.

What is the idea behind cognitive science and Artificial Intelligence?

First, we would like to know what intelligence is. Second, we would like to know how intelligent we can make computers. And third, we would like to know what it is that makes human beings able to be sometimes intelligent. I like to define the fields in terms of the questions we want to answer.

What do you think about recent developments in Artificial Intelligence and cognitive science? I am thinking about connectionism here: you are famous in the field for the physical symbol system hypothesis. In some of the interviews, I heard strong criticism of this hypothesis. What would you say about this? Is there a challenge—can the two be combined? Would you say that connectionism is also a part of cognitive science?

I think that extreme connectionists, those who want to do without symbols or rules, are wrong. Let's go back to the human mind: there are certain aspects of human mental operation that are obviously parallel. We can prove it. We know that the retina is a parallel device. Light can hit different parts of it and the impulse is sent up to the brain. We know that in areas 17 and 18 of the cortex there are essentially topological images of the retina. There are reasons to believe that most of the feature extraction from that is done in parallel.

On the other hand, there is equally overwhelming evidence that important things in human thinking are serial. I once made a man who was driving a car steer into a snowbank out in Iowa by asking him to perform a task that required visualization while he was driving the car. And he did not succeed in doing that very well. We know that most of the time, for any task requiring attention, human beings can do only one thing at a time, or one and a half.

What do you think about practice? When one is practicing and not even thinking about what one is doing, for example, when I am used to driving a car, it is almost automatic.

Not at all. I just have to get you into thick traffic, and the conversation will die down immediately. Either that, or you will be in a wreck. We can do a certain amount of time sharing when the demands for reaction are not frequent. But the idea that the human being is, at the level of conscious thinking, very parallel flies in the face of everyday common facts. It was said about President Ford that he could not walk and chew gum at the same time. That might be a little extreme. But the evidence points to a bottleneck of attention. We can have marginal quarrels about how narrow or how broad this bottleneck is.

We have to explain how a system that is highly parallel on its input devices and maybe on some of its output devices is highly serial at some place in between. It is primarily in the serial part of the operation, and maybe in part of the parallel, too, that the thing seems also obviously symbolic. I welcome the connectionists, and I wish they would put a lot of their attention into organs like the retina—how it works, how we extract features, and how we go from the nonsymbolic to the symbolic mode of representation. I get very bored with them when they try to go all the way and forget about all these other evidences.

In your opinion, then, one could combine the two: at the lower level, like pattern or speech recognition, connectionist models could be useful; but on the higher level—speech production, understanding, reading—we need symbolic models?

Yes, that has been my view long before connectionism. In fact, connectionism goes way back—to Hebb, and to Rosenblatt's perceptrons.[10] We have had that alternative model from the very beginning.

But only during the last few years it has become more powerful.

It is the first time that they have the chance to really do it, that they have computers with which you can do anything interesting. You need a lot of computing power. My kind of Artificial Intelligence you can almost do on a PC.

What do you think about challenges like the one from Searle, articulated in his Chinese Room Argument—that you cannot get to semantics from syntax?

The mistake is in the first premise: that what computers do is somehow rather syntactical as distinguished from semantic. Suppose computers have input organs. And suppose the input organ on a particular computer is such that when you point at a tape recorder it says, "Tape recorder," and it does not say that when you point at something quite different. I don't know of any criterion of understanding that would say that *I* understand what a tape recorder does, but the machine does not. It might not know what makes it run—well, I don't know, either.

What I took from Searle's Gedankenexperiment *is that you cannot explain it very well from a third-person view but that you should imagine you are inside the machine.*

He described the wrong machine. He described a machine that got a Chinese sentence, looked it up in a dictionary, word by word, took the English equivalent and wrote that down. I can do that, too, with a language I don't know. I could probably read Turkish that way, but I would not understand Turkish—I agree so far. But all this proves is that Searle's machine does not understand Chinese. It does not prove that *my* machine does not understand Chinese, because my machine is not only able to say a Chinese sentence about a cup of water but also to say whether this is true of *this* cup of water by looking at it.

So it has a relation to the world.

What Searle's machine lacked was a semantics. But to say that it is impossible for a machine to have semantics is ridiculous, because human beings who are machines have semantics. Semantics simply means that you have definitions connecting things that come in through your senses with the stuff that is going on inside.

I cannot take it as a first premise that computers don't have intentions. Denotation is an intention, and designation is an intention. The minute I have a program that allows me to recognize this thing as a tape recorder, then I have intentions, and so does my computer.

What do you think about the so-called commonsense problem and the notion of a nonrepresentational background?

The latter is an assertion. But I don't know of any scientific evidence to show that it is true. I think it is false. There is a lot of evidence that the kind of thinking observed in human beings can be done with the kinds of mechanisms I observe in computers. This is one reason why we worked a lot on systems that do their thinking by recognition, because all the evidence I see about human intuitions are precisely the phenomena of recognition. We can reproduce recognition in computers without any trouble at all. A computer that does medical diagnosis does it by recognition of symptoms and by associating these symptoms with what it has learned about disease entities.

We have done all the work on scientific discovery to explore creativity.[11] Our whole strategy has been to take up tasks that, if you saw a human being perform them, would guarantee that the human being was using intuition or was being creative. The question is what it would take a computer to do that, so that if you apply the same criteria to the computer you would have to agree that the computer is behaving intuitively or creatively.

What about bodily movement? Humans don't have problems with this . . .

If you are saying that we have not progressed nearly as far with computer motion as with human motion, I would have to agree with you. But that does not mean that we should romanticize about what humans can do. I had a problem with my knee last year and went to see a physician. He told me to walk a straight line with my eyes closed: I could not do it. Let's not romanticize about what human beings can do!

Let's take typing. I know typing "by heart"—I learned it; I don't have to think about it.

We would say that you are executing productions.

Okay. Now I am using an English keyboard, and I know it. But still I try to hit an umlaut key where there is none. How would you explain that experience?

In a very simple way. We would have in the program for a simulation of human behavior like this a set of productions. A production is a mechanism—when certain conditions are satisfied, certain actions are taken. You are typing from a manuscript, and you have a set of productions in which you are well trained. As soon as the condition is satisfied—there goes the action. You have a well-trained production when you see an *ü*; down goes your finger. John Anderson's spreading activation models deal with this very specifically.[12]

So tacit knowledge or knowledge that you could not make explicit is not an argument against your theory?

Not at all. A friend comes down the street, and I recognize him as Joe. I cannot give you a veridical account of the criteria used to recognize him. My own model of that is a program called EPAM (elementary perceiver and memorizer),[13] which picks up the features my eyes present to me, searches them down a discrimination net, and arrives at Joe. In the work we have done on protocol analysis,[14] we have a lot of evidence on what you can and cannot report about your mental experiences. One thing the evidence is very consistent on: you can report who or what you recognize, but you cannot report the sequence of tests you perform in order to make the recognition. There is nothing about our EPAM program that says that what it does is conscious. Consciousness is a separate issue. What our program does assert is that what is going on is a series of differential tests on features that allow different things to be discriminated.

So one can report about the results but not about the underlying neurophysiology and process.

An awful lot of things are going on here that I cannot tell you about. You should not even try. That is where introspection at the turn of the century had its great failure. It did not merely try to use verbal reports as data, as we do nowadays, but proposed that you could "see" inside. That was futile, because a lot of that is not available to sight.

In your opinion, the so-called commonsense problem is not a challenge or a difficulty for your theory?

Not at all. I have heard the argument. Dreyfus has been working the field for thirty years now, and, honest to God, he does not understand what goes on in computer programs. Searle does, but Dreyfus does not. He is just not a serious person.

Why are there so many problems to solve the knowledge representation or commonsense problem? There are a lot of attempts—scripts, frames, and so on—but they never got around the problem that you have to incorporate a huge amount of data.

Until recently, there have not been the manpower and the large computers to do it. Douglas Lenat has a ten-year program of building a huge semantic memory (CYC).[15] Then we will see. All our studies about chess indicate that a chess grandmaster has stored at least fifty thousand chunks. When people start to build programs at that magnitude and they still cannot do what they are supposed to do, *then* we will start worrying.

That is like saying that we won't have a theory of DNA until we have deciphered the whole human genome. Why isn't the evidence that we can do it for certain task domains perfectly good evidence that we understand the processes?

So your view is that you are on the right way, that we just have to build better and faster machines.

What we need is enough money and time. We need people like Lenat who invest ten years of their life to build this semantic memory.

Does that resolve the problem with questions like "Do doctors wear underwear?" or "Are station wagons edible?"

You don't have to have that sentence in memory. All you have to have in memory is that most people in our culture wear underwear. Many of the answers can be arrived at by very simple one- or two-step inferences. The way I get at "Vehicles are not edible" is by looking at my schema for vehicles, where I see that they are made of metal, and by looking at my schema for foods, where I see that they are made of carbohydrates, fats, and proteins, and I do a match . . . no big deal!

Most of the programs that try to simulate semantic memory have in them a so-called inheritance property, so that you can make special inferences from the more general ones.

Searle uses these questions to express his disbelief that you do inferences every time you see a station wagon, sit down on a chair, step through a door, and so on. Because every time you want to sit down, you would have to make inferences to conclude whether this particular chair will support your weight.

Evidently Searle doesn't understand how production systems can be used to deal with default values. I enter his office. He says, "Please sit down." I notice an object that satisfies the recognition conditions for "chair." Nothing about the object evokes the production "If a sitting place looks fragile, test it before you sit on it," because there is no cue in the situation that causes recognition of "fragility" (a node in the EPAM net). Hence the condition of the production is not satisfied, and it is not fired.

When Old Mother Hubbard went to her cupboard, as the nursery rhyme says, she found it bare. Did she have to test, "No flour, no sugar, no canned peaches, . . ." ad infinitum? Of course not. She simply did not notice any objects sitting on the shelves. None of the productions for recognition of the nonexistent items was evoked.

What do you think about the relationship between theory and experiment? Do scientists in Artificial Intelligence make more experiments, whereas cognitive science is more theoretical?

You cannot slice it up that way. The theorists tend to be philosophers; that is all they know how to do. Linguists to some extent, although linguists have that peculiar form of experiment of taking single sentences as counterexamples to anything you say. That is highly empirical, but in a peculiar way. Their standard of empiricism is quite different. Psychologists are tremendously empirical, and, by and large, even cognitive psychologists are suspicious of theory. Even in this department, you can find colleagues who would say that there is no such thing as a theoretical psychologist.

In Artificial Intelligence the dominant tradition has been an experimental tradition, a tradition of building big systems and seeing what they can do. There is a group of people who are rather theoretical. John McCarthy is a member of that group; Nilsson at Stanford Research Institute (SRI) is another example. There are both kinds, and there is honest disagreement about how far you can go each way.

I happen to look around a lot in chemistry and physics, and I see that for all the power of their theories, most of the time most of the people, when they actually want to deal with real-world phenomena, wallow around in experiments. I don't see theory doing anything great for the people who are trying to do high-temperature superconductivity. I don't see theory doing anything great for the people who are trying to do cold fusion. It occurs to me that even in a highly developed science like physics, experimentation plays a tremendous role.

Most of my work in Artificial Intelligence and cognitive science has been not experimental but closely tied to experiments. My idea of a theory is a model of the phenomena, not a bunch of mathematical theo-

rems. Mathematics is very powerful for finding out about highly abstracted, conceptualized systems and has to be supplemented with a lot
of messier techniques to deal with the real world.

*What do you think about the Turing Test? Do you think it is a useful test
idea?*

As a broad conception—that you compare the behavior of the computer with the behavior of the human being—yes. It is a principled way
of finding out. You might almost say it is the method of science. If I have
a set of differential equations, and if I think that those differential equations are a good theory of what the planets are doing, then I better record
the movements of the planets and compare them with the movements
predicted by my differential equations.

A computer program, formally speaking, is simply a set of difference
equations, in the sense that the next state of the computer will be a function of its state right now, just one cycle later. So we build our theories
as computer programs, as difference equations. Then we compare them
with human data. We want to compare them as close as we can, so we
often use thinking-aloud protocols as a source of data. Our problem of
theory verification is no easier and no harder than in any other science.

*Some people argue that you cannot take the Turing Test seriously, be-
cause it comes from the behaviorist point of view, and cognitive science
has to look at the underlying process and not at the interpretation from
the outside.*

I don't know any way of "looking at the underlying process." The
neurophysiologists have not told us how yet, and it will be a few years
before they do. So all I can observe is behavior. All I know about you is
what I see from the outside. That is all we know about human behavior,
and we have the same problem as with all those planets. We have the
same problem of inferring what goes on inside the box from what is
going on outside.

That is not behaviorism. Behaviorism was a theory saying that we
should satisfy ourselves with laws of the form "the following input leads
to the following output" and that we should not inquire about what is
inside the box.

*But it would be possible, even for a very simple machine, to "cheat" on
the Turing Test, and succeed.*

Not at all. The test for this would be whether you could write a program for a simple machine that would in fact match all the laboratory
data. Just try it! The people who are skeptical about Artificial Intelligence flop from one side to the other. On the one hand, they say that the

human being is so complicated that nobody could build a machine capable of doing commonsense reasoning and of exhibiting the behavior of commonsense reasoning. In the next breath, they tell us, Well, if all you have to do is match the behavior, you just write a simple program. If that is true, then the first statement is false.

I refer to people who are trying to slice up the capacities: story telling, visualization, pattern recognition, and so on. They say that at the moment, we could build machines that could satisfy only this criterion or that, but not all of them.

I think they are quite right. Again, people who object to this have a completely unrealistic view of what science is about. They somehow seem to imagine that physics, for example, is a field with a great theory and that theory predicts everything. There is no such thing: physics is a patchwork of little theories. If I want to talk about elementary particles at the nuclear level, I use completely different theories from when I want to talk about elementary particles at the level of quarks.

There is some sort of underlying conceptual unity, as there is in computer science. But the idea that there is *one* program, *one* set of difference equations that cleans up the field, is nonsense. The people who talk that way cannot possibly have ever studied physics.

What do you think is, at this moment, the state of the art in Artificial Intelligence and cognitive science? How would you situate the different flavors and directions, and what are the major problems to overcome?

The strategy has been to take some human tasks and write programs for them. When you think you understand that class of human tasks, you take some tougher ones. So we started out with the Tower of Hanoi puzzle and things like that, and more recently we have been doing high school geometry and physics and scientific discovery. We ask what kinds of tasks we have not modeled yet, and we push forward.

Where is the excitement nowadays? Well, in a couple of places. First, can you build a program that can start out with a verbal or pictorial description of a problem, build its own representation out of it, and solve the problem? And can you do that for serious problems? The first big advance in this was a program now fourteen years old: Gordon Novak's ISAAC program,[16] which takes problems out of a physics book written in English, constructs an internal representation, draws a picture of it on the screen, writes the equations for the problems, and solves them. It does translation from a verbal to a pictorial to an algebraic representation.

There are other things that have not been done. We have just completed some work on the "mutilated checkerboard" problem, an old puzzle about how to cover part of a checkerboard with dominoes. People think it is a hard problem, because in order to solve it, you have to change the representation. A lot of people are now working on where representations come from and how they change.

A lot of people are also working on the related problem of what diagrammatic and pictorial representations are about. Why are they a good thing? Why do they facilitate thinking?

And a lot of people are working on the commonsense reasoning problem, and something related to it, called the qualitative reasoning problem. For example, if I tell you that yesterday there was a freeze in Texas, you would not be surprised if I told you next that the price of grapefruit is going to be high next winter. The way you reason about this is, Well, they grow grapefruit in Texas, and a freeze will reduce the grapefruit crop, so if you have a smaller crop and the consumer still wants grapefruit, then the price will go up. We have not used any numbers, we have just talked about more or less. That is qualitative reasoning.

Let me ask another question. You said that when you want to solve a task or problem, the most important thing is to try to program it, and you will understand the model. Now, one could imagine a machine that wins every game of chess by pure brute force. But we know that human beings don't think that way.

So that machine would not pass the Turing Test, because it would not fit the behavioral evidence, which shows that no human being looks at more than a hundred branches of the tree before making a move. If we want to have a theory of human chess playing, we would need a machine that also matches that piece of evidence.

You can find out from the behavior of human chess players—and people have done it—that they are not computing more than a hundred positions. So we have behavioral evidence. If a computer is to be a model of human chess playing, it has to behave in such a way that it is not looking at more than a hundred. In fact, there is such a program. It plays chess only in tactical situations.

So the program with brute force would not be Artificial Intelligence?

It would be Artificial Intelligence, but not cognitive science. Take the best chess programs—Deep Thought, for example: it has just played two games against an American grandmaster, lost one and won one. It does not tell anything about how a chess grandmaster thinks, or very little.

What is the practical outcome of Artificial Intelligence and cognitive science? Or, to put it in other words: is there any? I heard, for example, that expert systems did not succeed very well economically. What are the reasons for that?

I am not in touch with that field at all. I think that is an exaggeration. Expert systems became a sudden fad, and of course lots of people tried lots of silly things that would not work. But expert systems have been around since the fifties.

But they do not diffuse like computers.

How do you know they don't? Let me tell you about an expert system that has been around since about 1956. At Westinghouse, they design motors, generators, transformers. Some of these are big, fancy devices, but some of them are small, run-of-the-mill devices that are nevertheless not sold off the shelf but built on customer specifications. Since 1956, there have been programs at Westinghouse that will take a customer's specifications for an electric motor, look into its memory to find a design that is similar to that motor, pull out these designs, pull out a lot of data about heating, transmission, and so on, and then redesign that motor for the customer's order, following almost the same procedure that engineers would follow. That has been done every day for thirty-three years. No one ever thought of calling it an expert system.

So in your opinion there are machines doing Artificial Intelligence, but not in a spectacular way.

In every architecture or construction firm today there is computer hardware that does most of the mechanical drawings of the buildings and stores various kinds of information about the buildings in its memory. One reason that it is done by machines and not by people any more is that computers know where they have put a wire through a pipe, whereas people make the drawings all the time without knowing it. Not only that, when you change from a number 2 bolt to a number 3 bolt between two I beams, the program will remember that and change the purchase requirements for bolts.

What is your view of the next five years in Artificial Intelligence?

That we will do a lot more of what we have been doing.

There is no big obstacle or barrier? Everything is going smoothly?

It is like in most fields of science. You push ahead, there are little ideas, and you solve new problems. Sometimes, these gradually change things around, the way DNA research did. But most science consists of taking a basic set of ideas that has not yet been exploited. The basic ideas of Artificial Intelligence have been around for only thirty years, so there are still areas to exploit.

So you are optimistic?

Not optimistic—realistic, I call it. That is what has been happening.

*I have heard many times that the early rate of progress in Artificial Intel-
ligence has at least slowed down, if it has not stopped altogether.*

That is a myth that Dreyfus started with his first essay, which was
mostly reprinted in the *New Yorker* magazine in 1970 or 1971, and later
in his book.[17] It got into the literature, and people accepted it. I don't
know a shred of evidence for it.

Look at the publication rates and you will see, on the contrary, a slow
steady growth, which has mushroomed in the last five years. So where is
the evidence for stagnation? Look at the programs in the year 1990,
compare them with 1985 and with 1980, compare them with 1975, and
you will see that there are new things every five years that we could not
do before. The stagnation is a myth.

It is a field that could be compared with molecular biology. There are
a tremendous number of complex phenomena to be explained, so people
are off into different parts of the forest wandering around, and we know
a lot more each year than we knew the year before. That is why it is such
a fascinating field. I have had the greatest time in my life during the last
thirty-five years.

*A number of social scientists are concerned about the social issues
implied by Artificial Intelligence. What do you think about them?*

I think everybody in the field should give concern to this. Everybody
in a technical field should be concerned with where it leads. Artificial
Intelligence is a lot like the invention of writing, or the invention of the
steam engine, in the sense that it does not have just one particular narrow application but penetrates everything, just as energy does. That is
why the steam engine was so important—everything can make use of
energy.

Everything can make use of intelligence, everything can make use of
writing. For that reason it is obvious that Artificial Intelligence, like writing and like the steam engine, can be used for good and bad purposes.
Therefore the task is not to predict whether it will be good or bad but to
design and modify our institutions so that we get as much of the good
uses as we can and avoid as much of the bad uses as we can.

Let's take the question of privacy as an example. Artificial Intelligence
gives some opportunities, even at a rather stupid level, for invasion of
privacy. You can take a vast mass of messages and screen them for key-
words; you can filter them down and focus on the interesting things.

There are some good applications for this. For example, I would not
have to read the *New York Times* anymore, because I could get my com-

puter to filter it and just pick out the five things each morning that I want
to see. So I can use the same thing either to invade your privacy or to save
myself from being killed by the *New York Times*. Well, we have to design our institutions so that they use it one way but not the other. For
example, we may have to encrypt the messages that go through the wires
so that they cannot be filtered.

*So it is a matter not of the inherent structure of a machine but only of its
use. What about working conditions? People are afraid of the computer
taking away their jobs.*

The answer to that is what you think about the Industrial Revolution.
Would you rather it not have happened? I don't mean that there were
not a lot of horrors that went with it, because there were. But would you
rather have a world in which the Industrial Revolution never happened?
Most of us, faced with that question, prefer that it happened. We don't
want to go back to being sixteenth-century peasants.

The main thing computing does, if it is successful, is raise the level of
productivity. So you have to ask if you can have full employment with a
higher level of productivity. We already know the answer to that: we
multiplied productivity by twenty-five or more in the Industrial Revolution, and we still can employ everybody, provided we worry about our
governmental and monetary institutions. It is not technology that determines employment in a society.

There are all sorts of human wants in this world, even in the West,
that are not being met. The United States are having great struggles with
Congress to have the defense system we think we want, the medical system we think we need, and so forth. One of the richest countries in the
world feels poor. That means that we could use much more productivity
than we have. Not only that: we would have to change our productivity
so that it does not destroy our environment. That is going to be even
more costly.

The idea that increased productivity is a menace is a wrong idea. The
idea that we have to worry about who gets hurt while these changes take
place is another matter. If we keep employment fairly high, then most
people are not going to be hurt.

What has caused the greatest migration in the United States in the last
fifty years? Air conditioning! Air conditioning moved tens of millions
from one part of the country to the other. There was no Sun Belt before:
nobody in his right mind would have lived in Texas until air conditioning came along, much less in Arizona. Now, what hardship did it cause?
They all thought it was a great thing. The migration was completely
voluntary.

The fact that jobs change is not necessarily frightening for people. It is often an opportunity, not always. What we should be worrying about is introducing not technology but institutions that make it possible for people to adapt to change. If people think they are making the change themselves, so that they say "*I'm* going to Texas" or "*I'm* going to California," there is no social problem. Only if people think that change is being imposed on them does it become a problem.

Recommended Works by Herbert A. Simon

Reason in Human Affairs. Stanford: Stanford University Press, 1983.
The Sciences of the Artificial. Revised ed. Cambridge, Mass.: MIT Press, 1981.
(With K. A. Ericsson.) *Protocol Analysis: Verbal Reports as Data*. Cambridge, Mass.: MIT Press, 1984.

Notes

1. A. Newell, "The Chess Machine: An Example of Dealing with a Complex Task by Adaptation," *Proceedings of the 1955 Western Joint Computer Conference*, pp. 101–8; and A. Newell, J. C. Shaw, and H. A. Simon, "Chess-Playing Programs and the Problem of Complexity," *IBM Journal of Research and Development* 2 (October 1958): 320–25.

2. See A. Newell, J. C. Shaw, and H. A. Simon, "Empirical Explorations of the Logic Theory Machine: A Case Study in Heuristics," *Proceedings of the 1957 Western Joint Computer Conference*, pp. 218–30.

3. A. Newell, J. C. Shaw, and H. A. Simon, "Report on a General Problem-Solving Program" in *Proceedings of the International Conference on Information Processing* (Paris, 1960); and A. Newell and H. A. Simon, "GPS: A Program that Simulates Human Thought," in *Computers and Thought*, ed. E. A. Feigenbaum and J. Feldman (New York: McGraw-Hill, 1963).

4. See, e.g., G. A. Miller, "The Magical Number Seven, Plus or Minus Two: Some Limits on Our Capacity for Processing Information," *Psychological Review* 63 (1956): 81–97; and "Some Psychological Studies of Grammar," *American Psychologist* 17 (1962): 748–62.

5. E.g., D. E. Broadbent, "A Mechanical Model for Human Attention and Immediate Memory," *Psychological Review* 64 (1957): 205ff.

6. G. A. Miller, E. Galanter, and K. Pribram, *Plans and the Structure of Behavior* (New York: Holt, Rinehart, and Winston, 1960).

7. A. Newell, J. C. Shaw, and H. A. Simon, "Elements of a Theory of Human Problem Solving," *Psychological Review* 65 (1958): 151–66.

8. For these "landmarks" in the history of cognitive science, see H. Gardner, *The Mind's New Science: A History of the Cognitive Revolution* (New York: Basic Books, 1985), especially chap. 3.

9. U. Neisser, *Cognitive Psychology* (New York: Appleton-Century-Crofts, 1976).

10. D. O. Hebb, *Organization of Behavior* (New York: Wiley, 1949); and F. Rosenblatt, *Principles of Neurodynamics* (New York: Spartan, 1962).

11. See, e.g., P. Langley, H. A. Simon, G. L. Bradshaw, and J. M. Zytkow, *Scientific Discovery* (Cambridge, Mass.: MIT Press, 1987).

12. see J. R. Anderson, "Spreading Activation," in *Essays in Learning and Memory,* ed. J. R. Anderson and S. M. Kosslyn (New York: Freeman, 1984).

13. See H. A. Simon, *Models of Thought* (New Haven: Yale University Press, 1979).

14. K. A. Ericsson and H. A. Simon, *Protocol Analysis: Verbal Reports as Data* (Cambridge, Mass.: MIT Press, 1984).

15. D. B. Lenat, "When Will Machines Learn?" *Proceedings of the International Conference on Fifth Generation Computer Systems* (Tokyo: Institute for New Generation Computer Technology, 1988), vol. 3, pp. 1239–45; and D. B. Lenat and R. V. Guha, *Building Large Knowledge-Based Systems: Representations and Inference in the* CYC *Project* (Reading, Mass.: Addison-Wesley, 1990).

16. G. S. Novak, "Representation of Knowledge in a Program for Solving Physics Problems," *Proceedings of the Fifth International Joint Conference on Artificial Intelligence* (1977): 286–91.

17. H. L. Dreyfus, *What Computers Can't Do: The Limits of Artificial Intelligence* (New York: Harper and Row, 1972).

JOSEPH WEIZENBAUM

The Myth of the Last Metaphor

I was introduced to computers in the early fifties, at Wayne University (now Wayne State University) in Detroit. A mathematics professor there decided to build a computer for the university. So we did build one. I was a graduate student then. We had to do everything: soldering, all the logic design, and the assembler language. That is how I began with computers.

After a long while, I participated in the design of a computer system for the Bank of America. That was in the late fifties and early sixties. The actual site of the work was Palo Alto, California, because the headquarters of the Bank of America is in San Francisco, and the Stanford Research Institute had done some preliminary work that we had to know about. That was, of course, very near Stanford University. I already knew Ed Feigenbaum, who was then at Berkeley.

When the design of the bank system was over, the General Electric Company kept many of us who had worked on it on, although we really had nothing to do. I began to work on a list processing system, later called SLIP. I was very interested in the use of the computer in socially significant areas. I also got to know Kenneth Colby, a psychoanalyst in the Palo Alto area. He was working on a simulation of the psychoanalytic process. So we talked a lot about that. I did develop that list processing system, and I got an honorary appointment (which is to say, no pay and no parking place) at the Stanford Computation Center. The interaction with the people there, and my general interest—at the time, list processing was a new thing, and the people most interested in it were psychologists (Feigenbaum himself had done work in psychological modeling)—drew me into the field.

In 1963 I was called to MIT. The reason I was called there was to participate in a project then called MAC (machine aided cognition). There we developed a system for time-sharing computers, which means simply that a number of people—eventually a very large number of people—could have consoles attached to a single computer and could use

Joseph Weizenbaum was born in Berlin in 1923 and emigrated to the United States in 1936. He studied mathematics at Wayne University, Detroit, served in the United States Army Air Corps from 1942, and then resumed his study after the war. At Wayne State he was involved in designing and producing a computer before moving into industry in 1953. In 1963, he joined MIT as Professor of Computer Science, a position he held until his retirement in 1988.

the console as if they had the whole computer. That was the basic idea. This made it possible for the first time since the very beginning, when people actually *had* the whole computer to themselves, to interact with the computer conversationally—that is, you do a little programming and try it out, as opposed to having to write a program, punch a lot of cards, give the cards to a computing center, and get the results back perhaps the next day or later.

Under the influence of this conversational computing, I became interested in actual conversation, in natural language conversation with the computer. I began to work on that project, and it got me right into one of the key areas of Artificial Intelligence, which was then and is still the processing of natural language by the computer.

That was the time that you developed ELIZA?

That's right. I had developed a lot of machinery for looking at written text, and I was now ready to program some sort of conversational mechanism. It was clear to me that when two people talk to each other, they both share some common knowledge. So if we were to talk about commercial aviation, we both would have to know something about it. I was confronted with the problem that today might be called knowledge representation, which was not what I was interested in at that moment. I tried to think of a kind of conversation where one of the partners does not necessarily have to know anything. So I thought of cocktail party conversations, or the conversations people have with bartenders who simply say "Yes, yes" or "Oh, that's the saddest story I've ever heard" and things like that.

Then it came to me, first, that the psychiatrist does little talking and, second, when he or she asks a question that would be a sign of ignorance if someone else asked it—for example, the patient has just said, "I like tea," and the psychiatrist asks, "What, in your view, is tea?"—that is considered not a dumb question but a "searching" question. I relied on that and built a system that imitated—a much better word would be *parodied*—the psychiatrist in the first session. Actually, it is a parody of a Rogerian psychotherapy, where the psychiatrist very frequently reflects back what the patient says.

That became ELIZA[1]—ELIZA because of Eliza Doolittle in *Pygmalion*, who learned to speak very well, although it was never made clear whether she developed her intelligence and knew what she was talking about.

Looking back on your evolution from a participant in Artificial Intelligence toward a critique of Artificial Intelligence, do you think ELIZA played a key role?

First of all, I am not sure that I accept the characterization as a "critique of Artificial Intelligence": it is not broad enough. I look, with a critical eye, on the whole computer field and, in fact, on modern science and technology in general.

I don't want that misunderstood, either: it is not that I reject it. One thing I have learned in my life, which is now getting longer, is that I, as well as many other people, am very much tempted by binary thinking. I don't mean thinking in bit strings or if-then constructions; I mean the kind of thinking that sets very sharp alternatives. For example, when everything is fine between my wife and myself, the whole world is fine. If she is angry with me, it's the end of our marriage, the end of the world. I am vastly exaggerating, but this is what I mean by "binary thinking." I think it is wrong: love and hatred, good and bad coexist; it is not one *or* the other. Similarly, in modern science and technology, in the computer field and Artificial Intelligence, to be critical does not mean to throw everything out. I think I should caution against this kind of binary thinking, to which, as I said, I am tempted myself occasionally or more than occasionally.

As to the question of which role ELIZA played in "converting" me to look much more critically at my own work and field: what happened (this is a well-known story by now) is that psychiatrists and other people took this parody seriously. I was stunned by that. In particular, I was shocked that practicing psychiatrists—Ken Colby, for example—rushed into print to say that this was the beginning of automatic psychiatry. That seemed to me offensive to my sense of what a human being is and to my sense of what it means to help another person in crisis, quite independently of whether one is a psychiatrist—or especially when one is a psychiatrist.

This led me to the question of what I had done there, and to the larger question of what I was doing at all, and to the still larger question of what *we* were doing at all. It cannot be understood without also taking into consideration what time it was: the midsixties, approaching 1968, the famous year. The Vietnam War was on, students and others protested against the war, the Civil Rights Movement was in full swing, and MIT was, as an institution, right in the middle of the Vietnam War, to a certain extent as a think tank for weapons and for social science strategies and so on. The idea of the secure hamlet and the free-fire zone that were used in the devastation of Vietnam by the Americans—these ideas originated at MIT, for example.

So it was a time, anyway, when one was at least provoked to think about whether to go along with this or to oppose it in one's own institution. Another thing that comes into it, and very importantly—not by the way—is my own history.

I was born in Germany in 1923. I left Germany in 1936, when I was thirteen years old, and came to the United States. Thirteen is already old enough to remember a lot. I had lived as a Jewish boy in Hitler's Germany for three years, and I have remained, to this day, very interested in the history of fascism in Germany and throughout the world and in the political history of Europe. I knew the role, I might say the miserable role, that German scientists, including physicians, played in German military history, also during the first World War. I think of Fritz Haber here, who enthusiastically invented poison gas as a military tool, as a means to kill people.

I saw myself in the same situation as these German scientists before and during World War II. I had to ask myself: do I want to play that kind of role? Everything put together—the highly politicized time, the questions that ought to have been on everybody's mind anyway: what comes out is that either one should push all this sort of thinking down—that is *Verdrängung*, repression—or one has to look at it seriously. Out of such serious investigation of oneself comes a critical stance.

You mentioned that you were interested in socially significant areas. Why did you undertake to program ELIZA at all? What was the idea of simulating a conversation with a psychiatrist?

I have already explained that it was a way to avoid the difficult problem of knowledge representation. It really did not have to do much with social significance. When I mentioned social significance, I meant that when I first started mathematics, I wondered whether I ought to be doing that, whether I should not be in a field that helps people more. I think the events in Germany and what I saw when I first came to America reinforced my social conscience. All the events in which I participated as a Jewish boy in Berlin . . . and then I came to America, and for the first time I saw black people. It did not take very long at all before I began to see them pretty much in the role of the Jews in Germany at that time. Consequently, I began to sympathize with their aspirations. I had the idea, perhaps wrongly, that I understood their position better than kids who had been brought up here. That is where the social conscience comes from, to a certain extent.

ELIZA as such has no relationship to this, except that when I first started list processing, there was a language called IPL/5 developed by Newell, Simon, and Shaw. It was a list processing language, and the idea was that such languages would be especially useful to social scientists, psychologists, and so on, as opposed to the dominant languages of the FORTRAN kind, which were thought to be particularly useful to engineers and scientists. I did not think that I was making a direct contribution to social science or to psychology or psychiatry. I happened to be in a branch of the computer field that—rightly or wrongly—was identified with social science.

Some people in the field thought ELIZA to be a step in the wrong direction—that you had to bring in knowledge representation to get computers to understand.

Technically, ELIZA was an interpreter. There was the machinery that made it all go, the interpreter, and then there was a script that decomposed the sentences that were typed in and generated the responses. This script could be anything. The most famous script is the one where ELIZA plays the psychiatrist, but there were others. I remember an undergraduate at that time who is now full professor at MIT—Steve Ward by name—who wrote a script that was a parody of the way President Nixon talked. That was not hard to do: Nixon has a very characteristic way of speaking, like "Let me say this about that" and so on.

After I had the interpreter debugged and running, I did other experiments that I, in fact, published.[2] Specifically, I remember one script in which one could define things that the system would then "remember," for example, things like "The diameter of a circle is twice the radius of the circle." This was typed in English, and the system stored it. And "The area of the circle is π times the square of the radius of the circle." Then one could type in "A plate is a circle" and "The diameter of the plate is two inches" and ask later, "What is the area of the plate?" The thing would see that the plate is a circle, that it did not have a definition of the area in terms of the diameter, but that there was a relationship between the diameter and the radius. It would look that up and finally tell you what the area of the plate was. The examples could get much more complicated than that, including balls and spheres and cubes.

The interpreter served the same function that, for example, Danny Bobrow's thesis served. Danny Bobrow was a graduate student at that time and wrote a system[3] that took in data like "The distance between Boston and New York is 250 miles," "My car gets so many miles per gallon," "How many gallons would it take for me to drive from Boston to New York?"—things like that, and more complicated than that. ELIZA, as an interpreter, could also do such things.

Once I did a trick: a possible statement was "When I say A I mean B." I put in things like "When I say 'Ich sage' I mean 'I say' " and "When I say 'wann' I mean 'when' " and so on. In a few lines I taught it how to "understand" German input and deal with it in these problems having to do with shapes of volumes, for example. It is clear that I could not totally avoid the knowledge representation problem, and I got to what was then state of the art—talking about 1966. It was more or less contemporaneous with Danny Bobrow's doctoral dissertation at MIT.

But the very large problem of knowledge representation is, in a sense, still not solved. A lot of work has been done and continues to be done. Indeed, one of the important questions that ELIZA generated was the question of what it means for a machine to understand anything at all, and the larger question of what it means to understand at all, independent of whether you talk about machines or human beings. It was so easy to be lulled into the idea that a machine understands me. It is amazing what happened to people who tried this thing. After a few sentences of conversation backward and forward, typed on a typewriter and a console, they would say: yes, I know it is a machine, but it really does understand; I really feel understood. It raised very significant questions, which are not yet settled by any means. I think the entire discussion about Searle's Chinese Room *Gedankenexperiment* is addressed to the question of what it means to understand. It is clear that that question has not been answered altogether.

What could be the deeper reason why people believe that computers understand? For instance, in your book you wrote that your secretary knew that it was only a machine, but still she would not allow you to look over her shoulder.[4]

I have to say that I don't know. But what ELIZA showed up most clearly is the extent to which we make attributions to other people and even, as it turns out, to computers or to the author of the program. We attribute understanding, we attribute meaning. When the psychiatrist program says something like "You have not said anything about your father," they attribute what they think to be the human motivation for such a question to the computer.

There is something like this: we as human beings are simply forced to make inductions all the time. By that I mean that, in our daily life, we constantly have to go beyond the evidence that is presented to us. For example: suppose you have never seen a hand calculator, and now I show you one. I tell you that it can do additions: if you type in 12, then +, and then 15, 27 will come out. And now you type in five pairs of numbers, $12 + 15 = 27$, $1 + 2 = 3$, and so on. You make five experiments. You are literally incapable of believing that the machine you hold in your hands can correctly add exactly the numbers you've typed in and nothing else. You will believe that it will correctly add *any two numbers* you type in. Not exactly any two: if you type in the first hundred digits of π and then the first hundred digits of e, the hand calculator might not deal with that. You might find that out. But even if you try that, you still are incapable of believing that this thing will add exactly the five pairs of numbers that you have typed in and nothing else. We cannot help ourselves. That is also true in dealing with the everyday world.

If you are sitting next to someone in a plane, and you start to talk about horse races or sailing, you have to do this kind of induction with the person you are talking with: this person knows not just all the things he or she has told you but a lot more. A good teacher, for example, answers students' questions to less than the very depth of his own knowledge. A student attributes to the teacher a depth of knowledge that is far greater than that revealed by the question. But this can be abused: it may turn out that the professor has in fact nothing but superficial knowledge, and he uses the inductions people are forced to make to appear to have much more authority of knowledge than he actually has.

But in order to survive in a complex world, we have to make a lot of inductions. When I come in here and sit down on this chair, I assume that the chair will not collapse. I don't examine it.

It is necessary, and your example is a very good one. You come into a restaurant, and all the seats are taken. Just as you come in, a very skinny person gets up from a chair and leaves room for you. Then you sit down on that chair, and obviously, logically, you have no reason to assume that the chair will carry someone of your weight just because it carried someone of much less weight. You have a lot of experience, but when you sit down on that chair confidently, you have generalized beyond the set of your experience. We do that all the time, and it is necessary.

The child learns that. The game of peekaboo that parents play with their children gets generalized very deliberately to teach the child that "not seeing me" does not mean that I am gone from this earth. Finally, when this all works well, the parents can go to a movie, leaving a babysitter, and when they leave, the child does not believe that they are gone forever. The child has learned, beginning with this game of peekaboo, that parents come back. This idea that "invisible" does not mean "gone altogether" is an induction. By that I mean: drawing a conclusion beyond that strictly entitled by the evidence.

Let's take the Chinese Room Gedankenexperiment. People outside the room make the induction that the man inside understands Chinese. The Gedankenexperiment shows us that this is wrong. But that is why a lot of people, especially in the Artificial Intelligence community, fight the Chinese Room Argument. Why do people defend so anxiously the idea that computers could understand or could be made to understand?

I think the clearest formulation of the counterargument to Searle is by Dan Dennett, who says that ultimately it is not the man inside the room who understands but the system as a whole. I have tried very hard to put myself into this position, but I must confess that I don't understand it.

To me it is very clear that the man inside the Chinese Room does not understand Chinese. One example that I sometimes give is the following: suppose we had forgotten to feed this man, and he has become very hungry. Now comes a Chinese story of an emperor's feast, where lambs are roasted and potatoes prepared with delicious gravy and so on. The question is: does the man's mouth water? The answer is: of course not. This would also be Dennett's answer, but he would add that nothing depends on that, because the system understands. One would ask: the system's mouth does not water, either—the system does not have a mouth. Can the system really understand hunger when it does not have a mouth, a stomach, any biological needs? I must say I have never understood this.

But the question you asked was why people in the Artificial Intelligence community are so desperately insistent that Searle's argument is false—that there is understanding there, whether in the man or in the system or wherever.

I think it is a very deep question. I think the phenomenon we are seeing is fundamentally ideological. Indeed, in the inner circles of the Artificial Intelligence community, words such as *theology* are used. "This is a theological argument" is sometimes said. There is a very strong theological or ideological component. People are captured by this ideology and become dogmatic. When the dogma is attacked, emotions come into play, and the argument ceases to be based on evidence and other good reasons. Now: why that? I am sure the deep psychological roots and how one talks about them depends very much, I suppose, on one's training in history and on the kind of language uses that one has internalized.

I can imagine, for example, a Jungian saying that part of the arch-memory of people living today are the Greek myths—for example the Pygmalion myth, that is, the creation of life by artificial means, the power struggle among the gods, and so on. That comes up and expresses itself in another place—like the Golem myth among the Jews, where, again, life is created out of nonliving material and has great power, the Frankenstein story, and so on. . . . A Jungian might say that this is all part of the archmemory of people and we are seeing its expression here. A Jungian might express it this way, whereas there are other ways to express it.

I think it fundamentally has to do with power, and certainly with the power to make life. It seems to me inescapable now to see that, first of all, most of the people involved in this are men and that they have very idealistic fantasies about creating thinking robots that will—citing Moravec,[5] for example—reach a human level within the next forty years, quickly become superior to us, and finally realize that they can get along much better without us, and they will take over and we will disappear. This is a serious statement by a serious worker in the field. It could easily be the manifestation of a creation myth. I find it inescapable to believe that this is an expression of a uterus envy. Women are said to have a penis envy, but we are discovering that men have a uterus envy: there is something very creative that women can do and men cannot, that is, give birth to new life. We now see evidence that men are striving to do that. They want to create not only new life but better life, with fewer imperfections. Furthermore, given that it is digital and so on, it will be immortal, so they are even better than women. I think, very deep behind the scene, that this is what is going on.

Should we take this opinion of leading people in the Artificial Intelligence community seriously?

Let me rephrase your question this way: should we take these fantasies, presented in the form of scientific programs, seriously—should we pay attention to them or just laugh about them?

The question that is usually asked about ambitions of this kind is: can they be realized? There are a lot of good reasons why those systems will be very different from human beings, to say the least. Indeed, in my book *Computer Power and Human Reason* there is a paragraph where I point out that, technically, one can build devices that can see and feel and sense their environment and can modify their environment with hands and all that. And that I don't see any reason to think there is some sort of technical limit to the kind of "intelligence" one can achieve. But what I then insist on in this book is that however intelligent these things become, whatever power they have to come to understand their environment, to move in it and to modify it, their intelligence must always be very different from and alien to human intelligence. I still believe that.

From that point of view, you can say that these dreams will never be realized in the form the scientists of today dream them. But I don't think this answers the question: should we take such programs, such proposals, seriously, especially for their philosophical implications?

One very critical component of such programs is that they formulate a picture of what it means to be a human being. It is a very simple picture. It describes, in my view, a very small part of what it means to be a human being and says that this is the whole.

We have seen that such scientific ideas—speaking about modern science—enter the public consciousness very quickly and help to build a world picture, a *Weltanschauung*, of the general public, of people who have no idea where these things come from, and have very serious consequences in political and cultural life of the community. An example, which is not scientific, although it owes something to modern science:

the idea that some human beings are vermin and therefore not worthy of living as human beings. That idea made the Holocaust possible. It would have been impossible without such an idea. We, people in general, Americans included, find it necessary to have an image of the enemy in an animallike form, like the "Japs" in World War II.

What this worldview does is present a picture of what it means to be a human being, which allows us to deal with human beings in a way that I think we ought not to—to kill them, for example. About thirty years ago, in an article in *Life Magazine*—an interview, perhaps—Marvin Minsky made the now famous statement, which he has never taken back or modified (I heard him make it again last year; I think he is proud of it), that the brain is merely a meat machine. What is interesting, and what perhaps escapes readers in other languages, is that in English we have two words for what he calls "meat": meat is dead, can be burned or eaten, can be thrown away; whereas flesh is living flesh, and a certain sense of dignity is associated with it. We don't talk about eating flesh, and if we talk about burning flesh, it is a horror image. Why did he say "meat machine" and not "flesh machine"? It is a very deliberate choice of words that clearly testifies to a kind of disdain of the human being.

Not long ago, at a public meeting at Harvard University at which both Dan Dennett and Marvin Minsky were present, I raised exactly this point. Dan Dennett stood up—one might say in defense of Marvin—and said: "If we are to make further progress in Artificial Intelligence, we have to give up our awe of living things." An absolutely incredible statement, but not to the Artificial Intelligence community. Herbert Simon, I think, once called me a "carbon-based chauvinist"—carbon-based organic life as opposed to silicon-based, artificial life—and to defend the idea of carbon-based life is chauvinism, a kind of racism, as Dan Dennett would say.

What they mean by that is that it is illogical, prejudicial—that is, that it means to make a prejudgment. That is exactly right: of course it makes a prejudgment. That is part of the forces of induction I talked about. I consider myself a member of the human family, and I owe loyalty to this family that I don't share, to the same extent, with other families. I have values and judgments that I make about this family, to which I am related both in time and in space. I am related not only to all the people on earth but also to all the people who came before me and will come after me. I am especially related to my children, about which I make a prejudgment, whom I value more than other people. The drama of the sacrifice that Abraham was called on to make of his son, where he shows that he is able to do it and only in the last moment an angel stops him— the drama of that is precisely that it is not just anybody that Abraham is asked to kill, but his son, for whom he has special values.

Well, I've special values for the human family, and by extension for life, carbon-based life. So Simon called me "carbon-based chauvinist," and Dennett pointed out to me in just so many words: "If someone said the things that you say about life about the white race in your presence, you would accuse him of being a racist. Don't you see that you are a kind of racist?"

What I am trying to say is that with these world-encompassing projects that strive for superhuman intelligence and for the ultimate elimination of the human race and so on emerges a subtly induced world picture, expressed, for example, in Dan Dennett's statement that we have to give up our awe of living things. That enters our language and the public consciousness, and that makes possible a treatment of human beings vastly below the respect and dignity that they ought to have. In that sense, we should take such statements very seriously.

We have talked, until now, about the so-called "strong Artificial Intelligence" community, as Searle would say. How about the "weak Artificial Intelligence" community? Their argument goes like this: it is not important to believe that we could build machines that are like human beings. But if we try to do it, we get more insight about how the human brain works, so it is very important to do Artificial Intelligence research.

I do think that the idea of computation has served psychology very, very well. Initially, in the late fifties, it was the idea of computation that almost dealt a death blow to behaviorism. Behaviorism and Skinner's work say that we are not entitled to speak about what goes on inside the brain or the mind. These are fictions that have no scientific validity. Now comes the computer, and we can actually see internal logical operations take place and result in things that resemble speech and so on. The appearance of the computer was a very liberating event for psychology. Computers and computation continue to be very powerful, perhaps the most powerful metaphor with the help of which we can formulate understanding of the operations of the brain.

Some people claim that it will be the last metaphor.

As it happens, the latest technology has always claimed to be the last metaphor. It is only the last we know about, and it may well turn out that in fifty years—if we run that long—we will say how silly that was; now we have the God-knows-what, and that turns out to be the best way to think about it.

It is not necessary to say that it is the last metaphor. It is extremely useful. It yields an enormous amount of insight. The more we understand computation and the more we diversify how we do computation,

the richer the metaphor tends to become. So that, to take a current example, connectionism is a modification, an enlargement of the computational metaphor. And computing being such a powerful metaphor for
thinking about thinking, this enlargement becomes immediately sugges-
tive. That means it is a kind of fertilizer for scientific thinking. One might
add to this: "Yes, by a kind of fertilizer, you mean bullshit!" Well,
maybe so . . .

As to the question of whether it is the last metaphor: until a few years
ago, the last metaphor was the so-called physical symbol systems hypothesis of Newell and Simon. It suggested a picture of thinking as a
kind of production system, which is to say, a form of computation that
could be easily realized in computers as we knew them—something like
the IBM 704, but, of course, thousands of times faster with very much
more storage, but still basically a von Neumann machine. It turned out
to be, in some sense, inadequate. I don't know enough to make a final
judgment. But in any case, the Artificial Intelligence community has
turned against it in favor of hugely parallel computation. Now *this* is the
last metaphor. I don't know what will come next. Perhaps machines
with biological components. The beginning could be largely silicon-
based machines with a few biological components, then with more and
more biological components. Finally somebody could get the idea that a
man and a woman should come together and produce a thinking machine which is entirely biological, which is a baby . . .

*In the debate between connectionism and the physical symbol systems
hypothesis, the connectionists say that the underlying "hardware" is relevant, the biological structure. They claim that they are more concerned
with the human brain. The machine need not consist of neurons, but the
neurons have to be simulated.*

We have to make a distinction. We have now ordinary machines,
largely based on the von Neumann model: with a CPU and a storage, or
several CPUs and several levels of storage. It is possible to think of a
machine even like the Cray, the supercomputer, as fundamentally a von
Neumann machine. On the other hand, there is the connection machine
built by Danny Hillis and produced by the company Thinking Machines,
Inc.,[6] or other machines like it, with perhaps thousands of tiny processors, all operating simultaneously. For some applications, they would
not be programmed in the way we think of programming today but the
connections would be shaped by "experiences" given to them and so on.
Finally, another stage is the neuronal machine, which is a kind of connection machine, but whose connectivity is to be imitative of how the
neurons in the brain are connected to one another.

These are three classes. The second of these is still a Turing machine,

in my view. Strictly speaking, that means that any other Turing machine—a PC, a VAX, and so on—can be programmed to imitate this
machine precisely, in every detail, except for speed.

*But connectionists would say that this is a relevant restriction: speed is
critical to simulating human thinking.*

That may be. Certainly speed is very critical. I'll come back to that.

What I am suggesting is that this machine is imitable by any other
machine, so that the physical symbol system hypothesis is not falsified by
the appearance of the connection machine. That, in turn, means that
whatever limitation one sees in the physical symbol system hypothesis
continues to exist for the connection machine.

Let me now think of a neuronal connection machine, which, at the
moment, is merely a fantasy. No one has built one yet. Then, with the
exception of highly organized biological components, for example, the
idea of keeping a cat's brain alive and using it as part of the visual cortex—except for components like that, it is again a Turing machine. And
again the physical symbol system hypothesis is not invalidated, is not
falsified. I don't think that, in principle, it avoids the problems of this
hypothesis and that it is a step ahead.

Let me now come to the question of speed. One of the things that has
entered our language as a result of computers is the term "real time." If
someone had said "real time" as recently as fifty years ago, nobody
would have known what he could possibly mean. Now we use this term
to speak about responses of a machine that keep up with the rate at
which the machine is subjected to stimuli from its environment. For a
machine that is supposed to do calculations on the basis of prices of
current stock shares at Wall Street stock exchanges, real time means that
as soon as these prices are known, they enter the calculations of the
machine, not two hours and not ten days later.

It is perfectly obvious that when the frog's eye sees the little bug in
front of the frog's face move along and tells the brain that there is a bug
there, the brain tells the tongue to go out and get it—if this were to take
place at a rate slower than real time, the frog would starve to death. It
would be very hard to build a simulated frog out of a machine that oper-
ates at a very low speed. That would in fact be impossible, because it
would not be able to do this important task. However, in terms of a
theoretical explanation of (to quote a famous title) "What the Frog's Eye
Tells the Frog's Brain"[7] simulation in much less than real time is certainly sufficient. From a philosophical, as opposed to a practical, point
of view, real time and speed are not necessary.

That sort of consideration, among many others, enters into Dennett's
argument against the Chinese Room. He would argue that all this pass-

ing around little slips of paper would involve enormous amounts of time, and that it is hugely deficient in terms of practicality and therefore is a totally unrealistic fantasy. Whereas—and this is Searle's argument—it is the same task that we now do with computers, and it obviously can be done with computers, but it requires very fast computers. I think I understand that argument, but it is not crucial. It is an interesting argument, but it does not touch the issue.

Let's get back to these other classes of machines: they are not other classes from a Turing point of view. If, in fact, these machines are not Turing machines, that is, they are not imitable by huge VAXes, and if the Artificial Intelligence community insists that only these machines can achieve humanlike intelligence, then they must be saying that all the efforts and all the investment that we have made to achieve Artificial Intelligence in the machines we have had available for the last thirty years, and all or most of the thinking that was done, was simply wrong. If this were the case, that would not necessarily be devastating to anybody except maybe the Department of Defense and the military, and perhaps embarrassing for some people.

If one considers quantum mechanics and relativity, for example, to be finally the real representation of the real world, the final metaphor, then it would not be useful or correct to damn all these physicists who have done physics "on the wrong basis" all these years before it came along. One had to build these structures in order to come to the structure that we now see. This would be a fair argument. But at least we ought to stop talking the way we talk about the present results of Artificial Intelligence.

Recommended Works by Joseph Weizenbaum

Computer Power and Human Reason. San Francisco: Freeman, 1976.
Kurs auf den Eisberg. Zurich: Pendo, 1984.

Notes

1. J. Weizenbaum, "ELIZA: A Computer Program for the Study of Natural Language Communication between Man and Machine," Communications of the ACM 1 (1966): 36–45.

2. J. Weizenbaum, "Contextual Understanding by Computers," *Communications of the ACM* 8 (1967): 474–80.

3. D. G. Bobrow, "Natural Language Input for a Computer Problem Solving System" (Ph.D. diss., Massachusetts Institute of Technology, 1964). The system was called STUDENT.

4. J. Weizenbaum, *Computer Power and Human Reason* (San Francisco: Freeman, 1976).

5. H. Moravec, *Mind Children: The Future of Robot and Human Intelligence* (Cambridge, Mass.: Harvard University Press, 1988).

6. W. D. Hillis, *The Connection Machine* (Cambridge, Mass.: MIT Press, 1985).

7. J. Y. Lettvin, H. R. Maturana, W. S. McCulloch, W.S. and W. H. Pitts, "What the Frog's Eye Tells the Frog's Brain," in *Embodiment of Mind*, ed. W. S. McCulloch (Cambridge, Mass.: MIT Press, 1965).

Why Play the Philosophy Game?

When I was a graduate student, I became interested in Artificial Intelligence and cognitive science. From my point of view, there is not really a difference between cognitive science and Artificial Intelligence. So for me, becoming interested in Artificial Intelligence was in fact becoming interested in cognitive science.

Could you explain why there is no difference? What is Artificial Intelligence, and what is cognitive science?

It is not that there is no difference. It is that one discipline, in a sense, includes the other. Both Artificial Intelligence and cognitive science today—some of my colleagues may disagree with me—take a kind of information processing approach to cognition. This approach can be broadly described by the question, What is the brain doing as a computational device? What is interesting about what the brain does is that it does some interesting stuff with information. And as soon as we start talking about doing things with information we are talking about information processing, and a very good way to talk about information processing is to use the kind of languages and methods that computer scientists have been developing for a long time.

For me, that is what cognitive science is about, but there are obviously many ways to get at the kind of information processing that people do. One way is to build models of it, to try to simulate it—that is the Artificial Intelligence aspect of cognitive science. Another way is to find out what exactly are the methods the mind is using, why it is that these methods work, what it is about these methods that makes them accomplish what the mind is accomplishing. For that, we need to look at the other disciplines that have grown up alongside computer science or maybe in front of it: psychology and linguistics and so on. I view these as ways of getting more data about what it actually is that we are supposed to accomplish, and also of giving us hints about the kinds of algorithms that

Robert Wilensky was born in 1951. He studied mathematics and computer science at Yale College and Yale University. Thereafter he worked in the Department of Medical Computer Science at Yale University, where he developed a medical record system. As a research assistant at Yale, he began working in Artificial Intelligence. In 1978, he received his Ph.D. and moved to Berkeley, where he was Director of the Artificial Intelligence Research Project and of the cognitive science program from its start. Today he is Professor in the Department of Electrical Engineering and Computer Science at the University of California at Berkeley.

the mind actually applies. People are very good at doing certain kinds of things and very bad at others, and we should have some explanation of what should be easy or hard for us to do.

Do you think that the sciences you mentioned have the same status in cognitive science, or are some more important than others?

If I put it in information processing terms, Artificial Intelligence should seem central. The idea of looking at things in terms of how we describe information processes has got to be central. But I think we could take each of these fields and talk about their strengths.

Linguists know a tremendous amount about language phenomena, but they are not very good at telling you about the mechanisms—in fact, they are even misleading and controversial in how they describe their data. In computer science, we know about certain kinds of machines, but there tends to be a bias toward truly algorithmic solutions. People in computer science want to be able to prove properties of these systems, and they want to design systems that will always work. The mind is probably not going to be like that, and in some sense we all have to undo some of the prejudices of our fields in order to make progress.

How can one manage to know enough in one's own field and, at the same time, know something about the other disciplines involved in cognitive science?

One of the things you are suggesting, I guess, is, Isn't there too much that one needs to know in order to do cognitive science? I think this is right, but if you look at these disciplines, the answer is yes and no. Nobody in a discipline knows everything about it. In computer science, we know about compilers and theories and operating systems and a dozen other things, and then Artificial Intelligence. And if I go around in my department, nobody who is not in Artificial Intelligence knows about Artificial Intelligence. The same is true in linguistics: the people who do phonology don't know a whole lot about semantics, and the other way around. If you think about it, you have to ask: why are the departments packaged the way they are? Sometimes I feel that I would be only a little bit more out of place in a linguistics department than I am in a computer science department.

Superficially, there is a problem, but no bigger problem than the one we have in trying to be a computer scientist or a linguist. To answer your question more substantially: there are only a couple of fields you can know very well. I find in my own work that I have to know some things about computer science, some things about logic, some things about linguistics, and some things about philosophy. I cannot know everything that is going on in every field, so I have to choose. There are certainly things I don't know anything about.

Is cognitive science a unified discipline or just a mixture?

The way we divide things into fields is fairly arbitrary to begin with. Once we've divided them up we tend to assume that it is a natural division and that there are good reasons, for example, to have a psychology department. In many cases, the department structure is organized more by methodology than by contents: to be a psychologist, you make certain experiments, linguists look at sentences, and computer scientists play with computers . . . the methodology does not say much about what we are interested in. And if we are interested in how the mind works, we might conceivably want to use methodology from linguistics and psychology and computer science, preferably all together.

One way to look at a university is to realize that there are many ways to organize it. Corporations, for example, are organized in departments, and for projects you draw from different departments. Maybe that is not a bad metaphor for doing cognitive science. We want to know how the mind works, and therefore we have to draw on different people with different methodologies and expertise.

So you think it is not a disadvantage that there are as yet hardly any cognitive science departments?

You may know that here (at the University of California at Berkeley) we are moving now in the direction of institutionalizing an official cognitive science program on both the graduate and the undergraduate levels. I am the director of the cognitive science program here, and I was very cautious about moving forward for exactly these reservations I mentioned. What does it mean to have a cognitive science degree? I am convinced now that it is not a bad idea. Up until this point I thought it was important to have a degree in X, with some specialty in cognitive science. On the graduate level, I still think it should be that way.

Our proposal for the graduate-level program right now is that the various departments participating in the program will have a track within them for cognitive science. So you would graduate in psychology, but within the cognitive science track. There are still some questions about how to proceed, but this is the way it should be in science. We are going to see how it works and if cognitive science will really be an integrated science of the mind. How we structure it in the beginning is almost certainly going to be wrong, but that is how things happen.

What is the state of the art in Artificial Intelligence? What are the current challenges and problems the field has to deal with?

My view of the discipline is that we have been making very slow but steady progress for many years without a whole lot of change in programs. What changes is the perception by the public and the amount of attention from the media. The media got very excited about expert systems, which represented very little in terms of Artificial Intelligence technology, and they became important in the business world. My feeling is that there is a certain consensus within Artificial Intelligence about some problems and certain general ideas about how to solve them.

What most Artificial Intelligence researchers today would call the biggest current problem is the representation and application of large amounts of world knowledge. It took a long time for people to recognize this as the main problem. In terms of what we are doing about it, we are still very limited. It is not clear at this point whether we are still limited because there is some big thing that we don't understand or whether it is a matter of mucking along for years and making incrementally more and more progress. I tend to suspect it is the latter. There is not any elixir or magical answer. We are on the right path and still have lots of work to do.

Some people—mostly philosophers—claim that Artificial Intelligence has been famous for its failures during the last years and that earlier claims were exaggerated.

Philosophers are a funny group. I was interviewed by someone from a newspaper about what the philosophers at Berkeley said. I said: "Look, I have seen a guy talking about photomicroscopy. When the physicists made these developments in photomicroscopy, did they go ask the philosophers whether it was possible? Of course not. Why should they think that philosophers know any more about this?"

It is probably fair to say that some of my colleagues in the field have been unduly optimistic about how hard it was to do things. But it is silly to jump on an individual in his field, however prominent, who says that in ten years all the problems of that field would be solved and to generalize it by saying that *the* people in the field believe that. If I said, "Philosophers believe that . . . ," you would immediately recognize that it was silly to quantify over such diversity. Why should we think that what one philosopher said fifteen years ago has ever since dominated the field?

Years ago Hubert Dreyfus said that a computer would never beat him at chess.[1] Then a computer beat him fairly shortly afterward. He said that computers would never play master-level chess. Now computers do this routinely. Then he said that a computer would never be world champion of chess. It looks like even this is right on the horizon.

Of all the people I have met in the field over the years, the number of those who had in my opinion unduly optimistic expectations about what was possible is extremely small. What is interesting is that some things have become well enough understood that commercial applications have come out of them—expert systems, for example. Artificial Intelligence

researchers say that this is not particularly interesting Artificial Intelligence; it has been around for fifteen years before it became commercial,
and it does not solve any of the fundamental problems we are all interested in. In some sense, it is not even a big commercial success. So how
are we going to measure the progress in the field?

What is interesting is that some of these same philosophers are very
excited about neural networks. Now, if you take virtually any problem
you would like to solve and ask which system will solve it better—a
neural network or a conventional Artificial Intelligence system, there is
no question: for virtually any problem you can think of, the conventional system is orders of magnitude better, and there is no suggestion
that this will change—except for a few tasks. That is not to say anything
negative about neural networks, but it is interesting that the same people
who blamed Artificial Intelligence for its exaggerated assumptions now
say enthusiastic things about neural networks that are much more irresponsible than anything an Artificial Intelligence person ever said,
because they happen to find them consistent with their philosophical
beliefs.

*What is the reason for the trend toward connectionism and neural
network models? You said that the outcome is poor so far.*

Responsible neural network people will admit this. But this doesn't
mean it's not important or not good science. In fact, I view this as being
a somewhat less radical change than other people do. Since the beginning of Artificial Intelligence, there has been an interest in parallel processing. An example would be the early work on semantic networks by
Quillian,[2] which was an intrinsically parallel system. If you did it today,
people would say it is a local connectionist model. The notion that certain things have to be done in a massively parallel way is partly not
distinct from some of traditional Artificial Intelligence. It is also the case
that if you look at low-level vision, the mainstream Artificial Intelligence
doctrine of how to do it is by massively parallel processing. So it has
been in the textbooks for twenty years that for vision you need a large
number of small computing units. That is not new.

What is new is that we should be able to do some of the higher-level
processes this way. Some of the power of thinking about things in a
massively parallel way is very important; that is right. It has been ne-
glected and is now getting its due. Fields like Artificial Intelligence tend
to be "faddish." The fad of the seventies was frames, and they had a
somewhat similar effect: everyone was doing frames and talking about
frames. Now it is neural networks, and it seems as if they were the solution for all the problems. But none of them work. It is not that there is no

value to them. It is just another contribution that gets somewhat over-played in the current situation. It has been known for years that per-ceptron-type systems were potentially very good at pattern recognition, and today's extension is not very surprising.

Some people are intrigued by the learning aspect of these systems. It is indeed intriguing, but it remains to be seen how it fits in. It is a valid, important idea that is getting its due, but there is also a lot of hocus-pocus going on.

People often like neural networks for the wrong reason or play an unsavory game. The game is "Since I am working with neural networks, my work is real science, because obviously the brain works with neu-rons. Since you are not doing neural networks, your work is not science, but some sort of engineering." When you ask these persons if they make an empirical claim that neurons in the brain work this way, they will say, "No, no, no. What I am really saying is that this is an interesting kind of computation to study." Well, then you are not putting forward a much more scientific claim than I am.

People want to have it both ways: if you are not doing this, you are not scientific; you are just hacking, being an engineer. But if you try to pin them down on scientific claims, they will deny that they are making them. You cannot have it both ways; things have to settle down. What the right way is to think about these systems has not quite emerged yet.

When you have a neural network that does certain tasks in a certain way, you find that you can characterize it without recourse to neural networks. For example, what back propagation does is optimize in an error space using gradient descent. Indeed, people working on neural networks talk about them in these terms. Many of the techniques are well known outside of neural networks; you use them all the time in engineering, operations research, and so forth, for optimization. What they are then trying to say is that the mind might use some kind of gradient descent to do certain versions of learning. That is a thing worth saying, but note that you are saying it abstracted away from the imple-mentation.

For me, that is what Artificial Intelligence has been about all along. Artificial Intelligence has been saying, among other things, Here is a de-scription of what the mind is doing in terms of information processing, of how it is behaving, that is independent of physiological details. If the answer is that the mind is computing in parallel minimization of errors by gradient descent, that is a nice computer science/mathematical disser-tation. That is what computer science has been doing all along. Once you have understood that that is what it is doing, you can look for better ways to do it or talk about the particular way in which the brain does it.

I understand that a critique of conventional Artificial Intelligence is that it does not consider the hardware. It is based on a purely abstract notion—the physical symbol systems hypothesis—and does not consider the realization in neurons. Connectionists claim that they have a better approach.

But if you ask them about their claims, there will be two answers. Most of them—the responsible ones—would say that they would like to make those claims but cannot. For example: does the mind work by back propagation? The answer is no. They would be making scientific claims with a theory that is known to be wrong.

And the claims of conventional Artificial Intelligence?

The question is at what level the claim is being made. Probably nobody could make physiological claims, except maybe the vision people in certain cases. They talk about things like computing edge detection by Gaussian convolution. It turns out that there are apparently real receptors that work by Gaussian convolution. You can identify the neurons and make physiological claims. Whether this is traditional Artificial Intelligence vision or whether you would call this neural network—to me they do not seem very different.

But with the exception of some special cases like this, the only claims that we can make are very general ones. An analogy might be that the muscles and bones in the arm work like a lever. Does this description in mechanical terms have biological authenticity to it? In some very abstract way, yes; in a detailed way, it is underspecified.

My idea of conventional Artificial Intelligence models is that they will be this kind of models. This is not a novel position. Many people believe so and even articulated it. If I say the mind does natural language processing by doing X—for example, by integrating different knowledge sources to make certain decisions in a certain order—is that a theory of how people do it? Yes, but it is of course underspecified with regard to the details. You can view some of the connectionist theories as being a somewhat finer grain–level specification, but even then nobody can map these things down to physiology. Neurophysiologists will tell you that these neural networks are extremely simplistic with respect to what we know about how neurons actually function. It is not obvious that talking about things in these terms should give you immediately more scientific credibility. You cannot say that you have understood a neural network model until you have said what it computes. Those descriptions are not in terms of a neural network language but are descriptions of some overall property of the computational process.

So you say that there is similarity or analogy: that you can gain insights about how the brain works with certain programs, certain algorithms. Is this only an analogy, or are we, in certain very restricted cases and in some features only, creating something like the real mind?

It is mainly a question of what level of abstraction you want to make your claims on. One way is this: here is a mathematical function, an abstract characterization of what actually gets done. That is the highest-level description of the human system. The human visual system does X, and we describe X, if we are lucky, in a completely abstract way from how it actually gets computed. At some level we might talk about the algorithm; at some other level we might talk about how these algorithms are actually implemented in terms of neurons. Nobody—except maybe some of the vision people—will say that they are actually making theories on the implementation level. Jeff Hinton,[3] for example, is very careful and says that he works on the algorithmic level, because he knows that on the neurophysiological level, the details are completely implausible.

To come back to the question: to what degree do we ascribe cognitive significance to an Artificial Intelligence theory? If I write a computer program that does some natural language processing, there is not a claim that the mind—at the implementation level—works like a LISP interpreter. We all understand that that is not the claim that is being made, because the claims that are being made are at a more abstract level.

Why are computers good in some abstract tasks like inferencing and bad in tasks that are easy for humans, like pattern recognition?

I find it somewhat infelicitous to say computers are "good" at this or "bad" at that, and with people it is the other way around. The machines that we happen to have built happen to work that way. In some sense, this is not surprising, because we created computers to do very specific things. It is almost miraculous that they can do all those other things at all. We created computers to do ballistic calculations—that is what they were made for. All of a sudden, they were used for data processing as well. It is certainly the case that the kinds of things we first studied in computing in general were those where you arrived at a useful result by a tremendous amount of clever computations. That was also the early view of Artificial Intelligence: to beat a person in chess is to do clever computation with millions of positions better than the person does it.

But not all things that people are good at are things that are not amenable to computation per se; they are things we can do because we have lots of knowledge. It is often said that computers lack creativity. But people are not that creative, either: they are good at things because they have already done them before. People have not yet thought much about what it would be like to build different machines. We are just beginning to think about it. The reason why computers are not good at certain things is that we have not thought for a long time about how to make them good.

There is no in-principle problem with commonsense?

No one has been able to point out what it is. There are other areas of science where people have proven in principle that something cannot be done. If there is an in-principle problem with getting machines to do certain things, no one has even suggested what it is, let alone if it is true.

Let me be more specific. Searle, for instance, claims that we do not make inferences when we answer questions like "Do doctors wear underwear?"

To say that such statements are in some sense a possible challenge to the traditional Artificial Intelligence shows such a profound ignorance of what the enterprise is like that it is hard even to respond to them seriously.

The problem is that in our everyday practice we rely on a lot of background knowledge.

I find this particular attack to be almost scandalous, in the following way: if you go to an Artificial Intelligence researcher and ask him about the most important problem of Artificial Intelligence, he will tell you that it is how to get all this background knowledge into the machine. When did Artificial Intelligence people start to use the term *background knowledge*? About twenty years ago. For example, look at the work that was done on script understanding.[4] The knowledge that people have about a restaurant is exactly background knowledge. So the whole idea of representing knowledge in Artificial Intelligence has basically been the problem of how to represent background knowledge.

Now Searle comes along and says, "This cannot work because you need background knowledge!" Well, what is the new claim? This is exactly what people have been trying to do for twenty years. Searle says that the background is a mysterious thing and that it is not represented. You could go to a lot of detailed arguments: what exactly does it mean to be represented, and this and that. I think the arguments are crazy.

But the idea that the way a system is going to perform is by having somehow encoded in that system those things that we refer to with the notion of background—that is axiom number one in Artificial Intelligence. I don't know a single Artificial Intelligence researcher who would not believe that. Positions differ in how it should be encoded and what exactly it should look like, but the idea that you are operating by virtue of background information is axiomatic in Artificial Intelligence.

But Searle's argument is not that the background is not represented but rather that it is nonrepresentational.

I think there's a lot of confusion about the use of the terms *representation* and *inference*, and the confusion is primarily one of level. At one level, all we mean when we say that something is represented is that somehow there's some state of the brain that encodes something about the world. That's it. It's hard to imagine how the "background" could not be representational according to this usage. But philosophers seem to think that when Artificial Intelligence researchers say that something is represented, we are making a claim that this knowledge is "explicit." What they seem to mean by it is that a person can become consciously aware of this knowledge or can explicitly articulate it. But you will find that Artificial Intelligence literature *never* makes such a claim. Whether something is articulable or amenable to conscious introspection is a very complicated issue. But at least partly, such a question is an issue about *how* something is represented.

For example, do people know how far apart a person's eyes typically are? In one sense, sure: I will recognize as peculiar a picture that departs significantly from this norm. But this is not to say that I can tell the distance in inches. Now, philosophers seem to believe that the representation of the average distance of eyes implies that a person can state, "The average distance is 3.563 inches." But that is not the claim that Artificial Intelligence researchers are making.

Clearly, the form in which such knowledge is encoded is different from that of knowledge that we can readily articulate as English sentences. At an abstract, computational level, we might not care about this difference too much, and describe both facts the same way. But at an algorithmic or implementational level, it might be encoded in a fashion that enables matching but not introspection. There is nothing mysterious or antirepresentational about such an encoding. The point is that if you want to say automatic, nonconscious inference isn't inference, that's fine. You can call it something else, but that does not make it less amenable to computational analysis.

What is the reason for this ongoing argument?

Why pay attention to the philosophy game? Sure, it might be fun. But if you were doing physics, would you go ask a philosopher what he felt about the latest theory in particle physics? The first thing you have to do today in order to study physics is to get all your notions about time and space out of your mind. There are philosophers who work on time and space, but you would not ask *them* every time a physicist says something.

There are some things in Artificial Intelligence, I think, that are interesting for the question of how we think about people. Most of them are not even discussed by philosophers, unfortunately. The motivations are somewhat less savory. The philosophers see that people in Artificial Intelligence touch upon topics that are important to them. They seize on that and talk about it as if it were all the field is about. Searle talks about his Chinese Room Argument—one of the worst arguments that has ever been made in terms of technical structure, rhetorically beautiful, but just a lousy argument—as if the distinction between what he calls "strong Artificial Intelligence" and "weak Artificial Intelligence" was something any Artificial Intelligence researcher cared about. It is a totally artificial distinction, and uninteresting; it is not what Artificial Intelligence people think or care about.

If I asked, "Are you a strong biologist or a weak biologist? Do you believe this or that? Do you believe that if you created that thing in your laboratory, it would actually have those properties, or it would not actually have them?" A biologist would answer that as a scientist, he did not spend a lot of time thinking about that: "What would it be if I created a person in a laboratory—would it have a soul? . . . Well, I don't know . . ." Biologists don't worry about that too much, and the same thing is very much true in Artificial Intelligence. There are some researchers who would ask themselves whether they would attribute a certain property to a system with certain characteristics. But basically, why should you waste your time worrying about this when you want to figure out how to make systems that have these characteristics? Mostly it is like this: realizing that Artificial Intelligence gets a lot of attention, philosophers seize upon something that has a flavor of philosophy.

All the reasonable positions have been taken many years ago. What is left is to take some unreasonable position. If you say things that are sensible, no one will pay attention to you.

You said that information processing is central to Artificial Intelligence. Would you say that connectionism is also about information processing?

Absolutely. This is another point that has caused terrible confusion. The whole question of whether these systems actually have symbols or not is a minor argument. When does a system have symbols and when not? It is one of these philosophical questions. It does not matter. If you want to understand how a neurological system works and you ask a biologist about, let's say, the visual system, he will explain to you: "Well, this neuron takes this information from here, processes it, and passes it on to that neuron . . ." and so on. The neurophysiologist talks in terms of information processing, because this is in a sense what the mind is doing. It is not really the symbol processing paradigm that is being opposed. If I say that the mind does error minimization by gradient descent, that is a description in terms of information processing, and this is exactly how the connectionists talk about it.

If I ask you or Searle, What is interesting about what the brain does? Is it the way it metabolizes glucose? Is it the way chemicals move in and out? That is not unique to the brain. What is interesting is the information processing capability of the mind. It is so obvious in the following sense: You meet a guy from another planet and he acts just like you—he talks and smiles at the right moment and so on. Then you find out what is behind it. The answer is that whatever is behind it is secondary to the fact that you just had these interactions with him. What are these interactions? They are not simply physical; you don't want to describe them as sounds moving your membrane, but "He said and did things"—it is intrinsically an information processing description, there is no way around it.

Many people talk about levels of processes. David Marr[5] for example talked about the computational, the algorithmic and the implementational level. "Computational level" is a misleading name, but what he meant was the description of what is getting done. You have to say what the visual system accomplishes independent of how it is done. Again, what is interesting in what it is doing is not that it metabolizes, but the fact that something is interpreted as a line. That is intrinsically an information processing problem. Indeed, it is hard to understand what you could say about what the mind did without talking about it as an information processing entity. The brain is not special because it is made out of cells, since so is the liver. The question is, Why is your brain different from your liver? The answer is, presumably, that the brain does things that are very abstract. If I ask you something, you can give me the same answer by writing on the blackboard, or on a piece of paper, or by whispering, or by shouting. These are totally different forms of behavior. Why am I able to say that your answer is the same in every case? It is not there in any of those behaviors; it is an incredible abstraction of those behaviors. So, I would not even know what to say about the mind that is interesting if I did not talk in terms of information processing.

If you claim that the mind is, like the computer, an information processing device, where would you stop? I think it was McCarthy who said that a thermostat, too, is an information processing device, and we could ascribe it beliefs.[6]

I don't find anything problematic in saying that there are lots of things doing information processing. Searle said that if you think so, you must also assume that a thermostat has beliefs and so on. This is basically a silly argument. It is not quite a *reductio ad absurdum*. Why should we be at all uncomfortable if I said that my mind is doing information processing and your tape recorder is doing information processing, if the mind's information processing is so much more complex than what the tape recorder is doing?

So the difference is the complexity?

Among other things. Also the kinds of information processing. Just to give you an idea why the argument itself does not really work: Searle says that then you must attribute thermostats with beliefs. But if you think about it, almost every state that we want to attribute is problematic. What about dogs? Do dogs have beliefs? Does your dog desire, or hope, to get some meat? Well, you would say: "I guess it does, sort of." What about a mouse? What about a bee? Do they have hopes, fears, desires, and so on?

So you've got all those biological systems that can have experiences, and there you draw the line. You would like to understand that what a bee experiences as pain is sort of like what you experience as pain, but it is far less subtle. The point is that all those intentional states—beliefs, desires, fears, and so on—are endpoints of a spectrum, and we usually assume that they go together with what people are. But by looking at the biological world and asking the same question—for example, "Does the ant believe that it is carrying a piece of sugar?"—you realize that it is not a well-formed question. The answer would be "a sort of belief." So you discover that belief is not an all-or-none predicate. The further down you go in the biological world, the more problematic is the question.

Now, what about the thermostat? It is not clear whether it is a whole lot more sensible to ask if a thermostat has beliefs than to ask if a bee has them. That is what philosophers do all the time: they take a notion that should be a graded notion, a notion requiring a lot of subtle thinking in order to decide whether or not it is appropriate to apply, and they want a simple answer—yes or no. But the situation is much more complicated than that. If we say: humans do information processing, and thermostats do information processing—sure, but so what? Why should it bother us any more than if we say humans have beliefs and bees have beliefs?

What do you think about the Turing Test?

It is probably not useful, and nobody really worries about it. Building a system that duplicates a person in every detail is such a gargantuan task that the possibility is even beyond my own fairly optimistic expectations. It seems too hard. On the other hand, it seems to be perfectly reasonable that humans believe a system to be intelligent if it does a lot of things but does not pass that test. It is not a day-to-day issue in Artificial Intelligence, and if we think about it at all, we probably don't consider it as an interesting criterion.

What are the future prospects of Artificial Intelligence?

That is a good question. I am very much an optimist, but I have a funny view that I articulated from the beginning. I think things are getting better at a very slow pace; they are not going to get better much faster. Systems get a little bit smarter and can do a few more things. That is what it is going to be like for a very long time: very slow progress. It is not that because there are now neural networks we will have Artificial Intelligence tomorrow.

So you don't expect the great breakthrough?

I cannot believe it. I am no prophet, but I just cannot see it. Artificial Intelligence must be a hard problem, because if it were not, something other than us would worry about what the next step to take is. If neural networks were the magical answer, we would have known about that already. People working on the brain say that it is the most complex object in the universe, so Artificial Intelligence has to be difficult.

After all, we have been working on this problem for such a short time. There is an example that always impresses me: I remember that it was fifty years between Maxwell's equations and the first transmission of a radio wave. Today, fifty years seems like an enormous period in science, but really is not. If you worked on a field for a couple of thousand years and you've got the basic concepts right, you can expect reasonably fast progress. You see this in physics. Someone talks about cold fusion, and maybe six months later you know if he was right or not. But it took thousands of years to get the basic ideas down.

Artificial Intelligence was begun a very short time ago, and the number of people involved is very small. If you look at the people who are really doing Artificial Intelligence research—not expert system development but basic Artificial Intelligence research—there are maybe two hundred in the United States, maybe four hundred? . . . Anyway, it is a small number.

The basic paradigm—how to deal with knowledge—is about twenty years old. If you look at the people who were doing early Artificial Intelligence and at the tools that were available to them, it is quite laughable: they had to work on computers that were far smaller than your personal computer. I am still very optimistic, because I think we are just at the beginning. We are just beginning to know how to think about these problems, to lay down the concepts, and to think about what it would be like to have radically different computers, which is a perfectly reasonable thing to think about. One of my objections to my friend Hubert Dreyfus is that he has a lot of nerve to decide that it is not going to work, if we have only worked on it for a time that, in a historical perspective, corresponds to only five minutes.

And you think that the contribution of connectionism will be smaller than is now expected? What is the reason for its current boom?

The usual story is: people claim that it was mainly the influence of Minsky's and Papert's book *Perceptrons*,[7] where they said these are not powerful models and things would quickly get too complicated.

Let me give you another version: if you look at the people who are prominent in connectionism, like Geoffrey Hinton and Jerry Feldman—their background is largely in vision.[8] In vision they never stopped thinking about neural networks, because they always knew that the only way to talk about vision in a sensible way was to talk in terms of parallel processing on the neural level. No one ever tried to do anything else, really. If you pick up an Artificial Intelligence textbook, the one picture of a neuron you will find is in the chapter on vision.

Two other things: it might have taken twenty years to study these problems and to understand them well enough to think about them in terms of neural networks. The early attempts to put a neural network in a corner and let it learn just did not work. Now people ask how they could do some of the things of which they have gotten a certain understanding from work in more traditional Artificial Intelligence; for example, How could I do word sense disambiguation using a neural network–type system? Most of the understanding of what the problem is took twenty years of thinking to decide what you are working on. There really was a basis being built for doing neural networks properly, on top of the understanding that we have gained.

The other thing is simply a computational problem. Neural networks today are simulations on conventional computers—which is somehow ironic. To some degree, it would have been much more difficult to do even most of *these* experiments twenty years ago.

The technical criticism by Minsky and Papert still holds; it was correct. But people interpreted it in a much more negative way than they should have. The criticism was that certain styles of systems have limited processing capabilities. It is well understood that if you have systems that are more complicated than early perceptrons, they would not necessarily be limited in that way. But nobody knew how to do that; it seemed too hard. Most of the neural networks now are more complicated than perceptrons. Again, it is important not to underestimate the amount of effort that went into thinking about how you would do these things at

all, before you could think about how to do them with parallel computing systems.

I see the directions more as complementary and mutually beneficent
than as competing. The metaphor that "we took the wrong turn of way"
is bad, as John McCarthy pointed out: there are at least three ways in
which you could approach Artificial Intelligence—biological, psychological, and logical. There is no way to show why not all of them should
work. My conviction is that *none* of them will work without some subtle
interaction between them.

To give you an idea of this: Jitendra Malik, the vision person next
door, just came through with a wonderful result about texture computation.[9] The problem was always how to compute textures. There were
theories for years, nobody got it really right, and the classical theory was
recently disproved. He studied neurophysiology and described it in computational terms. What he has now—without making too strong
claims—is a plausible computational theory of texture that is at least
neurologically plausible if not correct. It may be incorrect in some details, but it could be that this is a correct characterization. And it is a
perfectly traditional Artificial Intelligence business. It refers to neurons
at some level, but you can talk about the computations independently of
them.

For me this is exemplary of what we would like in Artificial Intelligence theory: what the program is doing is a convolution, and a thresh-
olding, and then a rectification. This is a beautiful description that you
can give without talking about any of the details. We can speculate that
in the brain it is done by neurons like this and neurons like that, but we
don't know. That is what the answer has to be like: the visual system is
detecting texture by doing this particular kind of computation. And the
computation can be described quite independently of how the brain does
these computations. If I want even to describe what the neurophysiological level is doing, I have to describe it in these terms of computation.

These different approaches are not at all antagonistic to one another:
it is hard to understand how one could proceed without the other. We
cannot understand what the system is doing as long as we cannot describe it, and we will describe it not in terms of chemicals but in terms of
information processing.

We are going to have a lot of feedback between the different approaches. It is more of a race than a question of who is right and who is
wrong. Maybe the ones who look at logic will get there before those who
look at psychology or before those who look at biology, but it does not
mean that one of them is the wrong way to get there. They are all perfectly valid.

Recommended Works by Robert Wilensky

"Computability, Consciousness, and Algorithms." *Behavioral and Brain Sciences* (1990).

"Meta-Planning: Representing and Using Knowledge about Planning in Problem Solving and Natural Language Understanding." *Cognitive Science 5*, no. 3 (1981).

"Some Complexities of Goal Analysis." In *Theoretical Issues in Natural Language Processing*, ed. Y. Wilks. Hillsdale, N.J.: Lawrence Erlbaum, 1989.

"Toward a Theory of Stories." Reprinted in *Readings in Natural Language Processing*, ed. B. J. Grosz, K. S. Jones, and B. L. Webber. Los Altos: Morgan Kaufman, 1986.

Notes

1. H. L. Dreyfus, *What Computers Can't Do: The Limits of Artificial Intelligence* (New York: Harper and Row, 1972).

2. R. M. Quillian, "Semantic Memory," in *Semantic Information Processing*, ed. M. Minsky (Cambridge, Mass.: MIT Press, 1968), pp. 216–70.

3. Geoffrey Hinton contributed to D. Rumelhart, J. McClelland, and the PDP Research Group, *Parallel Distributed Processing: Explorations in the Microstructure of Cognition*, 2 vols. (Cambridge, Mass.: MIT Press, 1986).

4. See R. Schank and R. Abelson, *Scripts, Plans, Goals, and Understanding* (Hillsdale, N.J.: Lawrence Erlbaum, 1977).

5. D. Marr, *Vision* (San Francisco: Freeman, 1982).

6. J. McCarthy, "Ascribing Mental Qualities to Machines," in *Philosophical Perspectives in AI*, ed. M. Ringle (Brighton: Harvester Press, 1979), pp. 161–95.

7. M. Minksy and S. Papert, *Perceptrons* (Cambridge, Mass.: MIT Press, 1988).

8. See, e.g., J. Feldman, "A Connectionist Model of Visual Memory," in *Parallel Models of Associative Memory*, ed. G. E. Hinton and J. A. Anderson (Hillsdale, N.J.: Lawrence Erlbaum, 1981), pp. 49–81. For G. E. Hinton, see also Rumelhart, McClelland, and the PDP Research Group, *Parallel Distributed Processing*.

9. J. Malik and P. Perona, "Preattention Texture Discrimination with Early Vision Mechanisms," *Journal of the Optical Society of America 7*, no. 2 (n.d.): 923–32.

Computers and Social Values

Things for me began in college, with an interest in computers and language. In the very beginning, those were the two things that interested me. I actually spent a year studying linguistics in London before I went to MIT and got involved in computer science. So I came into my study of computing with an interest in combining language and computers. Then I found out that the work on Artificial Intelligence at MIT seemed to be an obvious match. I did my dissertation work at MIT, then I was on the faculty there for three years before coming here (to Stanford).

On reading your publications, one gets the impression that your view of what is possible in Artificial Intelligence has changed a little between your work on SHRDLU and your latest book.[1]
The shift was very gradual. When I was at MIT, I was part of the general enthusiasm that everything can be done in Artificial Intelligence, without much discussion about it—it was just the atmosphere. When I came to the West Coast and was working at Xerox PARC, the work there was very much in the same tradition.

But at the same time I began to have conversations with a broader group of people. There were a series of lunch meetings at Berkeley with Danny Bobrow, whom I was working with at Xerox PARC, and John Searle and Hubert Dreyfus. Bert (Dreyfus) had been in this field for years, and Searle was just beginning to formulate his arguments about Artificial Intelligence. Then Fernando Flores came to the area and I began to work with him, and he began to work with Dreyfus and Searle and a number of other people. And I began gradually to get a broader perspective on what I had been doing and to see where the limitations were that, from the inside, did not seem obvious. So this shift took place over a period of about ten years, from the midseventies to the mideighties.

Terry A. Winograd was born in 1946 in Maryland. He studied mathematics at Colorado College (B.A.) and linguistics at University College (London). He received his Ph.D. in applied mathematics from MIT in 1970. He was Assistant Professor of Electrical Engineering at MIT from 1971 to 1974, before joining Stanford University, where he has been first Assistant Professor and then Associate Professor of Computer Science and Linguistics since 1974.

In retrospect, could you explain this shift and its contents? What are your arguments now about Artificial Intelligence?

The assumption I began with—in an unreflected way—was the standard representational assumption of most of cognitive science and Artificial Intelligence, which is that all the phenomena observed in human thinking and language could be explained in terms of representational mechanisms. They were something like "LISP atoms" inside the head, something like pieces of programs in a generic sense. If you want to understand why in a particular setting a person used a certain kind of language in a certain way, that would ultimately be explained in terms of those representations and rules and their logic, even if their logic was not ordinary mathematical logic.

As I began to work on questions of meaning, it became clear that you had to account for things that were much fuzzier—not in Zadeh's sense of "fuzzy," but much less precise than the kind of things you had rules for, even precise fuzzy rules. I wrote a series of papers back in the early seventies, posing puzzles about language.[2] How is it that we understand a word?—Not because there are boundaries to a definition but rather because there are some clear prototypical examples that we apply to a shading of examples away from that. In that process, I was very aware of the kinds of problems being posed, and struggling to find out how they could be accomplished within this representational view.

What I was led to is this notion of what Bobrow and I called "resource limited processing":[3] some of the phenomena are the result of the fact that the symbol processor can operate only a certain finite amount to come up with what has to count as an answer. The phenomena of fuzziness had something to do with relevance; that is, if you process only a certain amount, you come up with only certain parts of the space. You may come up with an answer that, from the outside, looks as if it were imprecise. The problem in pushing this forward as a research agenda was to give any reasonable account of how the structures were organized so that you came up with the things that people came up with. The more we pushed into that, the more it was ad hoc, and the less it seemed satisfying. You could always model a particular phenomenon by a particular script or a particular program, but when you tried to generalize—asking if this mechanism, on a broad range of things, would do humanlike reasoning, it was always very unclear: you hoped so, but you could not be happy with that.

That was the time I began reading things like Heidegger[4] and Maturana[5] and recognizing that these people were challenging the basic assumption that I was trying to elaborate. They say that there may be some sort of phenomena that you would not look at as operations on representations. And I was already, in some sense, moving away from the more traditional logical views. Therefore I welcomed these insights.

In your last book you challenged the idea of representation in general—the idea that our mind works with representations or a language of thought when we are thinking. How can a person with these views continue to work in Artificial Intelligence?

This raises a different question: whether I still do Artificial Intelligence or not. There are two very different starting points you can have for doing Artificial Intelligence. One is the "science fiction starting point," which says, "We want to do whatever is necessary to make the machine think like a person thinks, to make it do the same things, pass the Turing Test, and so on." There is another view that says, "If we look at the range of things that computers can be programmed to do effectively, there are many things that go beyond traditional data processing, calculations, and so on. They are symbolic, they involve search, they involve those techniques that are developed by Artificial Intelligence, but they need not be followed as a replacement for the kind of thinking humans do but rather as a particular piece of calculation that is relevant to their work."

If you define an expert system as something meant to replace an expert, this is not the kind of work I am interested in. If you ask how the techniques of symbolic processing and heuristic processing could be used in a machine that was important and useful for an expert doing his task, like a doctor doing a diagnosis, then there is a whole interesting series of possibilities and questions, where you say, "Where do these techniques actually work and what can't they do?" So, in that sense I am interested in building "expert systems," but they are "expert" in the second sense, in that they interact with what people do in their thinking.

So, in a certain sense, you are building tools for humans, but you don't claim that they have any insight into how humans work?

I claim that *I* have insights, but I do not claim that these insights are embodied in the programs. The insights about how humans work are in a different domain, about the nature of interpretation and so on; they are not the kind of insights you could embody in a piece of code. They are insights that you can use to decide what to put into this code, in giving a background, a perspective in which to make sense of what the code does. The insights into human language or thought are shaping the design of our computer systems. On the other hand, the system design does not carry within it these insights.

*What is cognitive science about? What are the underlying assumptions—
what is the project of cognitive science?*

There is one possible definition, which is that cognitive science is any
reasonable and systematic study of language, thought, intelligence,
learning ... but that definition is so broad that it is not particularly
useful except maybe to distinguish it from physics. I think there is a
coherent paradigm in cognitive science that has dominated the journals
and the conferences.

I happen to tie this paradigm to what Newell and Simon call the physical symbol systems hypothesis.[6] They argue that all the observable
phenomena that we consider cognitive—language, problem solving, and
so on—can be explained by the functioning of what they call a physical
symbol system, that is, a structure of tokens that can be interpreted according to a set of well-defined rules, where "rule" is used in a broad
sense: an algorithm is a kind of rule, as is deductive logic or as are programs. Therefore any account of how people think and use language will
be equivalent to an account of how some machine based on symbolic
processing will do these things.

I think that general way of looking at things pervades cognitive psychology, a certain amount of philosophy—those philosophers who
would be happy to call themselves "cognitive scientists," certainly Artificial Intelligence—almost entirely. The current connectionism is an interesting branch away from that, at least partially.

*In your view, is symbol processing the main idea of cognitive science,
or are there other currents critical of this but still within cognitive
science?*

Maybe I drew the picture too narrowly with symbol processing, because if you look at people who do neural network modeling and, to
look at the extreme side of it, people who do modeling with continuous
differential equation solving—not the quasi-symbolic things that most
connectionists do—I could say they are cognitive scientists as well. What
ties them together is basically the way to approach questions like language and problem solving from a mind-mechanism point of view: what
is the causal structure of the mechanism that goes *clang, clang* when the
mind works? It may turn out to be continuous differential equations in
neural nets; it does not have to be symbolic.

Basically, cognitive science is the shift to study the phenomena from
the standpoint of the functioning of the individual mind. The major initial push in cognitive science came in linguistics, with Chomsky, shifting
away from "the language" as the object of study to "the working of the
mind," even in the abstract sense of working that he has of the individual
language user.

Would you think of yourself as a cognitive scientist?

To a large extent, I was raised in that tradition and can't move away from it easily. Much of my thinking is in that mood. I have a different way of approaching questions that is not cognitive science proper, a sort of "social construction" orientation, where the object of study is not the individual and what is going on in his head but rather the discourse that exists in the contexts of individuals. Every speech act, every piece of discourse in this context, is by individuals, but the relevant structure does not lie within the individual.

In your opinion, what is the relation of Artificial Intelligence and computer science to the other disciplines within cognitive science? What sort of cooperation is there, in particular with Artificial Intelligence?

Let me give two different answers to this. One is the conceptual abstraction, and the other is the political history. The conceptual abstraction: Artificial Intelligence goes hand in hand with cognitive science, to the degree that you mean by Artificial Intelligence the production of things that people do. On the assumption that people do a pretty good job in different domains and that you want to duplicate these—especially language and communication—then obviously any study of the mechanisms at work is relevant to the design of things that operate that way. In that sense, Artificial Intelligence would be an application of cognitive science, in the same sense that civil engineering is an application of physics.

Is it a relationship between theory and practice?

Yes, Artificial Intelligence is a practice. Cognitive science is the science of the underlying phenomena that determine how these machines should work and what they will do. But that was only one answer.

Historically, Artificial Intelligence was very highly funded and highly regarded, in some sense, as a way into the future. Fields like linguistics and philosophy have always been considered as irrelevant and unimportant, and therefore low-funded. Cognitive science was in some way created to capture some of the attention that Artificial Intelligence has for these more basic domains. And it went along with a kind of missionary zeal on the part of the Artificial Intelligence people: "The particular means that we are using in our programs are going to be the answers to these deeper questions. So if you are a psychologist or a linguist, we have built the machine that does general problem solving or whatever it is, and the particular way in which we have built it is going to tell you what to do." I think that has faded down over time, as it turned out that it was not so easy to take those things and make programs.

Let us now take a look at Artificial Intelligence in particular. How would you define Artificial Intelligence?

Let me take Artificial Intelligence here as a paradigm for doing work. Artificial Intelligence in a different sense is the general quest to build intelligent machines: today it could be with computers; tomorrow it could be with biocircuits or genetic engineering or anything that attempts to do this. That is such a general and unpredictable definition that it is hard to say anything about it, except in science fiction stories.

As a body of techniques, there has been a history going back to the work on cybernetics in the forties or before, which came to flower in the forties and fifties. It took as its basic paradigm the notion of control and feedback and these kinds of circuits. It was replaced in the late fifties— 1956 is the official date—by mainstream Artificial Intelligence, within which there were different subsubjects. What holds together mainstream Artificial Intelligence is basically the symbol processing hypothesis. Wiener's cybernetic circuits did not have symbols; they had feedback loops.[7] It is the explicit symbol processing that holds Artificial Intelligence together. Within that there are several very different schools of what symbols should stand for and how they should be processed. So you have the people who are very close to formal mathematical logic, other people who are close to standard programming, and still others who emphasize search. You can locate centers around the world or in this country that are associated with these views, but all within the same general paradigm. From the midfifties to the midseventies, over a twenty-years period, that was just taken for granted, only details were argued. Sure, there were huge arguments among these different approaches, but they all took for granted the symbol processing.

In the last few years, that has been called into question, mainly because it has not done what people thought it should have done by now. The reason for this is not that people suddenly found Heidegger more readable but that things were just not moving ahead the way that they should if you look at the predictions that had been made fifteen or twenty years ago. The funding agencies began to get nervous; there were shifts in the social-scientific phenomena that have led to what Sherry Turkle[8] calls "emergent Artificial Intelligence," that is, various new ways of looking at Artificial Intelligence where you don't program in symbols but where behavior somehow emerges out of lower levels with some symbolic computation—Douglas Hofstadter talked about this a lot.[9] This is called the connectionist school.

Hofstadter himself is on the border. There is one article by him, "Waking Up from the Boolean Dream," which appeared in one of his collections,[10] where he is basically saying that a straightforward traditional symbolic view does not work. What he proposes to replace it with—like the connectionists—is something that I call "quasi-symbolic." He is talking not about real neural networks with complicated signal patterns and this kind of thing but about a computing element that takes as threshold the sum of its inputs and so on.

But there is a sense that you have to develop intelligence instead of programming it in. Today that is still not the dominant view by far—it represents 20 percent or so. But it is challenging the dominant view to some extent for external reasons, for example, funding agencies want to fund something new and exciting instead of something old that does not work. DARPA (Defense Advanced Research Projects Agency), for example, announced that they want to take five hundred million dollars out of Artificial Intelligence and put them into neural nets.

What are the reasons for the rise of the connectionist movement? The first work by Rosenblatt,[11] for example, was in the late fifties and early sixties, but only within the last three or four years does everyone want to shift. Is it only because the funding situation has changed? Often the funding situation is an expression of tendencies, not their reason.

There are always different domains of explanation. One domain is: who decided which program should be funded where? But behind this is the question of why they decide it this year and not last year—the question of the general sentiment.

I think there is a tendency to abandon hope for quick riches. It has been obvious from the beginning when they tried to build machines that learn—Rosenblatt, for instance—that it was going to be a very slow and difficult process. On the other hand, when you came in from the symbolic end, that was the first thing you got to—for example, geometric theorem proving, which was presented as early as 1956.[12] So it appeared that the higher-level phenomena of language and formal thought would be very immediately accessible for the symbolic mathematics and only extremely distantly available for the so-called lower-level computation. The decision in the research centers was to try to go for it quickly.

Now, people see that it does work for some things but that there are a lot of things symbol processing is not doing—and it does not look as if it were just a matter of fine-tuning to do it. Maybe we do need to go back to our roots and work our way more slowly. My feeling is that they have to go back one step further eventually and look much more seriously at nervous systems.

In several ways the perception- and connectionist-type models are extremely simplified abstractions of what goes on in the nervous system. One is that an individual neuron is a very complex element; it is not a simple threshold. The second is that the structures of the nervous system are not homogeneous, uniform interconnections of simple elements but complex architectures of different kinds of cells with different kinds of cells, connections, layers, and so on. The properties of how we think is in some sense a function of that complexity.

The hope of "traditional" Artificial Intelligence was that at a functional level all those things were irrelevant because all that architecture could be described as a Turing machine, something that had a symbolic basis. If that is true, then you really do not care about the brain any more than you care about how the chips are arranged in my computer. I think that this view will be seen as an overoptimistic abstraction. How we actually think depends on the hardware, on the neural structure, in ways that cannot be simplified by saying that we are implementing only the higher level. And slowly and painfully—because I think it is very difficult—we must actually try to understand point by point how neural systems work and how brains work.

One could say that cognitive science has an interdisciplinary approach, with several sciences engaged in one research program. Was it difficult for you, with your background as a mathematician and computer scientist, to get involved in philosophy and psychology?

There are certain difficulties but much fewer than you might think. If you look at the philosophers and psychologists and so on, the segment of those disciplines that is engaged in cognitive science is a very particular segment. If I had to talk to Freudian psychologists, I would have to learn a huge amount of new material and perspectives. But Freudian psychologists are not interested in cognitive science. They see it as a sort of perversion of psychology. If I had to talk to philosophers in traditional philosophical fields—let's say moral philosophers—I would have to learn many things, but they are not the ones interested in cognitive science. The ones interested in cognitive science are the analytic philosophers of language, who were already very close to mathematics and formal systems. You are pulling from each discipline that corner of it that is concerned with these formalized systems.

What about Heidegger?

Heidegger has not been an influence in cognitive science. Heidegger was very difficult for me, but that was because I was coming from the outside. It was not my participation in cognitive science that led to my reading Heidegger; it was my distancing from cognitive science.

I tell you one gap that I see, which is frequent. Even when they are concerned with the same phenomena, the nature of a philosophical concern and that of an engineering concern are very different and often not understood. I have heard conferences, for example, about representations, where a philosophically oriented person gets up and talks about what representation is, the relationship between extension and intension, and so on. The computer scientist stands up and talks about what representation is: the relationship to theorem proving, semantic networks, logic, and so on. The two of them are talking totally at cross-purposes. The engineering questions and the ontological questions are linked only by somebody who really understands both. Often the engineers will not even recognize that there *are* ontological questions. The nature of the question is invisible to them, and they just act as if there were no problem. I read a paper by Brian Smith,[13] which is a critique of the work of Lenat and Feigenbaum. But Lenat and Feigenbaum just do not have ontological concerns, and they do not recognize them. When Brian gave his paper at a conference, Feigenbaum stood up and said that he had never heard anything so strange. Many people in Artificial Intelligence do not understand philosophical questions; that is why they often make naive statements.

Searle is a good example for this gap. Searle's notion of "strong Artificial Intelligence" versus "weak Artificial Intelligence" was misunderstood by almost everyone I know in the Artificial Intelligence community. Because for him, it is a philosophical, not a technical, distinction. He could imagine something passing the Turing Test, which is still only weak Artificial Intelligence—basically that is what the Chinese Room does. And everybody in Artificial Intelligence came to assume that by "strong" he meant something that is as good as a person and that by "weak" he means expert systems or something applied, which is not at all the distinction he is making. The distinction he is making is between a difference in ontology of "having" intentionality and "acting as though" you had intentionality, which are fundamentally different even if they might be behaviorally indistinguishable. This difference does not make sense to an engineer. What it does is what it is; there is no other question to be asked.

I happen to find Searle's distinction to be ultimately wrong—I mean nonsensical—maybe because I am an engineer at heart. Nevertheless, that is an example where there has been such a tremendous misunderstanding because of the different kinds of questions.

How could one overcome these misunderstandings and gaps?

In some idealized world—I am never talking about *real* computer science departments—I could imagine a curriculum that starts people off with a broad spectrum of courses oriented toward the nature of the paradigm in the discipline. Not what philosophers tell me about Artificial Intelligence as an Artificial Intelligence scientist, but what kind of questions philosophers want to ask and what style of answers have been thought of.

And the curriculum should do that in an open way, in an ethnographic way—like an anthropologist trying to understand this, not asking whether it is right or wrong, whether it is better or worse than Artificial Intelligence. Then we should go back and ask: what is the relevance of these questions to what we are doing? And the answer might often be: none.

Some of the questions that Searle is asking ultimately have no relevance to what someone who is building an Artificial Intelligence system needs to think about. Maybe what they should think about is whether what they have built successfully should be used. Philosophical questions do have relevance to practical questions, but they are not the immediate practical questions of Artificial Intelligence.

Do you think that philosophers generally understand the problems of Artificial Intelligence and cognitive science, or do they have a lack of information on their side?

They certainly have a lack of information in the sense that they don't know the differences. If you say to a philosopher in general (there are very, very few like Dennett, who are competent): "What is the relationship between, let's say, the notions of 'logic' or 'worlds' as taught by philosophers and what semantic networks do?" They have either some very naive view or no idea at all. To understand technically what computer scientists are doing and how it relates to things that philosophers think is important but involves in-depth-knowledge. And who can say whether these things will be relevant to them? So, once you have established that, say, semantic networks are the same as procedural logic, then the details are not interesting to your questions if things can be symbolic or not. But then you need to understand connectionist networks. Are they symbolic or not? The problem is that the Artificial Intelligence people don't slice their presentations in a way that focuses on these kinds of questions, because they are too engaged in the immediacies of "how does this work" and "how does that work." There are a few people who try to make those translations. Dreyfus has much more seriously learned about the technology than he had twenty years ago. He takes more seriously trying to understand what it is . . . with the help of his brother.

Another, more technical, question: what about the Turing Test, which you mentioned already? Is it a useful test?

It is a useful test; the problem is: useful for what? As a definition of intelligence, it is not a very interesting test. It simply says that intelligence is equivalence to human beings. So something could be in some other way considerably more intelligent, but because it was not just like a person, it fails the test. It is equating intelligence essentially with "identity with people," or falling within the degree of variation that you find among normal people. It is superficial. It seems that intelligence is a deep property; there is no reason to take as standard exactly what people do—except that people are obviously more intelligent in almost everything they do than anything we see around us. So when you get to this level, you are at least sure to have done something. As a definition for intelligence, it is very funny. On the other hand, as a pragmatic thing, it certainly would be something to succeed.

As a procedure, it is also funny, being a test of the ability to deceive rather than a test of the ability to achieve. ELIZA comes close to passing the Turing Test to some extent, with almost no intelligence by any standard that we have, because it is very cleverly designed to deceive within a certain context. So you could imagine somebody being so clever at deceiving with something that was not very intelligent that they managed to "pass the Turing Test." It is unlikely with intelligent examiners. But it is funny to define something as your test where the success does not come from achievement, because there is no kind of measure outside the deceiving.

It is also a highly unlikely test in the sense that we will have systems that are in some sense intelligent long before coming even close to passing the Turing Test, because why should they have the same understanding of, for example, bodily things if they are not embodied in human bodies? Why should they have the same kind of reactions as people have? Why would you program into a machine the entire range of experience that human beings have? It is not something that you would do, except for this fooling thing. It is not a practical test; it is not something like a benchmark test, because the only systems that come close to succeeding it are the ones especially programmed for that purpose. To summarize: the Turing Test is useful only as a kind of *Gedankenexperiment* to raise the question, How would you define what it means for something to be intelligent? It does not answer that; it just raises the question.

What role do theory and experiment play in cognitive science and especially in Artificial Intelligence?

Every science is going to have its own methods. Chemistry is an example. A lot of what we know about chemistry was initially that somebody happened to spill something, and something happened that was surprising. That led them to ask questions about the phenomena that led to the development of theory. The purpose of experiments is not only to verify theories to the extent they are verifiable but also to suggest directions for a theoretical explanation. It reveals phenomena that otherwise you may not even have noticed. I see this as the main benefit that can come from experimentation.

Now, what about Artificial Intelligence, which is a different kind of practical experimentation? What is experimented with are not human phenomena but rather phenomena that happen when you program a machine to do certain things. To the extent that the programs are simply demonstrating preexisting theoretical constructions, they are not very interesting. They are demonstrations rather than experiments. So, I say that people have scripts, and I write a program that uses scripts. It puts out these scripts and so demonstrates that scripts can put out scripts—a kind of circular thing.

The things that are going to be interesting with experiments are those where you honestly cannot predict what a certain computational technique is going to do. They may reveal new phenomena. The phenomena might have nothing to do with human thought; they may be artifacts or the way in which this particular kind of mechanism works, or they may be phenomena that are more general, and you have to go back to look for them and to find them.

I think the benefit of the experiments in Artificial Intelligence for cognitive science has been overdrawn. Much too much work is justified in the sense that it should be interesting to psychology. In principle that is not a bad argument, but it is a matter of having some realistic judgment of what the phenomena are.

In your opinion, what are currently the main problems that Artificial Intelligence must overcome? You said Artificial Intelligence did not achieve all that was predicted. What is the future outlook for Artificial Intelligence?

The main problem area is one that you can label in different ways: commonsense reasoning, background, context. It is the way in which thinking and language are shaped and affected by the sort of broader, imprecise context of what you know, what is going on, and what is relevant. This is not modeled in a straightforward way by rules or algorithms. For a while—in my earlier work—I thought that those problems would be handled by elaborating the representations. You have something as basic representations, and then you have something on top of that, which involves relevance or connection. That was a particular view that some people still pursue. The connectionists go another way and say that there is something not on top of the symbol structure but below the symbol level, in the way things are built.

The phenomenological critique is that the phenomenon is not something that you will understand by looking at the mechanism itself, because it is something where the structure comes out of interaction among people or organisms and the environment. Even if ultimately it is being caused by physical mechanisms in the individual brain, the relevant structure that you can understand as a scientist has to be looked for outside the individual, in the emergence of structure among a community, within coupling with the environment, to use Maturana's term.

In the opinion of the phenomenological tradition, Artificial Intelligence cannot overcome this problem of commonsense or background. They think that Artificial Intelligence will be a failure in the long run. Is that your opinion, too?

Well, the long run is long. If someone were to say to me, "Five hundred years from now it will be possible to build, cell by cell, something that is architected like the human brain, which goes through all the experiences people have, step by step. These are androids, which are more or less like persons, with the only difference that they are built in laboratories," I would have to say, "I am a skeptic and an agnostic, but I cannot say 'no.'" There is no a priori reason why you could not imagine artificially constructing something that functions like the human brain. If someone said to me that we are going to achieve that by programming, then I am much more than a skeptic: I do not think that can be done. It is going to require a completely different understanding of what it means to construct this artificial thing, what it means to give it the appropriate learning and experience and so on.

As president of the CPSR,[14] you are also concerned with the social impact of these technologies. What do you think about the outcome, the actual usefulness, of Artificial Intelligence?

I think there is a lot of practical work that can be done using Artificial Intelligence techniques such as heuristic search and explicit deduction. If we ask what we teach our students in Artificial Intelligence—not the indoctrination but the contents—then there is a body of programming techniques. Like all programming techniques, there are problems to which they are well suited and problems to which they are badly suited. We will get a much better understanding of the nature of the problems to which these techniques are well suited. But simultaneously we have to recognize those applications for which they are ill suited.

There is some danger in that, especially in the building of so-called expert systems. If they are not fully understood by the people who buy them and use them and sell them, then you can distort the nature of what is being done.

Take an obvious example: the financial industry. There are expert systems for deciding on loans. Those systems will not have the depth of understanding that a human would, talking to a person for a loan. This does not mean that they cannot be used in order to decide on who gets loans. It does not mean even that, statistically, the banks that use them will not end up with fewer defaults than the banks that do not use them. On the other hand, there are value judgments to make: whether the appropriate question for the bank to ask is simply "Should we use this thing that statistically gives us fewer defaults, or is there some other human value involved?" Making no loans within the so-called bad districts of a city will lead to better statistics—it is called "redlining," and it is what people do, but officially it is considered wrong. Once you put programs in, you begin to make this kind of phenomena invisible. A program able to "learn" could do the equivalent of redlining without anyone programming it in explicitly, it being simply a statistical outcome of the experience with which this program works. So you have certain social policies being set implicitly by the variables that happen to be in this machine. There is a real danger in the near term, with expert systems as a way of thinking.

The fact is that an Artificial Intelligence program is not the key in that problem, but because Artificial Intelligence is more sophisticated, it may be harder to spot and more tempting to do. The same thing happens to the question of privacy and surveillance. The fact that there are computer records all over the world about me and that they can be networked creates the possibility for somebody or some agency to keep track in a very intimate way of what I am doing. The possibility of having Artificial Intelligence programs that look for patterns in these data increases that danger. It makes it possible to find things on the basis not just of a particular act but of a sequence of acts that create a certain pattern considered suspicious by them. Artificial Intelligence does not create the problem, but it expands it.

What can be done to attenuate the impact on society?

The first generic topic is education. If people who use expert systems have a deeper understanding of what they really are or are not, they will make better judgments about using them. I think it is a matter of public education about what they can and cannot do.

Beyond that, it is a social question of having an organized way in which to push for certain values. To go back to the example of loans: what is the value that is being taken? If the value is purely profit for the person or the company, then you can end up with very different things from what you get if you say that it is a value to have housing for the population, or to improve the situation of the people who have the most trouble instead of giving more money to those who already have it. But those questions cannot be addressed as technical questions about the computer but only as questions about society: how do we articulate and then pursue these values? We as computer scientists cannot answer these questions.

Recommended Works by Terry A. Winograd

Language as a Cognitive Process. Vol. 1: *Syntax*. Reading, Mass.: Addison-Wesley, 1983.
Understanding Natural Language. New York: Academic Press, 1972.
(With F. Flores.) *Understanding Computers and Cognition: A New Foundation for Design*. Norwood, N.J.: Ablex, 1986. Reprint. Reading, Mass.: Addison-Wesley, 1987.

Notes

1. T. Winograd Terry and F. Flores, *Understanding Computers and Cognition: A New Foundation for Design* (Norwood, N.J.: Ablex, 1986; reprint, Reading, Mass.: Addison-Wesley, 1987).

2. T. Winograd, "Artificial Intelligence—When Will Computers Understand People?" *Psychology Today* 7, no. 12 (1974): 73–79; "Towards a Procedural Understanding of Semantics," *Revue Internationale de Philosophie* 3 (1976): 117–18; "On Primitives, Prototypes, and Other Semantic Anomalies," in *Proceedings of the Conference on Theoretical Issues in Natural Language Processing*, University of Illinois (1978): 25–32; "A Framework for Understanding Discourse," in *Cognitive Processes in Comprehension*, ed. M. Just and P. Carpenter (Hillsdale, N.J.: Lawrence Erlbaum, 1977), pp. 63–88; and "Frame Representations and the Procedural-Declarative Controversy," in *Representation and Understanding: Studies in Cognitive Science*, ed. D. Bobrow and A. Collins (New York: Academic Press, 1975), pp. 185–210.

3. D. Bobrow and T. Winograd, "An Overview of KRL: A Knowledge Representation Language," *Cognitive Science* 1, no. 1 (1977): 3–45; D. Bobrow, T. Winograd, and the KRL Research Group, "Experience with KRL-0: One cycle of a Knowledge Representation Language," in *Proceedings of the Fifth International Joint Conference on Artificial Intelligence* (1977): 213–22; D. Bobrow and T. Winograd, "KRL: Another Perspective," *Cognitive Science* 3, no. 1 (1979): 29–42; T. Winograd, "Extended Inference Modes in Reasoning by Computer Systems," *Artificial Intelligence* 13, no. 1 (1980): 5–26, reprinted in *Applications of Inductive Logic*, ed. J. Cohen and M. Hesse (Oxford: Clarendon Press, 1980), pp. 333–58.

4. M. Heidegger, *Being and Time*, trans. John Macquarrie and Edward Robinson (New York: Harper and Row, 1962), *What Is Called Thinking?* trans. Fred D. Wieck and J. Glenn Gray (New York: Harper and Row, 1968), *On the Way to Language*, trans. Peter Hertz (New York: Harper and Row, 1971), and *The Question Concerning Technology*, trans. William Lovitt (New York: Harper and Row, 1977).

5. H. Maturana, *Biology of Cognition* (n.p., 1970), "Neurophysiology of Cognition," in *Cognition: A Multiple View*, ed. P. Garvion (New York: Spartan Books, 1970): 3–23, "Biology of Language: The Epistemology of Reality," in *Psychology and Biology of Language and Thought: Essays in Honor of Eric Lenneberg*, ed. G. A. Miller and E. Lenneberg

(New York: Academic Press, 1978), pp. 27–64; and, with F. Varela, *Autopoesis and Cognition: The Realization of the Living* (Dordrecht: Reidel, 1980), and *The Tree of Knowledge* (Boston: New Science Library, 1987).

6. A. Newell and H. Simon, "Computer Science as Empirical Inquiry," in *Mind Design*, ed. J. Haugeland (Cambridge, Mass.: MIT Press, 1976); and A. Newell, "Physical Symbol Systems," *Cognitive Science* 4 (1980): 135–38.

7. N. Wiener, *Cybernetics, or Control and Communication in the Animal and the Machine*, 2d ed. (1948; reprint, Cambridge, Mass.: MIT Press, 1961).

8. S. Turkle, "Romantic Reactions: Paradoxical Responses to the Computer Presence," in *The Boundaries of Humanity: Humans, Animals, Machines*, ed. J. Sheehan and M. Sosna (Berkeley and Los Angeles: University of California Press, 1991), pp. 224–52.

9. D. Hofstadter, *Gödel, Escher, Bach: An Eternal Golden Braid* (New York: Basic Books, 1979).

10. D. Hofstadter, *Metamagical Thema: Questing for the Essence of Mind and Pattern* (New York: Basic Books, 1984).

11. F. Rosenblatt, "Two Theorems of Statistical Separability in the Perceptron," in *Proceedings of a Symposium on the Mechanization of Thought Processes* (London: Her Majesty's Stationary Office, 1959), pp. 421–56; and *Principles of Neurodynamics* (New York: Spartan Books, 1962).

12. See the papers collected in *Computers and Thought*, ed. E. Feigenbaum and J. Feldman (New York: McGraw Hill, 1963).

13. B. Smith, "The Owl and the Electric Encyclopedia," *Artificial Intelligence* 47, no. 1–3 (1991): 251–88.

14. Computer Professionals for Social Responsibility is a public-interest alliance of computer scientists and others interested in the impact of computer technology on society. Since its beginning as a small discussion group formed over a Palo Alto computer mail network in 1981, CPSR has grown into a national organization. Here is an extract from CPSR's statement of purpose: "We work to influence decisions regarding the development and use of computers because those decisions have far-reaching consequences and reflect basic values and priorities. As technical experts, CPSR members provide the public and policymakers with realistic assessments of the power, promise, and limitations of computer technology. As concerned citizens, we direct public attention to critical choices concerning the applications of computing and how those choices affect society."

Lotfi A. Zadeh studied electrical engineering at the University of Tehran as an undergraduate. He continued his studies in the United States, at MIT (S.M.E.E., 1946) and at Columbia University, where he received a Ph.D. in 1949. He taught at Columbia University and joined the Department of Electrical Engineering at the University of California at Berkeley in 1959, serving as the department's chairman from 1963 to 1968. He has also been a visiting member of the Institute for Advanced Study in Princeton, New Jersey; a visiting professor at MIT; a visiting scientist at IBM Research Laboratory in San Jose, California; and a visiting scholar at the AI Center, SRI International, and the Center for the Study of Language and Information, Stanford University. Currently he is Professor Emeritus of Electrical Engineering and Computer Science at Berkeley and the director of BISC (Berkeley Initiative in Soft Computing).

LOTFI A. ZADEH

The Albatross of Classical Logic

I have always been interested in the issue of machine thinking. Let me show you something that I wrote in a student magazine published at Columbia University, *Thinking Machines—A New Field of Electrical Engineering*. Here are some of the headlines: "Psychologists Report Memory Is Electrical," "Electronic Brain Able to Translate Foreign Languages is Being Built," "Electronic Brain Does Research," "Scientists Confirm on Electronic Brain." Behind these headlines are the questions: How will electronic brains or thinking machines work and affect our living? What is the role played by electrical engineers in the design of these devices? These are some of the questions I tried to answer in this article.

Now look at the date of the article: 1950. That was six years before the term *Artificial Intelligence* was coined. I want to underscore the headline "Electronic Brain Able to Translate Foreign Languages Is Being Built" published in 1950. Forty years later, we still do not have the kind of computer that can really translate foreign languages. It shows how easy it was and still is to overestimate the ability of machines to simulate human reasoning.

How would you outline cognitive science? What sciences are contributing to cognitive science?

When you talk about cognitive science you are not talking about something that represents a well-organized body of concepts and techniques. We are talking not about a discipline but about something in its embryonic form at this point, because human reasoning turns out to be much more complex than we thought in the past. And as we learn more about human reasoning, the more complex it becomes.

In my own experience, at some point I came to the conclusion that classical logic is not the right kind of logic for modeling human reasoning. So that is why I came to fuzzy logic. I have always been interested in these issues. But I was in an electrical engineering department, and my interest was in systems analysis and that kind of thing. And only in 1968 or so, when I was moving into computer science, did I reorient myself. But my first paper on fuzzy logic appeared in 1965.[1]

There are several sciences contributing to the field of cognitive science. What roles do they have, in your opinion?

I would list them as follows, not necessarily in the order of importance, but more or less: psychology, linguistics, certain aspects of philosophy, Artificial Intelligence, and to a lesser extent perhaps anthropology. These are some of the principal components of what is called cognitive science today. The meaning of the term *cognitive science* is changing with time. My perception at this point is that there is not as yet a kind of unity. You have an aggregation of disciplines, the ones I mentioned and others that play secondary roles. This collection does not yet have an identity of its own. There are some research programs, some departments in a few cases—UC San Diego is an example. In the near future, there might be some more programs and even some more departments.

I do not anticipate that cognitive science as a *field* will become an important field numerically. I don't expect that there will be many departments or many students. I don't expect that, because of the conservatism of the academic world. When employers want to hire somebody, they hire somebody with a degree in an established field. It will be difficult for people with a degree in cognitive science to get a job. Whether they apply to psychology, linguistics, or computer science departments, the answer will be that they are at the periphery and not in the center of the field. It will take many years for this to change.

What is the very idea of cognitive science?

Well, in many ways cognitive science is essentially part of psychology. It is within psychology to a greater extent than within any other field that you find the basic elements of cognitive science, because cognitive science is concerned with human thinking, human reasoning, human cognition. Artificial Intelligence plays a secondary role, because Artificial Intelligence has to do with the modeling of human reasoning using computers. Cognitive scientists have a more direct interest in human reasoning, whether or not it is modeled by computers.

Do you think that Artificial Intelligence plays only a minor role in the enterprise?

I said secondary, not minor. It is a relatively important role, but nevertheless it is a secondary role. You will find that the overlap between cognitive science and Artificial Intelligence is not that great. You will find that the attendance of Artificial Intelligence conferences by people who might call themselves cognitive scientists is relatively small. And the attendance of cognitive science conferences by Artificial Intelligence people is also relatively small. Furthermore, I do not think that the overlap is growing; it is likely to remain small.

You mentioned that Artificial Intelligence was overestimated from the beginning. How would you, looking back, see the history of Artificial Intelligence? Are there stages?

At the time the article mentioned was published, there was a big discussion on the ability of machines to think. The term *Artificial Intelligence* was not in existence at that time yet, but still there was already this discussion going on. Officially, the term *Artificial Intelligence* was coined at a conference in Dartmouth in 1956[2] by Marvin Minsky, John McCarthy, Oliver Selfridge, and a number of other Yale people. It was at that conference that the field was launched and Artificial Intelligence came into existence.

In the late fifties and early sixties, the concerns of Artificial Intelligence were centered on game playing and to a lesser extent on pattern recognition, problem solving, and so on. Some of the fields that are very important today—like expert systems, knowledge-based systems, computer vision, natural language processing—were not, at that time, of any importance. You could perhaps see that at some time they might become important, but they were not in terms of what was being discussed at that time.

I would say that in the late sixties and early seventies the attention was shifted to knowledge representation, meaning representation, and, in connection with that, natural language processing. In the mid-seventies, some expert systems made their appearance. The first one was DENDRAL by Feigenbaum and his people at Stanford.[3] It was followed by MYCIN, designed by Buchanan and Shortliffe, also at Stanford.[4] Then came PROSPECTOR by Hart, Nilsson,[5] and some others at the Stanford Research Institute. Once these systems had made their appearance, the interest in issues relating to the design and conception of such systems—that is, knowledge representation, meaning representation, management of uncertainty, inference, and so on—became more pronounced.

So a definite shift took place, as problems like game playing and pattern recognition receded in importance. Natural language continued to be important because of its relevance for interfaces for databases.

Then, in the early eighties, we witnessed the Japanese Fifth Generation Project,[6] which attracted new attention to the field. That infused new life into the whole thing in terms of money, people working in the field and so forth. Since then, many things have happened. Artificial Intelligence has become a big field. Some of the big conferences have more than five or six thousand participants.

It is a field that is undergoing rapid change, but right at this point there is a feeling that the expectations were exaggerated. Many of the expectations did not materialize. So there is also some sense of disappointment.

When did this feeling appear?

I would say three or four years ago. There were some articles critical of Artificial Intelligence that point to all sorts of things that were promised but not delivered. I see as a reason for the lack of accomplishments the almost total commitment of Artificial Intelligence to classical logic. Artificial Intelligence has become identified with symbol manipulation. I think that symbol manipulation is not sufficient when it comes to finding the solutions for the tasks Artificial Intelligence has set for itself in the past, like natural language, understanding, summarization of stories, speech recognition, and so forth. Many of these problems turn out to be much more difficult than was believed in the beginning.

Furthermore, my personal view is that symbol manipulation does not provide the adequate techniques for human reasoning. Most of the Artificial Intelligence people are likely to disagree with me there, but I am convinced that in the end they will be proved to be wrong. I refer to this commitment to classical logic as to the "albatross of Artificial Intelligence."

If I understand you correctly, the symbol manipulation approach cannot go far without introducing fuzzy logic. Could you explain why the Artificial Intelligence community, until now, did not accept this approach?

I think it is a matter of historical development. The pioneers of Artificial Intelligence were mainly mathematicians or logicians, people who felt very comfortable with classical logic. They felt that symbol manipulation was sufficient. Most of them still feel that way. They don't like numbers, they don't like probabilities, they don't like fuzzy logic, they don't like anything that is not symbol manipulation.

You said that three or four years ago there was some criticism of Artificial Intelligence. But there were critical voices even twenty years ago, for example, by Dreyfus.

In some things I agree with him; in some things I disagree with him. He was right as to the fact that some of the expectations that were built up by the "artificial intelligentsia" were exaggerated. Let me cite one example that I came across not long ago. Marvin Minsky said in 1973 that "in a few years from now, computers will be able to read and understand Shakespeare." To me, this is a prediction that is not likely to materialize within the next twenty or thirty years, if at any time. Predictions of this kind soured many people on Artificial Intelligence. In that respect Dreyfus and other people who criticized those exaggerations were right.

I part company with Dreyfus when it comes to the use of computers to solve many problems that can be solved effectively by humans. To cite a few examples: we have computers that can play chess very well, on the grandmaster level, essentially, and it is not impossible that in two or three years a computer will beat the world champion. This is not a very important application, of course. But we will have computers that do a lot of useful and interesting things: understand speech, control complex processes, recognize patterns, read handwriting.

When you say that they will understand speech, do you mean like human beings?

No, not like human beings. That is my point: this may never happen; or if it does, it may be many years from now. They will understand speech in a more restricted way, in the sense that you will be able to pick up a telephone and call a number and ask things like "When is United Airlines flight 325 going to arrive at Los Angeles?" There will be no human operator on the other side, but a machine that understands and analyzes your question, looks through its database, and, with a synthesized voice, gives you the answer, something like "Flight 325 has been delayed. It will arrive at Los Angeles International Airport at 5:58 P.M." Systems like that will be in wide use. Another example: at this point, when you want to program a VCR, you do it by pushing different buttons. In the future—and not in the remote future but maybe in two or three or five years from now—you will be able to give a command to the VCR using speech: "Start recording channel 5 at 4 P.M. and stop at 6 P.M." You will be able to ask questions like "If I want to do this and that, how do I do it?" The VCR may be able to tell you certain things. I think the voice input–voice output systems will have an enormous impact, because the most natural way of communicating with a machine may well be voice input–voice output.

I have great respect for Dreyfus, and I consider him my friend; nevertheless, such critiques as his pour too much cold water on the enterprise. Jules Verne said, at the turn of the century, that scientific progress is driven by exaggerated expectations. And if you pour too much cold water, you are going to realize less. The expectations may be exaggerated; nevertheless, they are worthy of support.

A good example is what happened in Great Britain ten or fifteen years ago. There was a debate between Donald Michie and Sir John Lighthill about Artificial Intelligence and robotics.[7] The point is this: I think that at that time Lighthill was right—there were exaggerated expectations. In any case, it was a mistake for the British government to stop its support of Artificial Intelligence. Now they are trying to get back into it, because the decision turned out to be a mistake.

Which kind of cooperation is there between Artificial Intelligence and cognitive science? Is it a relationship of theory and practice? Where do the fields overlap?

There is some overlap, but it is not getting stronger. I am talking about the United States now; in other countries the situation might be different. In the United States, there is a certain tendency on the part of the Artificial Intelligence people to have a superiority complex with regard to cognitive science. The Artificial Intelligence community view the cognitive science community as comprising for the most part people who are not capable of doing serious work insofar as Artificial Intelligence is concerned. I do not think that in the cognitive science community there is a similar superiority complex in relation to Artificial Intelligence. This feeling of superiority inhibits the growth of interaction. The Artificial Intelligence people think that they have nothing to learn from the others, so they don't go to their conferences.

But this kind of trouble can be seen often in interdisciplinary work, because of differences in language and socialization. Will these differences be overcome?

Frankly, I don't think so—not in the United States, at least, because the Artificial Intelligence community tends to be a young, self-confident, and also sort of self-contained community. There is the feeling that they don't need anybody—not just cognitive science, but also mathematics, economics, operations research. I think the cognitive science community would probably like to have more interaction with the Artificial Intelligence community, but, as I said, this feeling is not mutual. Within the cognitive science community, there is more interest in neural networks, in connectionism at this point. The attitude toward neural networks and connectionism within the Artificial Intelligence community is somewhat mixed, because the Artificial Intelligence community is, as I said earlier, committed to symbol manipulation. Connectionism is not symbol manipulation. That in itself constitutes a bar against the acceptance of the Artificial Intelligence community.

In Europe, things might be different. Artificial Intelligence people have perhaps a closer relationship with cognitive science, and that relationship may be likely to grow. Because I am not on the European scene, it is pretentious if I try to analyze the situation in a reliable way, but this is the feeling that I have. Another factor is that the pattern of support in Europe is different.

In the United States you have to sell research, and therefore you have to be a salesperson. You try to sell your field; you try to sell yourself. I am exaggerating a little bit, but the essence of what I say is true. For this reason, there tends to be a more competitive spirit in the United States. Different fields view one another as competitors for funding, attention, positions, and so forth. In Europe this is less pronounced because of the tradition of research funded by the government. But this is changing. The systems that are set up now in Europe are more similar to the system in the United States—for example, the ESPRIT program.

What do you think about the Turing Test?

The Turing Test is a useless test. I have always said this. It is a test that people talk about, but it does not make any sense to me.

All you need is the following test: the human or the machine to be tested is behind the curtain, and you ask it to summarize what you tell it. No machine will be able to pass this test. You need no other test, just ask the machine to summarize what you said or typed.

But there are some programs that claim to do summarizing?

Extremely rudimentary. These programs do not have the capability at this point, and they are not likely to have it in the future to summarize nonstereotypical stories of a length of one thousand words or so. They are able to summarize very short, stereotypical stories. At this point we have no idea whatsoever how to build a machine able to summarize. Summarization is much more difficult than machine translation because it requires understanding. Because it is a test of understanding, it requires the totality of human knowledge. We cannot implement this knowledge on our machines, and at this point we do not have the slightest idea how it could be done.

What are the major problems that Artificial Intelligence has to overcome to make progress?

One of the most important problems in Artificial Intelligence is the problem of commonsense reasoning. Many people realize that a number of other problems like speech recognition, meaning understanding, and so forth depend on commonsense reasoning. In trying to come to terms with commonsense reasoning, Artificial Intelligence uses classical logic or variations of it, like circumscription, nonmonotonous reasoning, default reasoning, and so on. But these techniques make no provision for imprecision, for fuzzy probabilities and various other things.

I think it is impossible to come to grips with commonsense reasoning within the framework of traditional Artificial Intelligence. We need fuzzy logic for this. Most people in the Artificial Intelligence community are not prepared to accept fuzzy logic, because they made investments in classical logic; they have spent years learning and using it. They are not prepared to throw it away for something they don't know. Human nature does not work that way. That is part of the reason for this refutation.

But commonsense reasoning is a prerequisite for the solution of many other problems, and it will not be an easy problem to solve.

What do you think about the Chinese Room Argument? In some ways your own idea of a test is similar to it. Is Searle right?

I think Searle is right but that the whole thing is much simpler than he presents it. Basically, what is involved is this: machines are symbol manipulating devices. Humans can understand because of the connection between these symbols and the world we live in. Computers, as they stand, do not provide this correspondence. If I say "red flag," to a computer this is nothing but the symbols "R-E-D-F-L-A-G," but to the humans it is something they can see and feel; they know what it means. As long as you merely manipulate symbols, you have a limited capability of understanding what goes on in the world.

You may be able to fool people: database systems and expert systems can do it. Database systems can answer all kinds of questions, not because they understand but because they manipulate symbols, and this is sufficient to answer questions. Such a system could answer the question "What is the best wine?" by looking it up in the database, but it cannot taste wine. What Searle basically points to is the limitation associated with symbol manipulating systems. But as I said earlier, it is not necessary to go into all these things that Searle has done to explain it. I feel that it makes the issue more complicated than it needs to be. To people in Artificial Intelligence, these are not really deep issues, although there have been debates going on for maybe forty years.

In your opinion, will social problems arise with the further development and application of Artificial Intelligence?

There may well be. There are certain issues, one of which is concerned not with Artificial Intelligence but with the capability of computers to store large volumes of information so that our lives can be monitored more closely. There is a danger that our personal freedom will be restricted because it will be easy to track our lives, to track what we have done twenty or more years ago.

But what is more important is that many decisions will be based on computers that use Artificial Intelligence. When you apply for credit—

this is done already—your application will be assessed by computer, and the computer may say no. Somebody bought a program that decided that you should not get credit. The same thing can happen in other fields—when it comes to promotion or whatever. Some machine will look at your qualifications and make a decision.

When the machine makes an error, it is very difficult to fight it. I have a collection of horrible stories about humans who got stuck with errors made by computers. There is a definite danger that this might happen. People like Professor Weizenbaum or Terry Winograd are worried about these issues and thereby perform a valuable service.

What could be done about these future problems?

It is very difficult to do much about them. For totalitarian governments it will be much easier to check what people are doing, what they have said and written and so forth. Even now, many of these intelligence agencies monitor telephone conversations, and if you have the capability of processing huge amounts of information, they can monitor everybody's telephone conversations. I don't think this will happen, but the capability is there. And once the capability is there, there will be the temptation to misuse it. It is not very likely that these dangers will become so pronounced as to lead to a revolt or to the destruction of computers. It will be more insidious, a gradual taking-over of decision-making processes of various kinds; and at some point people may feel that they have not much left to do.

To conclude: what are, in your opinion, the future prospects of Artificial Intelligence?

Artificial Intelligence may not maintain its unity. It is very possible that within the next decade we will see a fragmentation of Artificial Intelligence. It will be broken up into pieces. The most important piece will be knowledge-based systems and expert systems. This branch started within Artificial Intelligence, but it is like a child that has grown up and is leaving its parents. There are many people working in these fields who are not members of the Artificial Intelligence community. They come from operations research, from statistics, from all sorts of fields.

Another field that may well leave Artificial Intelligence is robotics, also a very important field that has equally grown inside Artificial Intelligence. It already has its own conferences, publications, and so forth. The two currently most important fields within Artificial Intelligence are going to leave Artificial Intelligence. Then there is the field of voice input–voice output, and that too is leaving Artificial Intelligence.

So the question is: what is going to remain? What is going to remain is not *that* important. Computer vision will go together with robotics.

The things that are going to leave Artificial Intelligence are those that do not lend themselves to symbol manipulation: robotics, computer vision, and knowledge-based systems have problems with symbol manipulation because of imprecision and so on. If you look at the programs of Artificial Intelligence conferences and subtract these things, you are left with things like game playing, and some Artificial Intelligence–oriented languages. These things will not be nearly as important as Artificial Intelligence was when knowledge-based systems were a part of it. Artificial Intelligence will be like parents left by their children, left to themselves. Artificial Intelligence, to me, is not going to remain a unified field.

What about connectionism?

I think there are exaggerated expectations at this point with regard to connectionism. I think it is certainly an interesting field, and it has been in existence for a long time. There will be some applications that will turn out to be useful and important. But at this point, people are expecting too much. They think that connectionism and neural networks will solve all kinds of problems that, in reality, will not be solved.

I myself am interested in this field, just as I have been interested in Artificial Intelligence for a long time, being aware of the exaggerations. The mere fact that the expectations are exaggerated does not mean that one should leave aside that field, that it should not be supported, that it should be criticized. It means merely that one should be cognizant of the fact that more is promised than is likely to be delivered.

Recommended Works by Lotfi A. Zadeh

"Fuzzy Logic." *Computer* 21, no. 4 (1988): 83–92.
"Fuzzy Sets." *Information and Control* 8 (1965): 338–53. (The first article on fuzzy logic.)
(Ed.) *Fuzzy Logic for the Management of Uncertainty*. New York: Wiley, 1992.

Notes

1. L. Zadeh, "Fuzzy Sets," *Information and Control* 8 (1965): 159–76.

2. Summer institute at Dartmouth College; for a historical overview of cognitive science see H. Gardner, *The Mind's New Science* (New York: Basic Books, 1985).

3. DENDRAL, a rule-based heuristic program for analysis of mass spectographs in organic chemistry. E. A. Feigenbaum, B. G. Buchanan, and J. Lederberg, "On Generality and Problem Solving: A Case Study Using the DENDRAL Program," *Machine Intelligence* (Edinburgh: Edinburgh University Press, 1971), vol. 6.

4. MYCIN, a medical diagnosis of meningitis and blood infections; see E. H. Shortliffe, *Computer-Based Medical Consultation: MYCIN* (New York: Elsevier, 1976); and

B .G. Buchanan and E. H. Shortliffe, eds., *Rule-Based Expert Systems* (Reading, Mass.: Addison-Wesley, 1984).

5. PROSPECTOR, an expert system for mineral exploration.

6. See E. A. Feigenbaum and P. McCorduck, *The Fifth Generation: Artificial Intelligence and Japan's Computer Challenge to the World* (Reading, Mass.: Addison-Wesley, 1983).

7. D. Michie reports this controversy in the book by D. Michie and R. Johnston, *The Creative Computer* (Harmondsworth: Viking, 1984).

GLOSSARY

Background   See COMMONSENSE KNOWLEDGE.

Back propagation   Also known as *generalized delta rule*. A learning algorithm in the connectionist paradigm (see CONNECTIONISM). The aim is to learn association pairs of patterns. The input is propagated through the network, producing a certain output that is due to the pattern of weights. For each output unit the difference between the desired ("target") and the actual output is computed. This value is called the error; it is recursively propagated via the hidden units to the input units. The weights are changed because of the (local) error they have contributed to the global error. The overall performance of the network will improve after this change of weights (for the given input-output pair).

Behaviorism   Behaviorism "held that mental states are identical with sets of actual and counterfactual overt behaviors and that inner states of the subject, though . . . causally implicated in such behaviors, were not theoretically important to understanding what it is to be in certain mental states" (Clark 1989, 22). Behaviorism was criticized by IDENTITY THEORIES, which stressed the importance of inner (brain) processes, identifying mental states with them. Behaviorism is connected with the names of J. B. Watson and B. F. Skinner and was the leading school of psychology in the first half of the twentieth century. The advance over earlier schools (relying mainly on introspection as a psychological research method) was the concentration on the study of behavior. Although methodological behaviorism is still an accepted foundation even of cognitive science, logical behaviorism (making ontological claims) was the school of thought against which the "cognitive revolution" turned its attacks.

Boltzmann machine   A connectionist network type that finds a global energy minimum by applying statistical learning methods. It solves constraint satisfaction problems and finds a global solution by using a kind of simulated annealing strategy for getting out of local minima. See also CONNECTIONISM.

Cartesian dualism   See DUALISM.

Character recognition   See PATTERN RECOGNITION.

Chinese Room   In his paper "Minds, Brains, and Programs" (1980), John Searle argues that computational theories in psychology are essentially useless. For this purpose, he presented the Chinese Room parable. He imagines himself locked in a room, in which there are various slips of paper with doodles on them, a slot through which people can pass slips of paper to him and through which he can pass them out; and a book of rules telling him how to respond to the doodles, which are identified by their shape. One rule, for example, instructs him that when squiggle-squiggle is passed in to him, he should pass squoggle-squoggle out. So far as the person in the room is concerned, the doodles are meaningless. But unbeknownst to him, they are Chinese characters, and the people outside the room, being Chinese, interpret them as such. When the rules happen to be such that the questions are paired with what the Chinese people outside recognize as a sensible answer, they will interpret the Chinese characters as meaningful answers. But the person inside the room knows nothing of this. He is instantiating a computer program—that is, he is performing purely formal manipulations of uninterpreted patterns; the program is all syntax and has no semantics. Searle claims, with this parable, that semantics (meaning, understanding) cannot arise from purely formal symbol manipulation and that the system does not have the causal power that humans have (INTENTIONALITY) to generate meaning. As a reaction to connectionism, John Searle restated the parable as a "Chinese gym hall" where a number of people instantiate a program, with the same result, namely, that they do not understand Chinese (Searle 1990, 42). Already in his 1980 article Searle mentioned a few possible counterarguments against his parable. Among the most frequent arguments are the so-called SYSTEMS REPLY and the ROBOT REPLY.

Commonsense knowledge The everyday, practical knowledge used by people in mastering real life situations. The "commonsense problem" in Artificial Intelligence is based on the realization that commonsense knowledge plays a role in solving most tasks, except where a domain of knowledge can be strictly defined and restricted (e.g., chess, microworlds). This has also often been called the "frame problem." The research on "naive physics" is part of the effort to solve the commonsense problem, concentrating on the representation of everyday physical capacities and knowledge (e.g., about inertia and gravity; see Hayes 1969 and 1985). The debate in Artificial Intelligence centers around the question of whether this problem is one of quantity (and therefore can be solved by huge knowledge bases) or one of quality, that is, that commonsense knowledge comprises practices, skills, and capacities that are not representational. In Searle's theory of intentionality, he defends the notion of such a nonrepresentational "background" for all intentional states (Searle 1983).

Connectionism The basic idea of connectionism can be characterized as follows: it turns out that conventional computers (based on the VON NEUMANN ARCHITECTURE) have a radically different structure from our nervous system. The aim of connectionism is no longer to simulate cognitive processes by means of symbols and symbol manipulation (see PHYSICAL SYMBOL SYSTEM) but to simulate the processes taking place in neural systems. These artificial neural networks represent, of course, only very (mathematically) abstract and rough models of their natural originals. The connectionist approach is not so much interested in the function of a single neuron as in the global behavior of simple processors being called "units." One unit represents a neuron in a very abstract manner. It is connected by "weights" (comparable to synaptic connections) to some other units and computes its own activation by summing up the product of the weights and the other units' activations. The output of the unit is computed by a sigmoid or linear function mapping this sum of products ("netto input") to a certain range of values. This output again acts as an input for other units doing exactly the same in parallel. In most cases a synchronous and discrete timing is assumed. A connectionist system (and its environment) can be considered to consist of the following parts:

1. a set of processing units (which can be compared to neurons). They are assumed to be an n-dimensional vector, where n is the number of the system's units. Three kinds of units can be distinguished: input units, by which external stimuli can enter the system; output units, connected to an output device decoding the activations to an output; and hidden units, internal units that interact not with the environment but only with other units;

2. a state of activations that is defined over the units, that is, at each time t each unit has a certain activation, $a(t)$, which can also be seen as an n-dimensional vector;

3. an activation rule that computes the new level of activation at time $t + i$, that is, $a(t + i)$, by using the current activation $a(t)$ and the inputs of the unit;

4. an output function that maps the activation to the output of the unit;

5. the pattern of connectivity. It determines the influence that one unit has on another unit by weighting the inputs. The weights can be compared to the synapses of a natural

neural system. A positive weight has an excitatory, and a negative weight an inhibitory, influence on the unit's activation (under the assumption of positive activations). The pattern of connectivity is also called the architecture of the network and can be understood as an $n \times n$ weight matrix. It is responsible for the dynamics of the system and plays an important role in the context of the knowledge that is represented by this system;

6. the learning rule is responsible for changing the pattern of connectivity; that is, by changing the weights, the knowledge being represented in the network, and its dynamics, are changed. Normally the weights are changed in small increments or decrements, which are computed by the learning rule. It is determined by the current pre- and post-"synaptic" activation, by the current weight, sometimes by an external target value, and by a linear learning factor (normally 1);

7. the environment with which the system is interacting;

8. the periphery—that is, the sensory and motor system—plays an important role, because it is responsible for coding and decoding "external" stimuli and neural activations. Much of the knowledge being represented in the whole system can be found here;

9. the context of interpretation or the semantic embedding is, as items 7 and 8, not considered, in most publications concerning (natural or artificial) neural networks or models of cognition, to be one of the most influential factors. From an epistemological perspective, however, the context of interpretation is of special interest, because it determines the kind of knowledge that is represented in the system.

Connectionist networks can have either feed-forward architectures (the flow of activations goes in one direction from the input units to the output units) or recurrent architectures (with feedback connections). The latter entails that the network has its own dynamics, because it is interacting not only with the stimuli coming from the environment but also with its own activations. Both types of networks show quite complex behavior, which is due to the large number of quite simple processes taking place in the single units, to the parallel structure of these processes and to the massive connectivity. An interesting characteristic of connectionist networks is their ability to improve their performance, which is called "learning." Learning has a strong influence on the way that knowledge is represented as it changes the network's weights. Learning is realized by slight changes of the weights, where small local changes can cause a change in the global behavior of the network.

Connection Machine   Brought to market in 1986 by Thinking Machines, Inc., a spin-off of the MIT Artificial Intelligence laboratory. Originally conceived in 1981 by W. Daniel Hillis. The first commercial version of the Connection Machine has up to sixty-four thousand processors and is capable of executing roughly one billion instructions per second. Its architecture is an implementation of massive parallel processing (see CONNECTIONISM), each processor being small and relatively simple but provided with its own memory (see Waldrop 1987, 114).

Conversational analysis   An empirical method to study coherence and sequential organization of discourse. Recorded, naturally occurring discourses are analyzed with inductive methods, whereby recurring patterns are revealed. It is strongly connected with ETHNOMETHODOLOGY by its history as well as by its methodological and theoretical approach (see Levinson 1983, 286f.).

Cybernetics   The word was introduced by the mathematician Norbert Wiener (1948), who defined cybernetics as "the art and science of control over the whole range of fields in which this notion is applicable." It is a theory of self-regulating (feedback) systems that is applicable to both machines and living systems.

Delta rule   See BACK PROPAGATION.

Dualism   The philosophical view that mind and body (or matter in general) are substantially different. Cartesian dualism (named after Descartes) holds that mind and body are both substances; but whereas the body is an extended, and so a material, substance, the mind is an unextended, or spiritual substance, subject to completely different principles of operation from the body. Dualist theories differ in the way they explain the interaction between mind and matter, that is, how the nonmaterial can have causal effects upon matter. Where dualism is referred to in this book, what is meant is, more specifically, the relationship between mind and brain. The philosophical explanations discussed in this book—behaviorism, identity theory, functionalism, and eliminative materialism—are attempts at nondualistic (materialistic) theories of the mind-body relationship.

Eliminative materialism   The commonsense and traditional philosophical views hold that propositional attitudes such as beliefs and desires are real inner causal states of people. Eliminative materialism criticizes that these mental categories arise from folk psychology and cannot serve as the bases for scientific explanations. "Eliminative materialism is the thesis that our common-sense conception of psychological phenomena constitutes a radically false theory, a theory so fundamentally defective that both the principles and the ontology of that theory will eventually be displaced, rather than smoothly reduced, by completed neuroscience" (P. M. Churchland 1990, 206).

Ethnomethodology   A conception of sociological research that studies the everyday methodology with which members of a group or society organize their practice of action and interaction and ascribe meaning to it. For ethnomethodology, social reality is the product of such meaning ascriptions. "Ethnomethodological studies analyze everyday activities as members' methods for making those same activities visibly-rational-reportable-for-all-practical-purposes, i.e., 'accountable', as organizations of commonplace everyday activities. . . . I use the term 'ethnomethodology' to refer to the investigation of the rational properties of indexical expressions and other practical actions as contingent ongoing accomplishments of organized artful practices of everyday life" (Garfinkel 1967, vii and 11).

Expert systems   An expert system is an Artificial Intelligence program incorporating a knowledge base and an inferencing system. It is a highly specialized software that attempts to duplicate the work of an expert in some field of expertise and is used as consultant in the domain—for example, for diagnosis in medicine (MYCIN), analysis of chemical compounds (DENDRAL), configuration of computer systems (XCON), and so on. The aim of expert systems research is to make knowledge of experts more widely accessible. The main problem with building expert systems is the elicitation and representation of experts' knowledge (knowledge engineering). The whole enterprise is based on the assumption that such knowledge can be formalized and expressed in terms of KNOWLEDGE REPRESENTATION methods. According to Dreyfus's and Dreyfus's critique (1987), this is a fundamental obstacle and due to the nature of expertise.

An expert system contains these four components:

1. knowledge base: representation of the heuristic knowledge of one or more experts. Most knowledge bases are made up of production rules (see PRODUCTION SYSTEM), but other forms of knowledge representation do appear.

2. inference engine: it performs the reasoning function using the rules, implements a search strategy and pattern-matching scheme to select rules to apply.

3. database: it holds information about the status of the system. It contains initial conditions or facts and, later, inferred facts.

4. user interface: this is the part that interacts with the user, asking for input and providing output; an important part of expert systems is the explanation component that makes the inference process transparent to the user.

Functionalism   Functionalism was originally the name of an American school of psychology at the turn of the century, based on the functional significance of adaptive behavior. Recent functionalist theories can be seen as a reaction against IDENTITY THEORIES. Mental states are identified not with the brain's physical states but with its functional states. Such states can be specified in formal terms (i.e., in terms of logical and computational processes) and are neutral with respect to their physical realization, which is the reason for the attractiveness of functionalist theories to Artificial Intelligence research and cognitive science. Putnam ("Minds and Machines," 1960) specifies the functional states as the states of the TURING MACHINE ("Turing machine functionalism") and suggests that mental states are functionally equivalent in brains and computers (see Gregory 1987, 280).

Graceful degradation   The capacity of a system to deal with shortcomings of hardware (damage, etc.) and partial or erroneous data. Connectionist models of cognitive processes display graceful degradation of both types and are, in this respect, more psychologically real than symbol processing models, because natural cognitive systems clearly have the sort of flexibility described by the term graceful degradation.

Hidden units   See CONNECTIONISM, PERCEPTRON.

Hopfield net   An interactive network developed by the physicist John Hopfield (1982) by analogy with a physical system known as a spin glass. A spin glass consists of a matrix of atoms that may be spinning either pointing up or pointing down. Each atom exerts a force on its neighbor, leading it to spin in the same or in the opposite direction. A spin glass is an instantiation of a matrix or lattice system that is capable of storing a variety of different spin patterns. In the analogous network, the atoms are represented by units, and the spin is represented by binary activation values that units might exhibit (0 or 1). The influence of units on their neighbors is represented by means of bidirectional connections (hence "interactive network"). Any unit can be connected to any other unit. In Hopfield networks, as in all interactive networks, activations are updated across multiple cycles of processing in accord with an activation rule (see Bechtel and Abrahamsen 1991, 38).

Identity theory   Also known as *mind-brain identity theory*. A materialist theory of the mind that identifies mental processes with purely physical (neuronal) processes in the central nervous system (hence the name *central-state materialism*). The consequence of this view is that only beings with identical neuronal structure can have the same mental states. Criticized by FUNCTIONALISM.

Intentionality   Intentionality is "aboutness," that is, the quality of certain things to be "about" other things. A belief is "about," for example, an iceberg, but a thing like the iceberg itself is not "about" anything. The technical term *intentionality* is often confused with the ordinary language notion of "intention" in the sense of a plan to do something.

The term was coined by the Scholastics and revived by Franz Brentano. In Brentano's definition, intentionality is the defining distinction between the mental and the physical. For him, all and only mental phenomena exhibit intentionality.

Searle (1983) formulates intentionality as "that property of many mental states and events by which they are directed at or about or of objects and states of affairs in the world" (p. 1), where every intentional state consists of an intentional (representational) content in a psychological mode—for example, belief or wish (p. 12).

Knowledge representation   In its most general sense, knowledge representation means that an intelligent system has descriptions or pictures that correspond in some salient way to the world or a state in the world. In Artificial Intelligence, these descriptions have to be such that an intelligent machine can come to new conclusions about its environment by formally manipulating them. In the symbol manipulation approach to Artificial Intelligence, one can distinguish propositional (sentencelike) representation, often in a network model (see SEMANTIC NETWORKS), from complex representation models (schemata), like frames or scripts, that allow also for the representation of more general, abstract information about objects, states, or events. A schema is a higher-order unit of knowledge representation that groups subunits (slots in frames, conditions, actions, etc., in scripts) and, by being activated as a whole, should give access to the relevant information pertaining to the situation and should therefore provide models of knowledge with more psychological plausibility (see Stillings et al. 1987, 25ff.). Knowledge representation research is confronted with problems of relevance and of COMMONSENSE KNOWLEDGE.

Language acquisition   The question of how a child acquires her native language has been an important research area of linguistics and psychology for a long time. The potential for spontaneous language learning is innate and in this form unique to human beings, but the discussion revolves around what and how much is innate. Language acquisition research studies language development in children. The development has discernible stages, leading the child from one-word, via two-word, to more complex utterances, finally approaching adult language at school age. Language development involves the development and use of specific grammars that allow the child to produce regular sentences by rules that are both overgeneralized and deviant from adult language. Whereas Chomskian linguistics concentrated on the linguistic competence of the child, language development is now seen to be connected with general cognitive development and, at the same time, with the development of social behavior.

LISP machine   A single-user computer or workstation designed primarily for the development of Artificial Intelligence programs. The name is derived from LISP (LISt Processor), a programming language especially useful for Artificial Intelligence programming and characterized by its list structures. LISP was developed by John McCarthy and his team at MIT and is now marketed in different "dialects."

Massively parallel processing   See CONNECTIONISM.

Methodological solipsism   Solipsism is the philosophical view that takes the subjective self with its consciousness for the only existing thing. On the contrary, methodological solipsism does not deny an external world but takes the solipsistic viewpoint as a starting point for studying the reality that exists outside the self. In his article "Methodological Solipsism Considered as a Research Strategy in Cognitive Psychology" (1980), Jerry Fodor shows that cognitive psychology subscribes to methodological solipsism insofar as it accepts a computational theory of the mind.

Naive physics   See COMMONSENSE KNOWLEDGE.

Neural network   Another name often used for connectionist models of cognition (see also CONNECTIONISM), inspired by neuroscience.

Neuroscience The common designation for disciplines concerned with the study of processes in animal and human nervous systems. Neurophysiology is the study of the functions of the nervous system; neuroanatomy is concerned with the structure of the nervous system; and neuropsychology studies the relation between neural and psychological functioning.

Pattern recognition The technology of automatic recognition and analysis of patterns by machines. The problem is to identify varying tokens of a pattern (e.g., a letter in character recognition, handwritten by different persons at different occasions) as an instance of a certain type of pattern (e.g., the letter *A*). From a series of detected features of a pattern, the system has to decide whether a feature is similar enough to the template to warrant its classification. This task has been notoriously difficult for "classical" (i.e., symbol manipulation) Artificial Intelligence, whereas network models were more successful from the very beginning (see also PERCEPTRON, CONNECTIONISM). Connectionist networks are particularly adapted to deal with the nature of the problem of pattern recognition, which encompasses a multitude of input features, incomplete and deviating input (see also GRACEFUL DEGRADATION), and computation of similarities (as weights) (Gregory 1987, 591f.).

Patterns of activation See CONNECTIONISM.

Perceptron Studying the problem of pattern recognition in networks, Frank Rosenblatt (1959, 1962) developed layered networks with binary units, which he called perceptrons. In a perceptron, one set of units receives inputs from outside and sends excitations and inhibitions to another set of units, which then send inputs to yet a third group, or back to earlier layers. He made the strengths (weights) of the connections between units continuous rather than binary and introduced procedures for changing these weights, enabling the networks to be trained to change their responses. Rosenblatt emphasized the difference between perceptrons and symbol manipulation (see Bechtel and Abrahamsen 1991, 5f., 14).

By the 1960s, substantial progress had been made with both network and symbolic approaches to machine intelligence, but this parallel research strategy was soon given up in favor of symbol manipulation approaches (see PERCEPTRONS).

Perceptrons The title of a book published in 1969 (reprinted in 1988) by M. Minsky and S. Papert. Their objective was to study both the potential and the limitations of network models. They analyzed mathematically what kinds of computation could or could not be performed with a two-layer perceptron. They demonstrated that certain functions cannot be evaluated by such a network, for example, the logical operation of XOR (exclusive or). In order for a network to compute XOR, it is necessary to include an additional layer of units (now referred to as hidden units) between the input and output units. Although Minsky and Papert recognized that XOR could be computed by such a multilayered network, they also raised further doubts about the usefulness of network models. Their central question was whether networks would scale well—that is, whether it would be possible to increase their size. Their intuitive judgment was that network research would not lead to important results (see Bechtel and Abrahamsen 1991, 14–16).

A more general criticism of network research was that it was based on psychological associationism, a school of thought that the new cognitive paradigm crusaded against.

Phoneme recognition An approach to SPEECH RECOGNITION for continuous speech. A phoneme is a distinctive sound of speech and thus the smallest distinguishable and distinguishing unit of speech. The number of phonemes in every natural language is rather small (e.g., forty in English). Speech input is matched against phoneme templates stored in memory. The system tries to reconstitute words from single recognized phonemes. This method of speech recognition is more difficult than word recognition but produces better results (see Frenzel 1987, 180).

Physical symbol system   A physical symbol system is any member of a general class of physically realizable systems meeting the following conditions:

1. it contains a set of symbols, which are physical patterns that can be strung together to yield a structure (or expression);

2. it contains a multitude of such symbol structures and a set of processes that operate on them (creating, modifying, reproducing, and destroying them according to instructions, themselves coded as symbol structures); and

3. it is located in a wider world of real objects and may be related to that world by designation (in which the behavior of the system affects or is otherwise consistently related to the behavior or state of the object) or interpretation (in which expressions in the system designate a process, and when the expression occurs, the system is able to carry out the process) (Newell and Simon 1976, 40ff.).

Physical symbol systems hypothesis   "The necessary and sufficient condition for a physical system to exhibit general intelligent action is that it be a physical symbol system." If a PHYSICAL SYMBOL SYSTEM is any system in which suitably manipulable tokens can be assigned arbitrary meanings and, by means of careful programming, be relied on to behave in ways consistent with this projected semantic content, and if (as Newell and Simon claim) the essence of thought and intelligence is this ability to manipulate symbols, then any physical symbol system (such as the computer) "can be organized . . . to exhibit general intelligent action" (Newell and Simon 1976; see also Clark 1989).

Printed character recognition   See PATTERN RECOGNITION.

Production system   In 1943, Emil Post proposed an "if-then" rule-based system that indicated how strings of symbols could be converted to other symbols. It was later shown that the Post system is formally equivalent to the TURING MACHINE. In the sixties, Newell and Simon developed this representation into a data-driven control structure at the basis of the present concept of a production system. Computation in a production system is the process of applying rules in an order determined by the data. A production system has three components:

1. working memory (data representing the current state of the "world");

2. production rules (a set of rules of the general form
IF condition in working memory
THEN action); and

3. rule interpreter (control module that carries out the matching process that determines which rule to activate—special control strategies are necessary, e.g., when the left side, or condition, of two or more rules is satisfied by the current data) (Newell 1973; see also Frenzel 1987, and Stillings et al. 1987).

Qualia   (Latin, plural of *quale*, meaning quality as opposed to amount, quantity, *quanta*.) To specify qualia is to say what something is like, the way in which things appear to the conscious subject. The sensible qualities of experiences (sight, touch, taste, smell and hearing)—for example, the "redness" or "softness" of an object—have a distinctive phenomenological character difficult to describe.

Reductionism holds that qualia can be fully explained in terms of neurophysiological events in the brain and its interactions with its environment. In the account of epiphenomenalism, qualia are causally dependent on brain events but cannot be identified with them. Dualism holds that qualia are independent of physics and belong to the autonomous, nonphysical realm of the mind (see Gregory 1987, 666).

Reductionism Assumption that a scientific explanation is found by reducing complex phenomena to separate simple elements. In psychology, reductionism looks for elementary events that, in combination, explain all forms of human behavior. Reductionism is criticized on the grounds that by reducing a phenomenon to its elements its specific characteristics are lost, so that an explanation on this level becomes void (see Gregory 1987).

Robot reply The robot reply to the CHINESE ROOM parable is resumed by Searle (1980, 420) as follows: "Suppose we put a computer inside a robot, and this computer would not just take in formal symbols as input and give out formal symbols as output, but rather would actually operate the robot in such a way that the robot does something very much like perceiving, walking, moving about, hammering nails, eating, drinking—anything you like. The robot would, for example, have a television camera attached to it that enabled it to 'see,' it would have arms and legs that enabled it to 'act,' and all of this would be controlled by its computer 'brain.' Such a robot would . . . have genuine understanding and other mental states."

Semantic networks A model for the representation of long-term knowledge—for example, concepts. The network consists of nodes that represent further concepts. They are connected by links of different types: class membership ("is a"), whole-part relation ("has"), functions, attributes, and so on. Semantic networks are inspired by the psychological theory of association of ideas but can also be hierarchical in their structure. The method of retrieval of information is called "spreading activation": when a source node is activated, the level of activation spreads through all its links to further nodes until a target node (requested information) is reached. Semantic networks are one of the earliest models of knowledge representation and basically parallel, because activation would have to spread simultaneously to all nodes and be controlled by the weights of the links (see Stillings et al. 1987, 26, 89).

Speech act theory In the definition by Austin (1965), by uttering a sentence one can be said to perform three simultaneous acts:

1. locutionary act: the utterance of a sentence with determinate sense and reference;
2. illocutionary act: the making of a statement, offer, promise, and so on in uttering a sentence, by virtue of the conventional force associated with it (or with its explicit performative paraphrase); and
3. perlocutionary act: the bringing about of effects on the audience by means of uttering the sentence, such effects being special to the circumstances of utterance (Levinson 1983, 236). The term *speech act* is generally used to refer to item 2. Speech act theory can be seen as a reaction against logical positivism, which held that a sentence that cannot be, at least in principle, tested for truth or falsity, is meaningless. This definition of meaning is replaced in speech act theory by one relying on felicity conditions that must be fulfilled for a speech act to succeed and is thus narrowly connected with the use of language.

Speech recognition Also known as voice recognition. An application of Artificial Intelligence with the aim of making computers recognize human speech, for example, for voice input of commands or text. The voice input to the microphone is digitized and compared to previously stored speech patterns. The recognized words can be passed on to a natural language processing system to reconstitute the recorded utterance as a whole.

One major problem of speech recognition lies in the wide variation from one to speaker to another. Another problem is the recognition of continuous speech, which demands powerful processing (in real time) and poses the problem of distinguishing individual words in the continuous flow of speech. One approach to this problem is PHONEME REC-OGNITION (see Frenzel 1987, 179f.).

Spreading activation   See SEMANTIC NETWORKS.

Symbolic interactionism   Goes back to the work of George Herbert Mead (1934). The name of this line of sociological and sociopsychological research was coined in 1938 by Herbert Blumer (1938). Its focus is processes of interaction—social action that is characterized by an immediately reciprocal orientation—and the investigations of these processes are based on a particular concept of interaction that stresses the symbolic character of social action (Joas 1987, 84). Blumer (1969, 2) defines three premises on which symbolic interactionism rests:

1. human beings act toward things on the basis of the meanings that the things have for them;

2. the meaning of things is derived from, or arises out of, the social interaction that one has with one's fellows; and

3. these meanings are handled in and modified through an interpretative process used by the person in dealing with the things he or she encounters.

System reply   The system reply to the CHINESE ROOM parable says that "while it is true that the individual person who is locked in the room does not understand the story, the fact is that he is merely part of a whole system, and the system does understand the story. The person has a large ledger in front of him in which are written the rules, he has a lot of scratch paper and pencils for doing calculations, he has 'data banks' of sets of Chinese symbols. Now, understanding is not being ascribed to the mere individual; rather it is being ascribed to this whole system of which he is a part" (Searle 1980, 419).

Theory dualism   In P. S. and P. M. Churchland's terms, a contemporary form of DUALISM that denies that psychological theory can be reduced to neuroscience or that a unified science of the mind-brain is possible (Churchland and Sejnowski 1990, 228; see also the interview with P. M. Churchland in this volume).

Turing machine   The British mathematician Alan M. Turing (1912—54) was a key figure in the conception of electronic digital computers and Artificial Intelligence. His fundamental contribution was the paper "On Computable Numbers" (1937), in which he showed that certain classes of mathematical problems cannot be proved by any fixed definite process or heuristic procedure. He proposed an automatic problem-solving machine, now called the Turing machine, which is abstract in the sense that its description defines all possible operations, though not all may be realizable in practice.

The Turing machine can be visualized as an indefinitely long tape of squares on which are numbers or a blank. The machine reads one square at a time, and it can move the tape to read other squares, forward or backward. It can print new symbols or erase symbols. Turing showed that this very simple machine can specify the steps required for the solution of any problem that can be solved by instructions, explicitly stated rules, or procedures (see Gregory 1987, 783f.; see also Weizenbaum 1976).

Turing Test   The Turing Test is the imitation game proposed by A. Turing (see TURING MACHINE) in his paper "Computing Machinery and Intelligence" (1950) in order to replace the question of whether a machine is intelligent or not. This is how Turing describes the game:

"It is played with three people, a man (A), a woman (B), and an interrogator (C), who may be of either sex. The interrogator stays in a room apart from the other two and communicates with them via teletype in order to avoid recognition by sight or voice. The object of the game for the interrogator is to determine which of the other two is the man and which is the woman." The interrogator is allowed to put questions to A and B, though not about their physical characteristics, and he does not hear their voices. He is allowed to experience or question only mental attributes. The next step of the game is the Turing Test, as it is understood now in Artificial Intelligence. Now one of the humans is replaced by a machine. The question is whether the interrogator can distinguish the remaining human from the machine (see Gregory 1987, 784; Turing's article has been reprinted in several collections, e.g. Feigenbaum and Feldman 1963, Hofstadter and Dennett 1981, and Boden 1990).

Vision   In cognitive science, the study of vision is in itself an interdisciplinary field, which should lead to an integrated theory of vision, perception, and knowledge through the work done in neuroscience, psychology of perception, and knowledge representation.

Computer vision is the automatic analysis of television pictures of natural scenes and other three-dimensional compositions. Computer vision may have as its source of data a single static monocular image, a stereo pair of images, or a sequence of images of moving objects. In computer vision, recognition (i.e., classification and assignment of names to objects; see PATTERN RECOGNITION) often appears less important than reconstructing the scene from one or more visual images.

Von Neumann architecture   Named after the mathematician John von Neumann, who was central to its development in the forties and fifties. It is still the most common set of design principles for computers today and has come to be synonymous with the one-CPU serial processing computer. It is also called "stored program design," because both the program and the data to be processed are stored in the computer's memory. The program is a sequence of instructions implementing the algorithm (a step-by-step procedure to solve a given problem). The instructions specify ways of moving data in and out of memory and tell the computer to perform arithmetic and logic operations. The "von Neumann bottleneck" and the limits of this architecture lie not primarily in the sequential processing of the instructions but in the high volume of data transfer between memory and CPU (instructions and data) (see Frenzel 1987, 289).

Weight   See BACK PROPAGATION.

BIBLIOGRAPHY

Recommended Introductory Reading

The following titles represent only a small and subjective selection from among the numerous collections of articles and textbooks that have been published on Artificial Intelligence and cognitive science. Some of the titles listed in the "Recommended Works" sections following each interview are also conceived as introductory works or texts but are not repeated here.

Bechtel, W., and A. Abrahamsen. 1991. *Connectionism and the Mind: An Introduction to Parallel Processing in Networks*. Oxford: Blackwell.

Gardner, H. 1985. *The Mind's New Science: A History of the Cognitive Revolution*. New York: Basic Books.

Lycan, William G., ed. 1990. *Mind and Cognition: A Reader*. Oxford: Blackwell.

Stillings, N. A., M. H. Feinstein, J. L. Garfield, E. L. Rissland, D. A. Rosenbaum, S. E. Weisler, and L. Baker-Ward. 1987. *Cognitive Science: An Introduction*. Cambridge, Mass.: MIT Press.

References

Titles cited in the "Recommended Works" sections following each interview are not included here. This list consists of the titles mentioned in the notes to the interviews and in the introduction and glossary.

Anderson, A. R. 1964. *Minds and Machines*. Englewood Cliffs, N.J.: Prentice-Hall.

Anderson, J. R. 1984. "Spreading Activation." In *Essays in Learning and Memory*, ed. J. R. Anderson and S. M. Kosslyn. New York: Freeman.

———, and G. H. Bower. 1973. *Human Associative Memory*. New York: Holt.

Austin, J. L. 1965. *How To Do Things with Words*. Oxford: Oxford University Press.

Bacon, F. 1960. *The New Organon, and Related Writings*. Ed. and with an intro. by F. H. Anderson. New York: Liberal Arts Press.

Bate, W. J. 1946. *From Classic to Romantic*. London: Harper Torchbooks.

Bateson, G. 1972. *Steps to an Ecology of Mind*. New York: Ballantine Books.

Baumgartner, P. 1993. *Der Hintergrund des Wissens. Vorarbeiten zu einer Kritik der programmierbaren Vernunft*. Klagenfurt: Kärntner Druck- und Verlagsgesellschaft.

———, and S. Payr. 1991. "Körper, Kontext und Kultur. Explorationen in den Hintergrund des Wissens." *Informatik Forum* 4, no. 2: 62–74.

———. 1994. *Lernen mit Software*. Innsbruck: Österreichischer Studienverlag.

Bechtel, W., and A. Abrahamsen. 1991. *Connectionism and the Mind: An Introduction to Parallel Processing in Networks*. Oxford: Blackwell.

Berlin, B. 1974. *Principles of Tzeltal Plant Classification*. New York: Academic.

———, and P. Kay. 1969. *Basic Color Terms*. Berkeley and Los Angeles: University of California Press.

Block, N., ed. 1981. *Readings in Philosophy of Psychology*. Vol. 2. Cambridge, Mass.: Harvard University Press.

Blumer, H. 1969. *Symbolic Interactionism: Perspective and Method.* Berkeley and Los Angeles: University of California Press.

Bobrow, D. G. 1964. "Natural Language Input for a Computer Problem Solving System." Ph.D. diss., Massachusetts Institute of Technology.

———, and A. Collins, eds. 1975. *Representation and Understanding: Studies in Cognitive Science.* New York: Academic Press.

———, and D. A. Norman. 1975. "Some Principles of Memory Schemata." In *Representation and Understanding,* ed. D. G. Bobrow and A. M. Collins, pp. 131–50. New York: Academic Press.

———, and T. Winograd. 1977. "An Overview of KRL: A Knowledge Representation Language." *Cognitive Science* 1, no. 1: 3–45.

———. 1979. "KRL: Another perspective." *Cognitive Science* 3, no. 1: 29–42.

———, and the KRL Research Group. 1977. "Experience with KRL-0: One Cycle of a Knowledge Representation Language." *Proceedings of the Fifth International Joint Conference on Artificial Intelligence,* pp. 213–22.

Boden, M., ed. 1990. *The Philosophy of Artificial Intelligence.* Oxford: Oxford University Press.

Boring, E. G. 1950. *A History of Experimental Psychology.* New York: Appleton-Century-Crofts.

Borst, C. V. 1970. *The Mind-Brain Identity Theory.* London.

Bourdieu, P. 1979. *La distinction. Critique sociale du jugement.* Paris: Editions de minuit.

———. 1980. *Le sens pratique.* Paris: Editions de minuit.

Brachman, R. J., and H. J. Levesque, eds. 1985. *Readings in Knowledge Representation.* San Mateo, Calif.: Morgan Kaufmann.

Broadbent, D. E. 1957. "A Mechanical Model for Human Attention and Immediate Memory." *Psychological Review* 64: 205ff.

Brown, R. 1973. *A First Language: The Early Stages.* Cambridge, Mass.: Harvard University Press.

———, and U. Bellugi. 1964. "Three Processes in the Child's Acquisition of Syntax." *Harvard Educational Review* 34: 133–51.

Bruner, J. S., J. Goodnow, and G. Austin. 1956. *A Study of Thinking.* New York: Wiley.

Buchanan, B. G., and E. H. Shortliffe, eds. 1984. *Rule-Based Expert Systems.* Reading, Mass.: Addison-Wesley.

Chomsky, N. 1965. *Aspects of a Theory of Syntax.* Cambridge, Mass.: MIT Press.

———. 1981. *Lectures on Government and Binding.* Dordrecht: Foris.

———. 1959. Review of Skinner's "Verbal Behavior," *Language* 35: 26–58.

———. 1957. *Syntactic Structures.* The Hague: Mouton.

———. 1956. "Three Models for the Description of Language." *IRE Transactions on Information Theory,* vol. IT-2, p. 3.

———, and Halle, M. 1968. *The Sound Pattern of English.* New York: Harper and Row.

Churchland, P. M. 1984. *Matter and Consciousness: A Contemporary Introduction to the Philosophy of Mind.* Cambridge, Mass.: MIT Press.

———. 1990. "Eliminative Materialism and the Propositional Attitudes." In *Mind and Cognition,* ed. W. G. Lycan, pp. 206–23. Oxford: Blackwell.

Churchland, P. S., and T. Sejnowski. 1992. *The Computational Brain.* Cambridge, Mass.: MIT Press.

———. 1990. "Neural Representation and Neural Computation." In *Mind and Cognition,* ed. W. G. Lycan, pp. 224–52. Oxford: Blackwell.

Cicourel, A. V. 1964. *Method and Measurement in Sociology.* New York: Free Press.

———, and R. Boese, R. 1972. "Sign Language Acquisition and the Teaching of Deaf Children." In *The Functions of Language: An Anthropological and Psychological Approach,* ed. D. Hymes, C. Cazden, and V. John. New York: Teachers College Press.

————. 1972. "The Acquisition of Manual Sign Language and Generative Semantics." *Semiotica* 3: 225–56.

Clark, A. 1989. *Microcognition: Philosophy, Cognitive Science, and Parallel Distributed Processing*. Cambridge, Mass.: MIT Press.

Collins, A. M., and M. R. Quillian. 1972. "Experiments on Semantic Memory and Language Comprehension." In *Cognition in Learning and Memory*, ed. L. W. Gregg, pp. 117–37. New York: Wiley.

Crick, F.H.C., and C. Asanuma. 1986. "Certain Aspects of the Anatomy and Physiology of the Cerebral Cortex." In *Parallel Distributed Processing: Explorations in the Microstructure of Cognition*, vol. 2: *Psychological and Biological Models*, ed. D. E. Rumelhart, J. L. McClelland, and the PDP Research Group, pp. 333–71. Cambridge, Mass.: MIT Press.

D'Andrade, R. 1981. "The Cultural Part of Cognition." *Cognitive Science* 5: 179–95.

————. 1987. "A Folk Model of the Mind." In *Cultural Models in Language and Thought*, ed. D. Holland and N. Quinn, pp. 112–50. Cambridge: Cambridge University Press.

Dennett, D. C. 1991. *Consciousness Explained*. Boston: Little Brown.

————. 1987. *The Intentional Stance*. Cambridge, Mass.: MIT Press.

————. 1981. *Brainstorms*. Cambridge, Mass.: MIT Press.

————. 1968. "Machine Traces and Protocol Statements." *Behavioral Science* 13: 155–61.

Derrida, J. 1972. *La Voix et le Phénomène*. Paris: PUF. Published in English as *Speech and Phenomena, and Other Essays on Husserl's Theory of Signs*. Evanston, Ill.: Northwestern University Press, 1973.

Dretske, F. 1981. *Knowledge and the Flow of Information*. Cambridge, Mass.: MIT Press.

Dreyfus, H. L. 1965. "Alchemy and Artificial Intelligence." RAND Corporation, memo.

————. 1985. "From Micro-Worlds to Knowledge Representation: AI at an Impasse." In *Readings in Knowledge Representation*, ed. R. J. Brachman and H. J. Levesque, pp. 71–94. San Mateo, Calif.: Morgan Kaufmann.

————. 1972. *What Computers Can't Do: The Limits of Artificial Intelligence*. New York: Harper and Row.

————, and S. E. Dreyfus. 1986. *Mind over Machine*. New York: Free Press.

————. 1988. "Making a Mind versus Modeling the Brain: Artificial Intelligence Back at a Branchpoint." In *The Artificial Intelligence Debate*, ed. S. Graubard, pp. 15–44. Cambridge, Mass.: MIT Press.

Edelman, G. M. 1987. *Neural Darwinism*. New York: Basic Books.

Elman, J. L. 1990. "Finding Structure in Time." *Cognitive Science* 14, no. 2: 179–212.

Ericsson, K. A., and H. A. Simon. 1984. *Protocol Analysis: Verbal Reports as Data*. Cambridge, Mass.: MIT Press.

Fauconnier, G. 1985. *Mental Spaces*. Cambridge, Mass.: MIT Press.

Feigenbaum, E. A. 1959. *An Information-Processing Theory of Verbal Learning*. Ph.D. diss., Carnegie Institute of Technology.

————, B. G. Buchanan, and J. Lederberg. 1971. "On Generality and Problem Solving: A Case Study Using the DENDRAL Program." In *Machine Intelligence*, vol. 6. Edinburgh: Edinburgh University Press.

————, and J. Feldman, eds. 1963. *Computers and Thought*. New York: McGraw-Hill.

————, and P. McCorduck. 1983. *The Fifth Generation: Artificial Intelligence and Japan's Computer Challenge to the World*. Reading, Mass.: Addison-Wesley.

Feldman, J. A. 1981. "A Connectionist Model of Visual Memory." In *Parallel Models of Associative Memory*, ed. G. E. Hinton and J. A. Anderson, pp. 49–81. Hillsdale, N.J.: Lawrence Erlbaum.

Feldman, J. A., and D. H. Ballard. 1982. "Connectionist Models and Their Properties." *Cognititive Science* 6: 205–54.

Fillmore, C. 1968. "The Case for Case." In *Universals in Linguistic Theory*, ed. E. Bach and R. Harms, pp. 1–90. Chicago: Holt, Rinehart, and Winston.

———. 1982. "Frame Semantics." In *Linguistics in the Morning Calm*, ed. Linguistic Society of Korea, pp. 111–38. Seoul: Hanshin.

———. 1985. "Frames and the Semantics of Understanding." *Quaderni di Semantica* 6, no. 2: 222–53.

Fodor, J. A. 1975. *The Language of Thought*. Cambridge, Mass.: Harvard University Press.

———. 1980. "Methodological Solipsism Considered as a Research Strategy in Cognitive Psychology." *Behavioral and Brain Sciences* 3: 63–109.

———. 1983. *The Modularity of Mind*. Cambridge, Mass.: MIT Press.

———. 1990. *"A Theory of Content" and Other Essays*. Cambridge, Mass.: MIT Press.

———, and J. Katz. 1964. *The Structure of Language: Readings in the Philosophy of Language*. Englewood Cliffs, N.J.: Prentice-Hall.

———, and Z. W. Pylyshyn. 1988. "Connectionism and Cognitive Architecture: A Critical Analysis." *Cognition* 28: 3–72.

Frege, G. 1985. "On Sense and Meaning." In *The Philosophy of Language*, ed. A. P. Martinich, pp. 200–212. Oxford: Oxford University Press.

Frenzel, L. E. 1987. *Crash Course in Artificial Intelligence and Expert Systems*. Indianapolis: Howard W. Sams.

Fuchs, W., R. Klima, et al., eds. 1975. *Lexikon zur Soziologie*. 3 vols. Reinbek: Rowohlt.

Fuster, J. M. 1989. *The Prefrontal Cortex: Anatomy, Physiology and Neuropsychology of the Frontal Lobe*. 2d ed. New York: Raven Press.

Gardner, H. 1985. *The Mind's New Science: A History of the Cognitive Revolution*. New York: Basic Books.

Garfinkel, H. 1967. *Studies in Ethnomethodology*. Englewood Cliffs, N.J.: Prentice-Hall.

Goldman-Rakic, P. S. 1986. "Circuitry of Primate Prefrontal Cortex and Regulation of Behavior by Representational Memory." In V. B. Mountcastle, E. Bloom, and S. R. Geiger, *Handbook of Physiology: The Nervous System*. Bethesda, Md.: American Physiological Society.

Goldsmith, J. 1990. *The Last Phonological Rule*. Chicago: University of Chicago Press.

Gregory, R. L. 1966. *Eye and Brain*. London: McGraw-Hill.

———, ed. 1987. *The Oxford Companion to the Mind*. Oxford: Oxford University Press.

Grice, H. P. 1975. "Logic and Conversation." Reprinted in A. P. Martinich, *The Philosophy of Language*, pp. 159–70. New York: Holt, Rinehart, and Winston, 1985.

Habermas, J. 1988. *Der philosophische Diskurs der Moderne*. Frankfurt: Suhrkamp.

———. 1984. *The Theory of Communicative Action (Theorie des kommunikativen Handelns)*. Trans. T. McCarthy, 2 vols. Boston: Beacon Press.

Halle, M. 1959. *The Sound Pattern of Russian*. The Hague: Mouton.

Haugeland, J. 1985. "The Nature and Plausibility of Cognitivism." In *Mind Design*, 3d ed., ed. J. Haugeland, pp. 243–81. Cambridge, Mass.: MIT Press.

Hayes, P. 1969. "The Naive Physics Manifesto." In *Expert Systems in the Electronic Age*, ed. D. Michie, pp. 463–502. Edinburgh: Edinburgh University Press.

———. 1985. "The Second Naive Physics Manifesto." In *Formal Theories of the Commonsense World*, ed. J. R. Hobbs and R. C. Moore, pp. 1–36. Norwood, N.J.: Ablex.

Hebb, D. O. 1949. *Organization of Behavior*. New York: Wiley.

Heidegger, M. 1962. *Being and Time*. Trans. J. Macquarrie and E. Robinson. New York: Harper and Row.

————. 1971. *On the Way to Language*. Trans. P. Hertz. New York: Harper and Row.

————. 1977. *The Question Concerning Technology*. Trans. W. Lovitt. New York: Harper and Row.

————. 1927. "Sein und Zeit." In *Jahrbuch für Philosophie und phänomenologische Forschung*, vol. 8, ed. E. Husserl. Halle.

————. 1968. *What Is Called Thinking?* Trans. F. D. Wieck and G. J. Gray. New York: Harper and Row.

Hillis, W. D. 1985. *The Connection Machine*. Cambridge. Mass.: MIT Press.

Hinton, G. E. 1981. "Implementing Semantic Networks in Parallel Hardware." In *Parallel Models of Associative Memory*, ed. G. E. Hinton and J. A. Anderson, pp. 161–88. Hillsdale, N.J.: Lawrence Erlbaum.

————. 1986. "Learning Distributed Representations of Concepts." *Proceedings of the Eighth Annual Conference of the Cognitive Science Society*, pp. 1–12. Hillsdale, N.J.: Lawrence Erlbaum.

————, and J. A. Anderson, eds. 1981. *Parallel Models of Associative Memory*. Hillsdale, N.J.: Lawrence Erlbaum.

————, and T. J. Sejnowski. 1986. "Learning and Relearning in Boltzmann Machines." In D. E. Rumelhart, J. L. McClelland, and the PDP Research Group, *Parallel Distributed Processing: Explorations in the Microstructure of Cognition*, vol. 1: *Foundations*, pp. 282–317. Cambridge, Mass.: MIT Press.

Hockett, C. F. 1958. *A Course in Modern Linguistics*. New York: Macmillan.

Hofstadter, D. 1979. *Gödel, Escher, Bach: An Eternal Golden Braid*. New York: Basic Books.

————. 1984. *Metamagical Thema. Questing for the Essence of Mind and Pattern*. New York: Basic Books.

————. 1981. "Reflections on John R. Searle: Minds, Brains, and Programs." In D. R. Hofstadter and D. C. Dennett, *The Mind's I*. New York: Basic Books.

Hofstadter, D. R., and D. C. Dennett. 1981. *The Mind's I*. New York: Basic Books.

Hopfield, J. J. 1982. "Neural Networks and Physical Systems with Emergent Collective Computational Abilities." In *Proceedings of the National Academy of Sciences*, pp. 2554–58. Reprinted in *Neurocomputing: Foundations of Research*, ed. J. A. Anderson and E. Rosenfeld. Cambridge, Mass.: MIT Press.

Hutchins, E. 1980. *Culture and Inference: A Trobriand Case Study*. Cambridge, Mass.: Harvard University Press.

————. 1987. "Myth and Experience in the Trobriand Islands." In *Cultural Models in Language and Thought*, ed. D. Holland and N. Quinn, pp. 269–89. Cambridge: Cambridge University Press.

Jakobson, R. 1968. *Child Language, Aphasia, and Phonological Universals*. The Hague: Mouton.

————, and M. Halle. 1956. *Fundamentals of Language*. The Hague: Mouton.

Joas, H. 1987. "Symbolic Interactionism." In *Social Theory Today*, ed. A. Giddens and J. Turner, pp. 82–115. Stanford: Stanford University Press.

Johnson, M. 1987. *The Body in the Mind*. Chicago: University of Chicago Press.

Katz, J., and J. Fodor. 1963. "The Structure of a Semantic Theory." *Language* 39: 170–210.

Keenan, E. L. 1975. *Formal Semantics of Natural Language*. Cambridge: Cambridge University Press.

Kleene, C. S. 1967. *Mathematical Logic*. New York: Wiley.

Köhler, W. 1929. *Gestalt Psychology*. New York: Liveright.

————. 1917. *Intelligenzprüfung an Menschenaffen*. Berlin: Springer.

————. 1925. *The Mentality of Apes*. New York: Humanities Press.

Köhler, W. 1969. *The Task of Gestalt Psychology.* Princeton: Princeton University Press.

Konishi, M. 1986. "Centrally Synthesized Maps of Sensory Space." *Trends in Neuroscience* 9: 163–68.

Kuhn, T. S. 1970. *The Structure of Scientific Revolutions.* Chicago: University of Chicago Press.

Laird, J. E., A. Newell, and P. S. Rosenbloom. 1986. *Soar: An Architecture for General Intelligence.* Research report, Carnegie-Mellon University.

Lakoff, G. 1988. "Cognitive Semantics." In *Meaning and Mental Representations,* ed. U. Eco, M. Santambrogio, and P. Violi, pp. 119–54. Bloomington: Indiana University Press.

———. 1973. "Hedges." *Journal of Philosophical Logic* 2: 459–508.

———. 1971. *On Syntactic Irregularity.* New York: Holt, Rinehart, and Winston.

———. 1987. *Women, Fire, and Dangerous Things: What Categories Reveal About the Mind.* Chicago: Chicago University Press.

Langacker, R. 1987. *Foundations of Cognitive Grammar.* Vol. 1. Stanford: Stanford University Press.

———. 1982. "Space Grammar, Analysability, and the English Passive." *Language* 58, no. 1: 22–80.

Langley, P., H. A. Simon, G. L. Bradshaw, and J. M. Zytkow. 1987. *Scientific Discovery.* Cambridge, Mass.: MIT Press.

Lazzaro, J., and C. Mead. 1989. "A Silicon Model of Auditory Localization." *Neural Computation* 1: 47–57.

Lenat, D. B. 1988. "When Will Machines Learn?" *Proceedings of the International Conference on Fifth Generation Computer Systems,* vol. 3, pp. 1239–45. Tokyo: Institute for New Generation Computer Technology.

———, and R. V. Guha. 1990. *Building Large Knowledge-Based Systems: Representations and Inference in the CYC Project.* Reading, Mass.: Addison-Wesley.

Lettvin, J. Y., H. R. Maturana, W. S. McCulloch, and W. H. Pitts. 1965. "What the Frog's Eye Tells the Frog's Brain." In *Embodiment of Mind,* ed. W. S. McCulloch. Cambridge, Mass.: MIT Press.

Levinson, S. H. 1983. *Pragmatics.* Cambridge: Cambridge University Press.

Malik, J., and P. Perona. 1982. "Preattention Texture Discrimination with Early Vision Mechanisms." *Journal of the Optical Society of America* 7, no. 2: 923–32.

Marr, D. 1982. *Vision.* San Francisco: Freeman.

Martinich, A. P., ed. 1985. *The Philosophy of Language.* New York: Holt, Rinehart, and Winston.

Maturana, H. 1978. "Biology of Language: The Epistemology of Reality." In *Psychology and Biology of Language and Thought: Essays in Honor of Eric Lenneberg,* ed. G. A. Miller and E. Lenneberg, pp. 27–64. New York: Academic Press.

———. 1970. "Neurophysiology of Cognition." In *Cognition: A Multiple View,* ed. P. Garvion, pp. 3–23. New York: Spartan Books.

———. 1987. *The Tree of Knowledge.* Boston: New Science Library.

———, and F. Varela. 1980. *Autopoiesis and Cognition: The Realization of the Living.* Dordrecht: Reidel.

McCarthy, J. 1979. "Ascribing Mental Qualities to Machines." In *Philosophical Perspectives in AI,* ed. M. Ringle, pp. 161–95. Brighton: Harvester Press.

———. 1968. "Programs with Common Sense." In *Semantic Information Processing,* ed. M. Minsky, pp. 403–18. Cambridge, Mass.: MIT Press.

———, and P. J. Hayes. 1969. "Some Philosophical Problems from the Standpoint of Artificial Intelligence." In *Machine Intelligence,* ed. B. Meltzer and D. Michie, vol. 4, pp. 463–502. Edinburgh: Edinburgh University Press.

McCawley, J. D. 1968. "The Role of Semantics in a Grammar." In *Universals in Linguistic Theory*, ed. E. Bach and R. Harms, pp. 125–70. New York: Holt, Rinehart, and Winston.

McClelland, J. L. 1979. "On the Time Relations of Mental Processes: An Examination of Systems of Processes in Cascade." *Psychological Review* 86: 287–330.

———, and J. L. Elman. 1986. "Interactive Processes in Speech Perception: The TRACE Model." In D. E. Rumelhart, J. L. McClelland, and the PDP Research Group, *Parallel Distributed Processing: Explorations in the Microstructure of Cognition*, vol. 2: *Psychological and Biological Models*, pp. 58–121. Cambridge, Mass.: MIT Press.

———, and D. E. Rumelhart. 1988. *Explorations in Parallel Distributed Processing: A Handbook of Models, Programs, and Exercises*. Cambridge, Mass.: MIT Press.

Mead, C. 1988. *Analog VLSI and Neural Systems*. Reading, Mass.: Addison-Wesley.

Mead, G. H. 1934. *Mind, Self, and Society: From the Standpoint of a Social Behaviorist*. Chicago: University of Chicago Press.

Merleau-Ponty, M. 1945. *Phénomenologie de la perception*. Paris: Gallimard. Published in English as *Phenomenology of Perception*. London: Routledge and Kegan Paul, 1962.

Mervis, C., and E. Rosch. 1981. "Categorization of Natural Objects." *Annual Review of Psychology* 32: 89–115.

Michalski, R., J. Carbonell, and T. Mitchell. 1983. *Machine Learning: An Artificial Intelligence Approach*. Palo Alto, Calif.: Tioga.

Michie, D., and R. Johnston. 1984. *The Creative Computer*. Harmondsworth: Viking.

Michon, J., and A. Anureyk. 1991. *Perspectives on Soar*. Norwell, Mass.: Kluwer.

Mill, J. S. 1843. *A System of Logic Ratiocinative and Inductive, Being a Connected View of the Principles of Evidence, and the Methods of Scientific Investigation*. 2 vols. London: J. W. Parker.

Miller, G. A. 1956. "The Magical Number Seven, Plus or Minus Two: Some Limits on Our Capacity for Processing Information." *Psychological Review* 63: 81–97.

———. 1962. "Some Psychological Studies of Grammar." *American Psychologist* 17: 748–62.

———, E. Galanter, and K. Pribram. 1960. *Plans and the Structure of Behavior*. New York: Holt, Rinehart, and Winston.

Minsky, M. 1968. *Semantic Information Processing*. Cambridge, Mass.: MIT Press.

———. 1985. *The Society of Mind*. New York: Simon and Schuster.

———, and S. A. Papert. 1969. *Perceptrons: An Introduction to Computational Geometry*. Cambridge, Mass.: MIT Press. Reprint. *Perceptrons*. Expanded ed. Cambridge, Mass.: MIT Press, 1988.

Mishkoff, H. C. 1988. *Understanding Artificial Intelligence*. 2d ed. Indianapolis: Howard W. Sams.

Moore, R. C. 1985. "The Role of Logic in Knowledge Representation and Commonsense Reasoning." In *Readings in Knowledge Representation*, ed. R. J. Brachman and H. J. Levesque, pp. 336–41. San Mateo, Calif.: Morgan Kaufmann.

Moravec, H. 1988. *Mind Children: The Future of Robot and Human Intelligence*. Cambridge, Mass.: Harvard University Press.

Neisser, U. 1976. *Cognitive Psychology*. New York: Appleton-Century-Crofts.

Newell, A. 1955. "The Chess Machine: An Example of Dealing with a Complex Task by Adaptation." *Proceedings of the 1955 Western Joint Computer Conference*, pp. 101–8.

———. 1980. "Physical Symbol Systems." *Cognitive Science* 4: 135–38.

———. 1973. "Production Systems: Models of Control Structures." In *Visual Information Processing*, ed. W. C. Chase. New York: Academic Press.

———, J. C. Shaw, and H. A. Simon. 1958. "Chess Playing Programs and the Problem of Complexity." *IBM Journal of Research and Development* 2: 320–25.

Newell, A. 1958. "Elements of a Theory of Human Problem Solving." *Psychological Review* 65: 151–66.

———. 1963. "Empirical Explorations with the Logic Theory Machine: A Case Study in Heuristics." In *Computers and Thought*, ed. E. A. Feigenbaum and J. Feldman, pp. 109–33. New York: McGraw-Hill.

———. 1960. "Report on a General Problem-Solving Program." *Proceedings of the International Conference on Information Processing*. Paris.

Newell, A., and H. Simon. 1976. "Computer Science as Empirical Inquiry." Reprinted in *Mind Design*, ed. J. Haugeland, pp. 35–66. Cambridge, Mass.: MIT Press.

———. 1963. "GPS: A Program That Simulates Human Thought." In *Computers and Thought*, ed. E. A. Feigenbaum and J. Feldman, pp. 279–96. New York: McGraw-Hill.

———. 1956. "The Logic Theory Machine." *IRE Transactions on Information Theory* IT-2, no. 3.

Norman, D. A. 1969. *Memory and Attention*. New York: Wiley.

———. 1988. *The Psychology of Everyday Things*. New York: Basic Books.

———, D. E. Rumelhart, and the LNR Research Group. 1975. *Explorations in Cognition*. San Francisco: Freeman.

Novak, G. S. 1977. "Representation of Knowledge in a Program for Solving Physics Problems." *Proceedings of the Fifth International Joint Conference on Artificial Intelligence*, pp. 286–91.

Oevermann, U. 1976. "Programmatische Überlegungen zu einer Theorie der Bildungsprozesse und einer Strategie der Sozialisationsforschung." In *Sozialisation und Lebenslauf*, ed. K. Hurrelmann. Hamburg.

——— et al. 1979. "Die Methodologie einer objektiven Hermeneutik und ihre allgemeine forschungslogische Bedeutung in den Sozialwissnschaften." In *Interpretative Verfahren in den Sozialwissenschaften*, ed. H. G. Soeffner. Stuttgart: Metzler.

Osgood, C. 1953. *Method and Theory in Experimental Psychology*. New York: Oxford University Press.

———. 1963. "Psycholinguistics." In *Psychology: A Study of a Science*, ed. S. Koch, vol. 6, pp. 244–316. New York: McGraw-Hill.

Palmer, S. E., and R. Kimchi. 1985. "The Information Processing Approach to Cognition." In *Approaches to Cognition: Contrasts and Controversies*, ed. T. Knapp and L. Robertson. Hillsdale, N.J.: Lawrence Erlbaum.

Partee, B. H. 1976. *Montague Grammars*. New York: Academic Press.

Payr, S. 1992. "Wissensgestützte maschinelle Übersetzung: Anforderungen und Probleme am Beispiel der Anaphorik und Textrepräsentation." Ph.D. diss., University of Klagenfurt.

Peirce, C. S. (1931–35). *Collected Papers*. Cambridge, Mass.: Harvard University Press.

Pellionisz, A., and R. Llinas. 1982. "Space-Time Representation in the Brain: The Cerebellum as a Predictive Space-Time Metric Tensor." *Neuroscience* 7: 2249–2970.

Pinker, S., and A. Prince. 1988. "On Language and Connectionism: Analysis of a Parallel Distributed Processing Model of Language Acquisition." *Cognition* 28: 73–193.

Place, U. T. 1956. "Is Consciousness a Brain Process?" *British Journal of Psychology* 47: 44–50.

Postal, P. 1972. "Some Further Limitations of Interpretive Theories of Anaphora." *Linguistic Inquiry* 3: 349–71.

Propp, V. 1928. *Morfologija Skazki*. Leningrad: Akademija.

Putnam, H. 1975. "The Meaning of 'Meaning.'" Reprinted in *Mind, Language, and Reality: Philosophical Papers*, vol. 2, pp. 215–71. Cambridge: Cambridge University Press.

———. 1988. *Representation and Reality*. Cambridge, Mass.: MIT Press.

Pylyshyn, Z. W.. 1984. *Computation and Cognition: Toward a Foundation for Cognitive Science*. Cambridge, Mass.: MIT Press.

———. 1973. "What the Mind's Eye Tells the Mind's Brain." *Psychological Bulletin* 8: 1–14.

Quillian, M. R. 1968. "Semantic Memory." In *Semantic Information Processing*, ed. M. Minsky, pp. 216–70. Cambridge, Mass.: MIT Press.

Quine, W.V.O. 1960. *Word and Object*. Cambridge, Mass.: MIT Press.

Reddy, D. R., et al. 1973. "The HEARSAY Speech Understanding System." *Proceedings of the Third International Joint Conference on Artificial Intelligence*. Stanford.

Reddy, M. 1979. "The Conduit Metaphor." In *Metaphor and Thought*, ed. A. Ortonyi, pp. 284–324. Cambridge: Cambridge University Press.

Robson, J. M., ed. (1973–74). *Collected Works of John Stuart Mill*. Vols. 7 and 8. Toronto: University of Toronto Press.

Rosch, E. 1977. "Human Categorization." In *Advances in Cross-Cultural Psychology*, ed. N. Warren. New York: Academic Press.

———. 1978. "Principles of Categorization." In *Cognition and Categorization*, ed. E. Rosch and B. B. Lloyd, pp. 27–48. Hillsdale, N.J.: Lawrence Erlbaum.

Rosenblatt, F. 1962. *Principles of Neurodynamics*. New York: Spartan.

———. 1959. "Two Theorems of Statistical Separability in the Perceptron." In *Proceedings of a Symposium on the Mechanization of Thought Processes*, pp. 421–56. London: Her Majesty's Stationary Office.

Ross, J. (1967), *Constraints on Variables in Syntax*. Ph.D. diss., Massachusetts Institute of Technology.

Rumelhart, D. E. 1975. "Notes on a Schema for Stories." In *Representation and Understanding: Studies in Cognitive Science*, ed. D. G. Bobrow and A. Collins, pp. 211–36. New York: Academic Press.

———. (1986). "Schemata and Sequential Thought Processes in PDP Models." In D. E. Rumelhart, J. L. McClelland, and the PDP Research Group, *Parallel Distributed Processing: Explorations in the Microstructure of Cognition*, vol. 2, pp. 7–57. Cambridge, Mass.: MIT Press.

———, P. M. Lindsay, and D. A. Norman. 1972. "A Process Model for Long-Term Memory." In *Organization of Memory*, ed. E. Tulving and W. Donaldson, pp. 197–246. New York: Academic Press.

———, and J. L. McClelland. 1981. "Interactive Processing through Spreading Activation." In *Interactive Processes in Reading*, ed. C. Perfetti and A. Lesgold. Hillsdale, N.J.: Lawrence Erlbaum.

———, and the PDP Research Group. 1986. *Parallel Distributed Processing: Explorations in the Microstructure of Cognition*. 2 vols. Cambridge, Mass.: MIT Press.

Rumelhart, D. E., and D. A. Norman. 1973. "Active Semantic Networks as a Model of Human Memory." *Proceedings of the Third International Joint Conference on Artificial Intelligence*, pp. 450–57.

Sapir, E. 1921. *Language*. San Diego: Harcourt Brace Jovanovich.

Schank, R. C. 1972. "Conceptual Dependency: A Theory of Natural Language Understanding." *Cognitive Psychology* 3: 552–631.

———. 1975. *Conceptual Information Processing*. Amsterdam: North-Holland.

———. 1982. *Dynamic Memory: A Theory of Learning in Computers and People*. Cambridge: Cambridge University Press.

———, and R. Abelson. 1977. *Scripts, Plans, Goals, and Understanding*. Hillsdale, N.J.: Lawrence Erlbaum.

Schutz, A. (1962, 1964, 1966). *The Collected Papers of Alfred Schutz*. Vol. 1: *The Problem of Social Reality*, ed. M. Natanson. Vol. 2: *Studies in Social Theory*, ed. A. Brodersen. Vol. 3: *Studies in Phenomenological Philosophy*, ed. I. Schutz. The Hague: Nijhoff.

Schutz, A. 1967. *The Phenomenology of the Social World*. Evanston, Ill.: Northwestern University Press.

Searle, J. R. 1983. *Intentionality: An Essay in the Philosophy of Mind*. Cambridge: Cambridge University Press.

———. 1990. "Ist der menschliche Geist ein Computerprogramm?" *Spektrum der Wissenschaft* (March): 40–47.

———. 1980. "Minds, Brains, and Programs." *Behavioral and Brain Sciences* 3: 417–57.

———. 1984. *Minds, Brains, and Science: The 1984 Reith Lectures*. London: Penguin Books.

———. 1992. *The Rediscovery of the Mind*. Cambridge, Mass.: MIT Press.

Selfridge, O., and U. Neisser. 1960. "Pattern Recognition by Machine. *Scientific American* 203: 60–68.

Shortliffe, E. H. 1976. *Computer-Based Medical Consultation: MYCIN*. New York: Elsevier.

Simon, H. A. 1979. *Models of Thought*. New Haven: Yale University Press.

Sloan Foundation. 1976. *Proposed Particular Program in Cognitive Science*. New York.

Smart, J.J.C. 1956. "Is Consciousness a Brain Process?" *British Journal of Psychology* 47: 44–50.

———. 1963. *Philosophy and Scientific Realism*. New York: Humanities Press.

———. 1959. "Sensations and Brain Processes." *Philosophical Review* 68: 141–56.

Smith, B. 1991. "The Owl and the Electric Encyclopedia." *Artificial Intelligence* 47: 1–3, 251–88.

Smolensky, P. 1986. "Information Processing in Dynamic Systems: Foundations of Harmony Theory." In D. E. Rumelhart, J. L. McClelland, and the PDP Research Group, *Parallel Distributed Processing*, vol. 1: *Foundations*, pp.194–281. Cambridge, Mass.: MIT Press.

———. 1988. "On the Proper Treatment of Connectionism." *Behavioral and Brain Sciences* 11: 1–74.

Sternberg, S. 1969. "Memory Scanning: Mental Processes Revealed by Reaction-Time Experiments." *American Scientist* 57: 421–57.

Strawson, P. F. 1950. "On Referring." Reprinted in A. P. Martinich, *The Philosophy of Language*, pp. 220–35. New York: Holt, Rinehart, and Winston, 1985.

Suchman, L. 1987. *Plans and Situated Action*. Cambridge: Cambridge University Press.

Talmy, L. 1985. "Force Dynamics in Language and Thought." In *Papers from the Parasession on Causatives and Agentivity*, ed. Chicago Linguistic Society.

———. 1983. "How Language Structures Space." In *Spatial Orientation: Theory, Typology, and Syntactic Description*, ed. H. Pick and L. Acredolo, vol. 3. Cambridge: Cambridge University Press.

Thomason, R., ed. 1974. *Formal Philosophy: Selected Papers of Richard Montague*. New Haven: Yale University Press.

Tulving, E., and W. Donaldson, eds. 1972. *Organization of Memory*. New York: Academic Press.

Turing, A. 1950. "Computing Machinery and Intelligence." *Mind* 59: 433–60.

Turkle, S. 1991. "Romantic Reactions. Paradoxical Responses to the Computer Presence." In *The Boundaries of Humanity: Humans, Animals, Machines*, ed. J. Sheenan and M. Sosna, pp. 224–52. Berkeley and Los Angeles: University of California Press.

Wagner, H. R. 1973. *Alfred Schutz on Phenomenology and Social Relations: Selected Writings*. Chicago: University of Chicago Press.

Waldrop, M. M. 1987. *Man-Made Minds: The Promise of Artificial Intelligence*. New York: Walker.

Weizenbaum, J. 1976. *Computer Power and Human Reason*. San Francisco: Freeman.

————. 1966. "ELIZA: A Computer Program for the Study of Natural Language Communication between Man and Machine." *Communications of the ACM* 1: 36–45.

————. 1967. "Contextual Understanding by Computers." *Communications of the ACM* 8: 474–80.

Whorf, B. 1956. *Language, Thought, and Reality*. Cambridge, Mass.: MIT Press.

Wiener, N. 1961. *Cybernetics; or, Control and Communication in the Animal and the Machine*. 2d ed. Cambridge, Mass.: MIT Press.

Winograd, T. 1974. "Artificial Intelligence—When Will Computers Understand People?" *Psychology Today* 7, no. 12: 73–79.

————. 1980. "Extended Inference Modes in Reasoning by Computer Systems." *Artificial Intelligence* 13, no. 1: 5–26. Reprinted in *Applications of Inductive Logic*, ed. J. Cohen and M. Hesse, pp. 333–58. Oxford: Clarendon Press.

————. 1975. "Frame Representations and the Procedural-Declarative Controversy." In *Representation and Understanding: Studies in Cognitive Science*, ed. D. Bobrow and A. Collins, pp. 185–210. New York: Academic Press.

————. 1977. "A Framework for Understanding Discourse." In *Cognitive Processes in Comprehension*, ed. M. Just and P. Carpenter, pp. 63–88. Hillsdale, N.J.: Lawrence Erlbaum.

————. 1978. "On Primitives, Prototypes, and Other Semantic Anomalies." *Proceedings of the Conference on Theoretical Issues in Natural Language Processing*, University of Illinois, pp. 25–32.

————. 1976. "Towards a Procedural Understanding of Semantics." *Revue Internationale de Philosophie* 3: 117–18.

————. 1972. "Understanding Natural Language." *Cognitive Psychology* 3: 1–191.

————, and F. Flores. 1986. *Understanding Computers and Cognition: A New Foundation for Design*. Norwood, N.J.: Ablex.

Wittgenstein, L. 1953. *Philosophical Investigations*. Oxford: Blackwell.

————. 1922. *Tractatus logico-philosophicus*. London: Routledge and Kegan Paul.

Zadeh, L. 1965. "Fuzzy Sets." *Information and Control* 8: 159–76.

Zipser, D. 1986. "Biologically Plausible Models of Place Recognition and Goal Location." In D. E. Rumelhart, J. L. McClelland, and the PDP Research Group, *Parallel Distributed Processing: Explorations in the Microstructure of Cognition*, vol. 2: *Psychological and Biological Models*, pp. 432–70. Cambridge, Mass.: MIT Press.

INDEX

Names

Anderson, A., 59
Anderson, J. A., 196
Anderson, J. R., 22, 157, 238
Austin, G., 150
Austin, J. L., 49, 321

Bacon, F., 182
Ballard, D. H., 108
Bate, W. J., 186
Bateson, G., 17
Berlin, B., 119, 166
Bobrow, D., 157, 191, 254, 283, 285
Bourdieu, P., 52, 55
Bower, G., 157
Brentano, F., 317
Broadbent, D. E., 233
Brooks, R., 30
Brown, R., 47
Bruner, J. S., 150
Buchanan, B. G., 303

Carbonell, J., 147
Carnap, R., 180
Chomsky, N., 5, 47, 85, 115, 117, 120, 122, 150, 173, 183, 287
Churchland, P. M., 21, 24, 99
Churchland, P. S., 34, 35, 99
Cicourel, A., 157
Colby, K., 249, 252
Collins, A., 157, 191
Crick, F., 23, 26, 34
Cummins, R., 82

D'Andrade, R., 49, 157
Dennett, D., 92, 180, 256–57, 259–60, 262, 293
Descartes, R., 71–72, 87
Dretske, F., 99
Dreyfus, H., 18, 52, 59, 61–62, 64, 90–91, 95–99, 101, 105–7, 136, 158, 186, 239, 245, 269, 280, 283, 293, 304–5, 316
Dreyfus, S., 316
Durkheim, E., 52, 55

Edelman, G., 185
Elman, G., 143

Fauconnier, G., 122
Feigenbaum, E., 81, 150–51, 249, 292, 303
Feldman, Jerry, 108, 280
Feldman, Julian, 59, 151
Fillmore, C., 119–20, 126, 189
Flores, F., 283
Fodor, J., 22, 24–25, 72, 82, 112, 137, 180
Frege, G., 117, 189
Freud, S., 98
Fuster, J. M., 32

Galanter, E., 233
Garfinkel, H., 47
Goldman-Rakic, P. S., 26
Goldsmith, J., 123
Goodnow, J., 150
Gregory, R. L., 23
Grice, H. P., 49

Habermas, J., 51–52, 55, 183
Halle, M., 115
Haugeland, J., 61
Hayes, P., 61
Hebb, D., 226, 236
Heidegger, M., 71–72, 285, 289, 291
Hillis, W. D., 261, 315
Hinton, G., 35, 137, 159, 191–92, 196, 273, 280
Hofstadter, D., 65, 92, 289
Hoijer, H., 47
Hume, D., 73
Husserl, E., 71–72
Hutchins, E., 49, 54

Jakobson, R., 115
Johnson, M., 126

Kant, I., 71–72, 186
Kay, P., 119, 166
Keenan, E., 118

Subjects

(Page numbers in boldface refer to the glossary.)